Sensors and Measurement Systems for Marine Engineering Applications

Sensors and Measurement Systems for Marine Engineering Applications

Editor

Dimitrios-Nikolaos Pagonis

MDPI

Basel • Beijing • Wuhan • Barcelona • Belgrade • Novi Sad • Cluj • Manchester

Editor
Dimitrios-Nikolaos Pagonis
Department of Naval
Architecture
School of Engineering
University of West Attica
Athens
Greece

Editorial Office
MDPI AG
Grosspeteranlage 5
4052 Basel, Switzerland

This is a reprint of articles from the Special Issue published online in the open access journal *Applied Sciences* (ISSN 2076-3417) (available at: https://www.mdpi.com/journal/applsci/special_issues/Marine_Sen).

For citation purposes, cite each article independently as indicated on the article page online and as indicated below:

Lastname, A.A.; Lastname, B.B. Article Title. *Journal Name* **Year**, *Volume Number*, Page Range.

ISBN 978-3-7258-1549-4 (Hbk)
ISBN 978-3-7258-1550-0 (PDF)
doi.org/10.3390/books978-3-7258-1550-0

Contents

About the Editor

Dimitrios-Nikolaos Pagonis

Dimitrios-Nikolaos Pagonis received his Master of Engineering degree in Electronic and Electrical Engineering from the University of Loughborough, U.K., in 1999, and his Master of Science degree and Ph.D. in Microelectronics from the University of Athens, in 2001 and 2004, respectively. During 2005–2007, he held a Postdoctoral position at the National Centre of Scientific Research "Demokritos". He joined the University of West Attica in 2010 (after a three-year academic break, during which he was a Flag State Surveyor for the Greek Ministry of Mercantile Shipping), where he is currently an Associate Professor in the Department of Naval Architecture. He has participated in various research projects while authoring/co-authoring one PCT patent and over 60 papers in international journals and peer-reviewed conferences. He has served as a member of journal editorial boards and conference scientific committees and as a guest editor and reviewer for various academic journals and conferences. His current research interests include 3D/4D printing, measurement systems employing 3D/4D printing, screen-printing, new microsystem technologies, microsystem modeling and simulation, sensor interfacing, and sensors/actuators on flexible substrates.

applied sciences

MDPI

Editorial

Sensors and Measurement Systems for Marine Engineering Applications

Dimitrios Nikolaos Pagonis

Naval Architecture Department, School of Engineering, University of West Attica, 12243 Athens, Greece; d.n.pagonis@uniwa.gr

1. Introduction

In recent years, vast developments and applications of sensor technologies have been recorded in various industries, including shipbuilding. Therefore, the employment of novel sensors in marine environments has significantly progressed, as illustrated by some key examples. For example, wireless sensor networks (WSNs) have emerged as an efficient and cost-effective alternative for the real-time monitoring of the marine environment, used for oil spill detection and localization [1], offering significant advantages such as ease of deployment [2]. Optical-fiber-based sensors have attracted considerable attention for environmental applications such as the in situ measurement of seawater salinity [3] and structural health monitoring in marine applications [4,5] due to their durability under extreme temperature and pressure conditions, high sensitivity, and flexibility. In addition, piezoelectric energy harvesters have been proposed to eliminate batteries from future sensing devices [6,7], while piezoelectric sensors have already been employed to determine ocean wave height and period, as well as underwater objects [8]. Furthermore, nanofibrous grids utilizing piezoelectric fibers demonstrate high-resolution, self-powered tactile sensing capabilities, suggesting their use in real-time motion tracking and spatial sensing in marine environments [9], whereas monolithic printed sensors pave the way for on-demand, on-site fabrication [10]. These examples underscore the vast technological developments that have been pivotal in addressing the challenges related to marine research and environmental protection.

Moreover, over the past three decades, the miniaturization of sensors has led to the newly developed field of "microsensors"—an emerging field that has grown rapidly, building on the significant advancements of the semiconductor industry. Consequently, the sensors deployed on ships have followed the same trend, taking their role one step further by incorporating semiconductor-based technology into standard marine equipment and measuring systems. Typical types of sensors employed aboard are gas detection sensors, gas/air flow sensors, humidity sensors, temperature/pressure sensors, speed/acceleration sensors, strain sensors, IMUs, etc., which are all essential with regard to vessel safety since the proper operation of all modern vessels relies heavily on the information provided by the onboard measuring devices that quantify critical performance parameters. As a common example, a typical Ro-Ro vessel has four medium-power 4-stroke engines installed onboard, which are required for the operation of the corresponding generator sets [11]. Failure to detect a possible malfunction by the appropriate sensing system in any of the installed engines, even one, can lead to immediate safety compromises, as outlined by the corresponding safety regulations [12], with severe consequences. In addition, modern sensing devices play a vital role in advancing marine technology by enabling the integration of new technologies, such as the Internet of Things (IoT), big data, and cloud computing [13]. These technologies rely heavily on sensing devices for functionality and also contribute to their rapid development.

This Special Issue encompasses the diversity of nine relevant studies, spanning from the design of novel sensors and energy-harvesting solutions for the maritime industry

Citation: Pagonis, D.N. Sensors and Measurement Systems for Marine Engineering Applications. *Appl. Sci.* **2024**, *14*, 3761. https://doi.org/10.3390/app14093761

Received: 23 April 2024
Accepted: 26 April 2024
Published: 28 April 2024

to advanced engine-monitoring systems. Specifically, it comprises eight articles and one review, which are briefly outlined in the next section. It is important to note that the purpose of this editorial is not to elaborate on each of the papers presented but rather to encourage readers to explore them further.

2. An Overview of the Published Articles

In the first article, Bardakas et al. (contribution 1) present a novel rotational–linear triboelectric nanogenerator (RL-TENG) that converts rotational motion to linear motion to harvest rotational energy. The device employs ZnO nanoparticles as a triboelectric material, demonstrating a modular design for various applications. The RL-TENG design offers several advantages, including reduced wear and increased temperature during operation, compared to traditional rotational tribogenerators. This technology has significant implications for maritime applications, particularly in wind energy harvesting for powering remote monitoring systems or other low-power devices.

In the second article, Pagonis et al. (contribution 2) introduce a mass air flow sensor designed for low- and medium-power internal combustion engines in the marine industry. The sensor fabrication process is based on additive manufacturing and PCB technology. The complex design of the sensing element housing geometry, which is manufactured through 3D printing, is based on standard airfoil geometry, while it is derived through suitable CFD simulations. The key features of the proposed device are its low cost, fast on-site manufacturing, robustness, and simplicity, suggesting numerous potential applications in marine engineering.

In the third article, Yu et al. (contribution 3) present a case study of super-long steel pipelines floating offshore, focusing on controlling deformation and avoiding resonance for safety. Their study employs wireless communication equipment and aerial photography technology to monitor strain and vibration during construction. This monitoring method prevents excessive deformation, resonance, and the destruction of anticorrosive coatings during floating transportation. The obtained results provide a detailed strain and modal analysis as well as effective monitoring technology for offshore steel pipeline transportation safety.

In the fourth article (contribution 4), Jiao et al. elaborate on a novel MEMS piezoresistive pressure sensor designed to operate under harsh environmental conditions. The specific sensor demonstrates adequate linearity and sensitivity over a wide operating range, while a digital temperature compensation system impedes linearity drift due to temperature variation. The specific technology has significant implications for maritime applications, particularly in harsh environments, where robust and accurate pressure sensing is required.

In the fifth article, Yang et al. (contribution 5) propose a novel Principal Component Analysis integrating a Long Short-Term Memory Network (PCA-LSTM) model for predicting the productivity of cutter suction dredgers. The developed model involves operational parameters that are based on mechanism analysis, while it is considered a deep learning-based approach capable of dealing with operation series data with a special memory mechanism. The specific methodology has the potential to improve the efficiency and productivity of dredging operations, leading to cost savings and reduced environmental impacts.

In the sixth article, Li et al. (contribution 6) present a novel method for measuring ocean surface currents in the Kuroshio region using Gaofen-3 SAR data. The developed method, which combines sub-aperture processing and least-squares (LS) technology to measure current vectors, demonstrates the capability of the Gaofen-3 SAR to accurately retrieve the ocean surface current field in the Kuroshio region. The specific methodology has significant implications for climate studies and maritime navigation in dynamic oceanic environments.

In the seventh article, Tsitsilonis et al. (contribution 7) present a method for identifying engine malfunctioning through instantaneous crankshaft torque (ICT) measurement analysis. Briefly, this study demonstrates the usefulness of engine ICT as a nonintrusive diagnostic measurement, allowing for quick and less resource-intensive identification of engine malfunctions. The developed methodology has significant implications for maritime applications, particularly for maintaining the reliability and safety of marine engines.

In the eighth article, Daud et al. (contribution 8) propose a novel Gaussian-process-based inversion methodology for Seabed Logging (SBL) for detecting potential hydrocarbon-saturated reservoirs underneath the seabed by employing electromagnetic waves (EM). The specific method allows for greater flexibility in modeling a variety of EM responses, while the obtained results indicate that it can efficiently predict the hydrocarbon depth in seabed logging, having significant implications for offshore oil and gas exploration and production.

The last article in this Special Issue, a review by Prabowo et al. (contribution 9), provides a comprehensive survey of the developed sensor technology for maritime applications, covering various aspects, such as logistics, shipping activities, the hydrodynamic characterization of new design hulls, advanced machinery performance, arctic-based field observations, vibration-based damage detection, corrosion control and monitoring, and the measurement of explosions on critical maritime infrastructures. The specific review highlights the importance of sensors in maritime-based industries and research, as well as the potential for further advancements in the field.

3. Conclusions

This Special Issue aims to contribute to the exploration of significant advancements in sensor and measurement system technologies. Through a brief overview of the selected research papers, this editorial highlights the innovative approaches and potential impacts of sensor-driven solutions in addressing the challenges faced by the maritime sector. The articles presented in this Special Issue provide valuable insights into the implications for future research in the maritime industry, which can be summarized as follows.

The transformative impact of sensor technologies on maritime engineering is underscored by the convergence of innovation, interdisciplinary collaboration, and the continuous pursuit of safety, efficiency, robustness, and low costs. For instance, additive manufacturing has emerged as a potential enabler that offers flexibility, speed, cost-effectiveness, and adaptability to diverse operational environments for sensor fabrication; the work by Pagonis et al. illustrates this paradigm shift, showcasing the seamless integration of additive manufacturing and standard airfoil geometry in the design of mass airflow sensors for marine engineering applications. In addition, the development of micro-pressure sensors tailored to harsh environments by Jiao et al. showed the need for robust sensor technologies in marine applications; most probably, future research will explore the integration of advanced materials to improve sensor performance and longevity in challenging marine conditions. Furthermore, the selected research papers highlighted the multifaceted roles of sensor/measurement system technologies in enhancing safety, efficiency, and predictive capabilities across various maritime domains. From the development of rotational–linear triboelectric nanogenerators by Bardakas et al., offering renewable energy solutions for maritime applications, to the advanced monitoring of offshore steel pipelines by Yu et al., leveraging wireless communication and aerial photography to mitigate transportation risks, each study underscores the transformative potential of sensor-driven solutions.

Additionally, the deployment of novel analytic techniques, as demonstrated by Yang et al., Tsitsilonis, and Theotokatos, empowers maritime stakeholders with predictive insights, real-time diagnostic capabilities, and better productivity prediction, thereby optimizing resource utilization and mitigating operational risks. Future research could explore the application of advanced radar technologies for precise ocean current measurements, aiding climate studies and maritime navigation in dynamic ocean environments, the potential of non-intrusive diagnostic measurements for quick and less resource-intensive identification of engine malfunctions (e.g., employing engine Instantaneous Crankshaft Torque), and de-risking hydrocarbon exploration in deep marine environments by accurately predicting hydrocarbon depths, leading to more cost-effective exploration campaigns.

The future trajectory of sensor technologies in the maritime industry is promising, focusing on novel sensors and measuring systems that can enhance safety, efficiency, sustainability, and technological advancements to meet the continuously evolving needs of marine applications. Thus, the current studies presented in this Special Issue should be seen not only as the research results of investigations carried out by the respective researchers but also as key starting points, inviting readers to continue with new studies on the themes explored.

In light of the aforementioned considerations, it is the editor's belief that future marine sensing devices will increasingly incorporate advances from other scientific disciplines, revolutionizing their manufacturing and operation, similar to the miniaturization that occurred due to the vast advances in the semiconductor industry in the past. Consequently, many of the current drawbacks associated with employing almost solely semiconductor technology as the foundation for modern sensing devices, such as complicated manufacturing processes, expensive sensor packaging, indirect electrical communication between the sensing integrated circuit and the necessary readout circuitry, and high prototype fabrication cost, will generally be overcome.

Funding: This research received no external funding.

Conflicts of Interest: The author declares no conflict of interest.

List of Contributions:

1. Bardakas, A.; Segkos, A.; Tsamis, C. Zinc Oxide-Based Rotational–Linear Triboelectric Nanogenerator. *Appl. Sci.* **2024**, *14*, 2396. https://doi.org/10.3390/app14062396.
2. Pagonis, D.; Benaki, V.; Kaltsas, G.; Pagonis, A. Design of a Mass Air Flow Sensor Employing Additive Manufacturing and Standard Airfoil Geometry. *Appl. Sci.* **2021**, *11*, 11579. https://doi.org/10.3390/app112411579.
3. Yu, J.; Ren, C.; Cai, Y.; Chen, J.; Wang, Y.; Chen, W. A Case Study of Floating Offshore Super-Long Steel Pipeline Combing with Field Monitoring. *Appl. Sci.* **2021**, *11*, 10186. https://doi.org/10.3390/app112110186.
4. Jiao, M.; Wang, M.; Fan, Y.; Guo, B.; Ji, B.; Cheng, Y.; Wang, G. Temperature Compensated Wide-Range Micro Pressure Sensor with Polyimide Anticorrosive Coating for Harsh Environment Applications. *Appl. Sci.* **2021**, *11*, 9012. https://doi.org/10.3390/app11199012.
5. Yang, K.; Yuan, J.; Xiong, T.; Wang, B.; Fan, S. A Novel Principal Component Analysis Integrating Long Short-Term Memory Network and Its Application in Productivity Prediction of Cutter Suction Dredgers. *Appl. Sci.* **2021**, *11*, 8159. https://doi.org/10.3390/app11178159.
6. Li, Y.; Chong, J.; Sun, K.; Zhao, Y.; Yang, X. Measuring Ocean Surface Current in the Kuroshio Region Using Gaofen-3 SAR Data. *Appl. Sci.* **2021**, *11*, 7656. https://doi.org/10.3390/app11167656.
7. Tsitsilonis, K.; Theotokatos, G. Engine Malfunctioning Conditions Identification through Instantaneous Crankshaft Torque Measurement Analysis. *Appl. Sci.* **2021**, *11*, 3522. https://doi.org/10.3390/app11083522.
8. Daud, H.; Mohd Aris, M.; Mohd Noh, K.; Dass, S. A Novel Methodology for Hydrocarbon Depth Prediction in Seabed Logging: Gaussian Process-Based Inverse Modeling of Electromagnetic Data. *Appl. Sci.* **2021**, *11*, 1492. https://doi.org/10.3390/app11041492.
9. Prabowo, A.; Tuswan, T.; Ridwan, R. Advanced Development of Sensors' Roles in Maritime-Based Industry and Research: From Field Monitoring to High-Risk Phenomenon Measurement. *Appl. Sci.* **2021**, *11*, 3954. https://doi.org/10.3390/app11093954.

References

1. Tabella, G.; Paltrinieri, N.; Cozzani, V.; Rossi, P.S. Wireless Sensor Networks for Detection and Localization of Subsea Oil Leakages. *IEEE Sens. J.* **2021**, *21*, 10890–10904. [CrossRef]
2. Xu, G.; Shen, W.; Wang, X. Applications of Wireless Sensor Networks in Marine Environment Monitoring: A Survey. *Sensors* **2014**, *14*, 16932–16954. [CrossRef] [PubMed]
3. Li, G.; Wang, Y.; Shi, A.; Liu, Y.; Li, F. Review of Seawater Fiber Optic Salinity Sensors Based on the Refractive Index Detection Principle. *Sensors* **2023**, *23*, 2187. [CrossRef] [PubMed]
4. Min, R.; Liu, Z.; Pereira, L.; Yang, C.; Sui, Q.; Marques, C. Optical fiber sensing for marine environment and marine structural health monitoring: A review. *Opt. Laser Technol.* **2021**, *140*, 107082. [CrossRef]

5. Chen, S.; Wang, J.; Zhang, C.; Li, M.; Li, N.; Wu, H.; Liu, Y.; Peng, W.; Song, Y. Marine Structural Health Monitoring with Optical Fiber Sensors: A Review. *Sensors* **2023**, *23*, 1877. [CrossRef]
6. Kargar, S.M.; Hao, G. A Drifter-Based Self-Powered Piezoelectric Sensor for Ocean Wave Measurements. *Sensors* **2022**, *22*, 5050. [CrossRef] [PubMed]
7. Karga, S.M.; Hao, G. An Atlas of Piezoelectric Energy Harvesters in Oceanic Applications. *Sensors* **2022**, *22*, 1949. [CrossRef] [PubMed]
8. Asadnia, M.; Kottapalli, A.G.P.; Shen, Z.; Miao, J.; Triantafyllou, M. Flexible and surface-mountable piezoelectric sensor arrays for underwater sensing in marine vehicles. *IEEE Sens. J.* **2013**, *13*, 3918–3925. [CrossRef]
9. Liu, Q.; Jin, L.; Zhang, P.; Zhang, B.; Li, Y.; Xie, S.; Li, X. Nanofibrous Grids Assembled Orthogonally from Direct-Written Piezoelectric Fibers as Self-Powered Tactile Sensors. *ACS Appl. Mater. Interfaces* **2021**, *13*, 10623–10631. [CrossRef] [PubMed]
10. Pagonis, D.N.; Matsoukas, I.; Kaltsas, G.; Pilatis, A. A Flow Sensing Device Formed Exclusively by Employing Additive Manufacturing for On-Site Fabrication Aboard a Ship. *Sensors* **2023**, *23*, 8481. [CrossRef] [PubMed]
11. Livanos, G.A.; Theotokatos, G.; Pagonis, D.N. Techno-economic investigation of alternative propulsion plants for Ferries and RoRo ships. *Energy Convers. Manag.* **2014**, *79*, 640–651. [CrossRef]
12. International Convention for the Safety of Life at Sea (SOLAS), 1974. Available online: https://www.imo.org/en/About/Conventions/Pages/International-Convention-for-the-Safety-of-Life-at-Sea-(SOLAS),-1974.aspx (accessed on 11 April 2024).
13. Kamolov, A.; Park, S. An IoT-Based Ship Berthing Method Using a Set of Ultrasonic Sensors. *Sensors* **2019**, *19*, 5181. [CrossRef] [PubMed]

applied sciences

MDPI

Article

Zinc Oxide-Based Rotational–Linear Triboelectric Nanogenerator

Achilleas Bardakas *, Apostolos Segkos and Christos Tsamis *

Institute of Nanoscience and Nanotechnology (INN), National Centre for Scientific Research "Demokritos", Patr. Gregoriou E & 27 Neapoleos Str., Aghia Paraskevi, 15310 Athens, Greece; a.segkos@inn.demokritos.gr
* Correspondence: a.bardakas@inn.demokritos.gr (A.B.); c.tsamis@inn.demokritos.gr (C.T.)

Abstract: In this study, we evaluate a prototype harvesting device that converts rotational motion to linear motion for harvesting rotational energy. Triboelectric materials are attached to the parts of the device that execute linear motion, resulting in a contact-separation mode of operation of triboelectric generators. As a triboelectric material, thin layers of ZnO nanoparticles deposited on Kapton films are evaluated. The design of the rotational–linear triboelectric nanogenerator (RL-TENG) exhibits several advantages since it does not suffer from the issues related to rotational tribogenerators such as wear and increased temperature during operation. Moreover, our approach can result in the modular design of energy-harvesting devices for a variety of applications. As a demonstrator, cups were attached to the rotating axis of the RL-TENG to harvest wind energy that is suitable for maritime applications.

Keywords: triboelectric nanogenerators; rotational energy; triboelectrification; TENGs; wind harvesting

1. Introduction

Triboelectric harvesting has been identified as a very promising technology for the conversion of mechanical energy to electrical energy due to its low cost, abundance of available materials and high conversion efficiency even at low operating frequencies [1–5]. The operation of triboelectric generators (TENGs) is based on the charge exchange between two surfaces that are in contact (triboelectrification) and the subsequent electrostatic induction when these surfaces are set in relative motion. A major category of TENGs focuses on the conversion of rotational energy that could be induced by air and water flow, car wheels, etc. In general, rotating TENGs can be classified, according to their design, into two categories: (a) disk type and (b) cylindrical type. Further classification can be performed based on triboelectric materials (dielectric–dielectric or dielectric–conductor) and the operation mode (sliding, contact separation, non-contact). One of the first approaches of a segmented disk TENG was presented by Lin et al. [6]. The tribogenerator was based on sliding triboelectrification and periodic overlapping of the segmented disks due to rotation. Large output currents were obtained at a high frequency (117.6 μA and 29.0 mA/cm^2 of 66.7 Hz at a rotating speed of 1000 rpm). An updated version of the disk generator was later presented by the same group [7] in order to improve the operational characteristics, the longevity and the stability of the device. The new design was based on a freestanding triboelectric layer on the rotational disk, while the electrodes were located onto the stationary disk. Zhang et al. [8] presented a direct-current triboelectric nanogenerator (DC-TENG) based on a rotating-disk design. It consisted of two disks and two pairs of carbon fiber flexible brushes that contacted the electrodes. Due to its design, the tribogenerator was serving as a constant current source for direct driving electronic devices.

Bai et al. [9] demonstrated a cylindrical rotating triboelectric nanogenerator (TENG) based on sliding electrification. The TENG was based on a core–shell structure with an alternative strip on the surface. During rotation, the relative sliding between the contact surfaces of the core and the shell results in an "in-plane" lateral polarization that drives

Citation: Bardakas, A.; Segkos, A.; Tsamis, C. Zinc Oxide-Based Rotational–Linear Triboelectric Nanogenerator. *Appl. Sci.* **2024**, *14*, 2396. https://doi.org/10.3390/app14062396

Academic Editor: Huey Hoon Hng

Received: 12 February 2024
Revised: 7 March 2024
Accepted: 11 March 2024
Published: 12 March 2024

the external current. A power density of 36.9 W/m^2 was obtained by a rotating TENG with 8 strip units at a rotation rate of 1000 rpm. Zhang et al. [10] presented a non-contact cylindrical rotating TENG for harvesting mechanical energy from water flow such as that in pipes and sewers. The generator was operated in freestanding mode between two rotating interfaces to minimize abrasion of the electrodes. When placed in water flow, the output voltage and the current of the TENG reached 1670 V and 13.4 μA, respectively. An eccentric-type TENG, consisting of a fixed metal electrode, a rotor-independent layer and a base, was proposed by Qu et al. [11]. Due to its design, the TENG could effectively reduce the wear of the friction layer and improve the durability of devices. A rolling friction contact-separation mode triboelectric generator was presented by Yang et al. [12] that was fabricated for harvesting vertical rotation energy by utilizing the integrated cylindrical surface with the conjunction of rolling contact electrification and electrostatic induction. Several rotational TENGs have also been presented for harvesting wind energy [13]. For instance, Park et al. [14] developed a continuously rotating TEG, aiming to operate at marginal wind velocities by maximizing the efficiency of the sliding contact, which was a one-dimensional fiber rather than a 2D contact area. A ring-type TENG for rotational energy harvesting was presented by Xin et al. [15]. The device could also be used as a self-powered rotational speed sensor. These are only some examples of the various designs that have been presented for harvesting rotational energy. A detailed presentation of rotational triboelectric generators can be found in [16].

An alternative design for the conversion of rotational energy was demonstrated by Mo et al. [17]. They developed a self-powered air filter for the removal of particulate matter from the environment based on a radial piston TENG. In this study, we propose a novel design for harvesting rotational energy that is based on the transformation of rotational motion to linear motion, allowing for TENGs to operate at contact-separation mode and thus minimizing the inherent issues of rotating TENGs, such as wear and the development of high temperatures.

2. Materials and Methods

2.1. Design and Fabrication of the Rotational–Linear Tribogenerators (RL-TENG)

For the design of the RL-TENG, Autodesk Fusion360 was used, while the fabrication of the device was performed using a 3D printer (Creality Ender 7). The main printing parameters are as follows:

1. Material: PETG (polyethylene terephthalate glycol);
2. Layer height: 0.2 mm;
3. Perimeters: 4;
4. Infill: 40% (triangles).

The design of the RL-TENG is shown in Figure 1a. It consists of (a) 3 pairs of contact-separation type triboelectric generators, i.e., T1, T2, T3 (Figure 1b), (b) the circular base upon which the TENGs are mounted and (c) a circular cam that is attached to the axis of rotation and is perpendicular to the base plane.

In Figure 1c, we can see a part of a triboelectric pair. The triboelectric material was placed on the surface of the component using double-sided aluminum tape. The area of the triboelectric surfaces that contacted each other is 1×1.67 cm^2. Figure 2 is a photograph of the RL-TENG. We can identify the three triboelectric pairs, i.e., T1, T2 and T3, as well as their wiring. In the central area of the tribogenerator, the cam is placed (not shown in the figure). Two different cams were used to control whether the triboelectric pairs worked in a synchronous or asynchronous way. The cam is mounted to the rotating axis (vertical to the plane, not shown in figure) to provide the rotational motion.

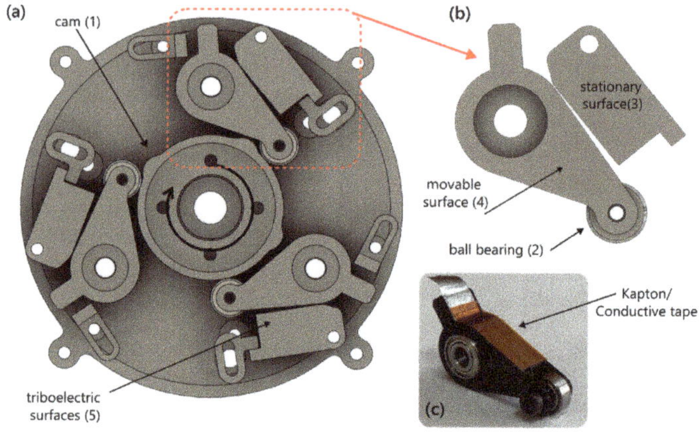

Figure 1. (**a**) Schematic of the energy-harvesting device, (**b**) a magnified view of a triboelectric pair showing the movable and stationary triboelectric surfaces and (**c**) a part of a triboelectric pair.

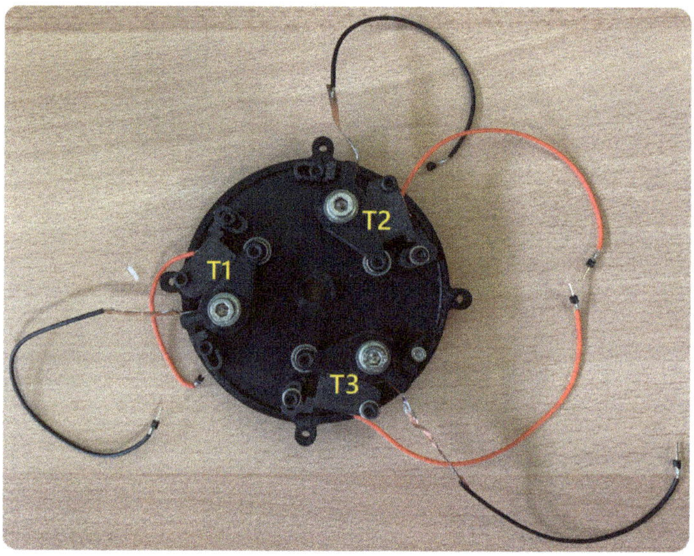

Figure 2. Photograph of the RL-TENG. We can identify the three triboelectric pairs (T1, T2 and T3).

For the triboelectric layers, Kapton® (DuPont™, Kingston, ON, Canada) and Kapton® with a ZnO thin film deposited by sol–gel on the surface were used. The thickness of both Kapton® layers was 75 µm. Briefly, a sol–gel solution was prepared by dissolving the appropriate amount of Zinc Acetate Dihydrate ($C_4H_{10}O_6Zn$, Merck, Germany, and urea ($CO(NH_2)_2$, The Science Company, Lakewood, CO, USA) in ethanol (C_2H_5OH, Carlo Erba Reagents, Emmendingen, Germany) using magnetic stirring, at 60 °C for 30 min, in order to end up with a 40 mM solution containing 5w/w% of urea. The sol–gel was left to cool to room temperature prior to use for 24 h. ZnO films were prepared by ten successive spin coatings on Kapton films. After each deposition, the film was annealed at 300 °C in air. A detailed description of the ZnO thin-film fabrication and deposition process as well as its performance as a triboelectric layer can be found in [18].

2.2. Characterization Setup

For the characterization of the RL-TENG, the setup shown in Figure 3a was used. It consists of an aluminum base with threaded holes for ease of mounting the required components. A DC servo motor (PD4-CB59M024035-E-01) attached to an aluminum profile extrusion was used to provide rotational motion, powered by a 24 V power supply (350 W) and controlled via USB. The Plug&Drive Studio 2 computer UI (Nanotec Electronic GmbH & Co. KG, Feldkirchen, Germany) was utilized for tuning and controlling the rotational speed of the servo motor. The RL-TENG was mounted on the rotational axis of the servo motor using a 3D-printed mounting plate with the appropriate spacers, as shown in Figure 3b.

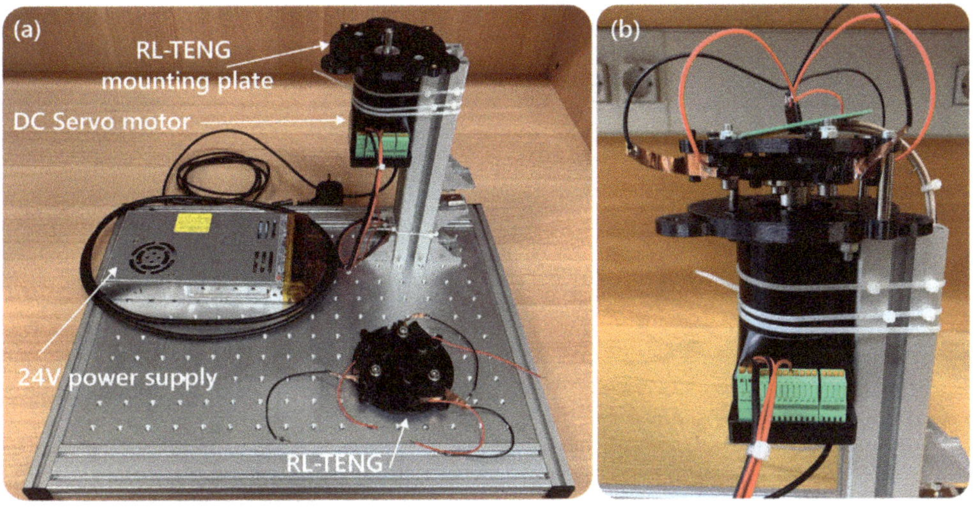

Figure 3. (**a**) The experimental setup used in the electrical characterization of the RL-TENG; (**b**) the RL-TENG mounted on top of the servo motor, aligned with the axis of rotation.

2.3. Electrical Measurements

Four types of measurements were performed: (a) time-dependent output voltage (transient), (b) capacitor charging, (c) open circuit voltage (V_{oc}) and (d) short circuit current (Isc). For the transient measurements, the RL-TENG was connected to an InfiniiVision DSO7104A oscilloscope (Agilent Technologies, Santa Clara, CA, USA) through a 10× probe, terminated at 1 MΩ input impedance. For the capacitor charging measurements, a full, quad bridge rectifier circuit was used. The capacitor voltage was monitored by a Keithley 617 electrometer (Keithley, Solon, OH, USA), with an input resistance of 200 TΩ. For the measurement of the short circuit current Isc, a Stanford Research Systems SR570 (Stanford Research Systems, Sunnyvale, CA, USA) low-noise current preamplifier was used. For the estimation of the power as a function of an external load, the methodology developed by Jayasvasti et al. [19] was adapted. In brief, the methodology consists of adding a small resistance (1 MΩ) in parallel with the oscilloscope probe (input resistance: 10 MΩ) to reduce the current flowing through the measuring device. It was demonstrated that this methodology is more accurate for the extraction of the power output compared to the typical method of measuring the voltage drop across the load. From the same measurements, the open circuit voltage can be estimated. All experiments were performed at 25 ± 2 °C and 25 ± 3%RH (relative humidity).

3. Results and Discussion

3.1. Operating Principle of Tribogenerators in Contact-Separation Mode

As we have described above, the RL-TENG is based on the transformation of rotational motion to linear motion. The output voltage of RL-TENG is a combination of the output voltages of each triboelectric pair T1, T2 and T3 that operate in contact-separation mode. In this operation mode, the movable part (active TENG in Figure 4a) is moving in a vertical direction relative to the stationary part (base TENG in Figure 4a). When the two triboelectric surfaces (Kapton and ZnO-on-Kapton) are in contact, charges with different polarities appear on the two materials, due to contact electrification. The type of charge that appears on each material depends on the electronegativity of the two materials. In our case, the surface of ZnO is charged positively, and the surface of Kapton is charged negatively, due to the differences in the electronegativities of the two materials. As the two triboelectric surfaces start moving away from each other, a charge build up will appear at the electrodes, located at the backside of the triboelectric materials, and a current will flow in the external circuit. A detailed theoretical analysis of the triboelectric nanogenerators in contact-separation mode has been presented by Niu et al. [20] al and Dharmasena et al. [21].

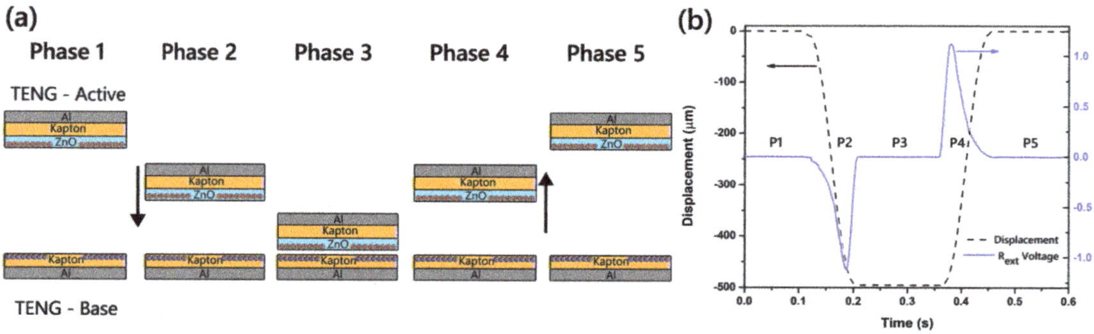

Figure 4. (**a**) Different phases of a triboelectric generator in contact-separation mode; (**b**) simulation results of the modeled tribogenerator, showing the output voltage during each phase.

Figure 4a illustrates schematically the TENG operation in contact-separation mode. Assuming that there are already charges developed on the active and the base surfaces of the tribogenerator, we can distinguish five different phases during an operation cycle.

- Phase 1: The active area is stationary at its balance position. No current flows through the external circuit.
- Phase 2: The active area begins moving until it reaches the position of maximum displacement. Current flows to the external circuit due to electrostatic induction. The current flow takes place for the duration of the movement of the active area.
- Phase 3: The active area remains still at the maximum displacement for a short amount of time. No current flows through the external circuit.
- Phase 4: The active area starts moving backwards towards its balance position. Current flows to the external circuit due to electrostatic induction, but in the opposite direction relative to phase P2.
- Phase 5: The active area reaches its maximum position and stops moving. No current flows through the external circuit.

During all the phases described above, the reference surface (TENG-Base) remains in the same position.

To analyze the electrical performance of the tribogenerator during these phases, we have modeled and simulated the device in COMSOL 5.0. For the simulation, the following parameters were used: thickness of Kapton: 50 μm; relative permittivity of Kapton: 3; and surface charge density: 50 $\mu C/m^2$. ZnO layer was omitted in the simulation since it is several orders of magnitude smaller compared to Kapton. Figure 4b shows the output voltage of the tribogenerator, connected to an external load resistance R_{ext} (=1 MΩ), during the various phases. As expected, voltage develops between the electrodes, due to electrostatic induction, when the two electrodes are in relative motion.

3.2. Characterization of the RL-TENG

The triboelectric pairs were measured for various rotational velocities using the setup shown in Figure 1. Figure 5 shows the transient signal of triboelectric pair T1 as a function of time for rotational velocities ranging from 60 rpm to 300 rpm. We notice that, as the rotational velocity increases, the triboelectric signal increases. In fact, the output voltage signal increases from 0.2 V for 60 rpm to 2 V for 300 rpm.

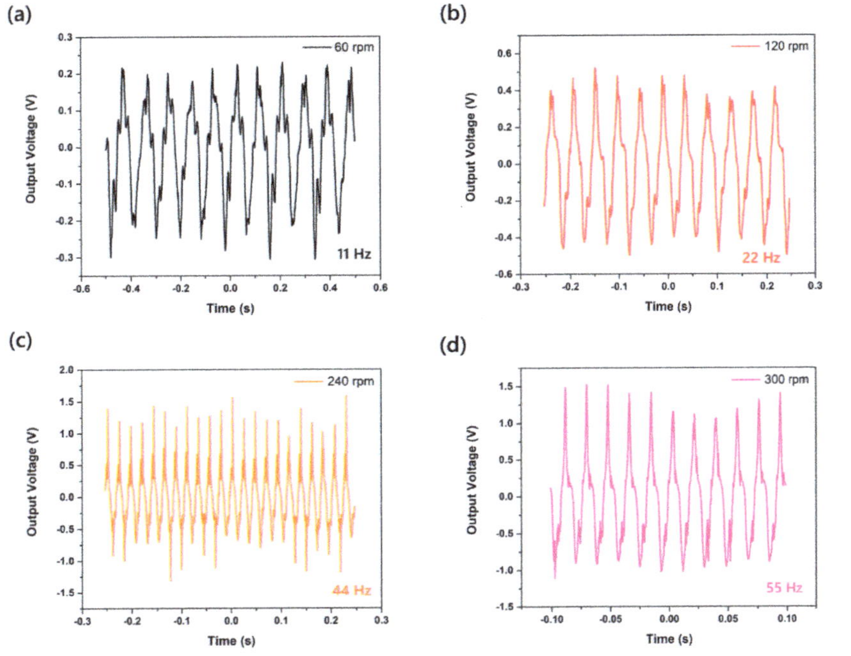

Figure 5. Output signal of T1 triboelectric pair of RL-TENG for various rotational velocities: (**a**) 60 rpm, (**b**) 120 rpm, (**c**) 240 rpm and (**d**) 300 rpm. (Output voltage frequency is a multiple of the number of cams located on each disk, in this case 11).

Figure 6a,b show the output current and voltage of triboelectric pair T1 as a function of the external load for various rotational velocities. The short circuit current increases from 133 nA to 267 nA, and the open circuit voltage increases from 2.24 V to 3 V as the rotational velocity of the RLTENG increases from 100 rpm to 300 rpm. Figure 6c shows the output power as a function of the rotational speed. The output power increases from 0.1 μW to 0.23 μW for the same rotational velocities. Maximum output power is obtained when the resistance value of the external load is 20 MΩ. Similar results are obtained from triboelectric pairs T2 and T3, although small differences are observed due to variations in the manufacturing and assembly process.

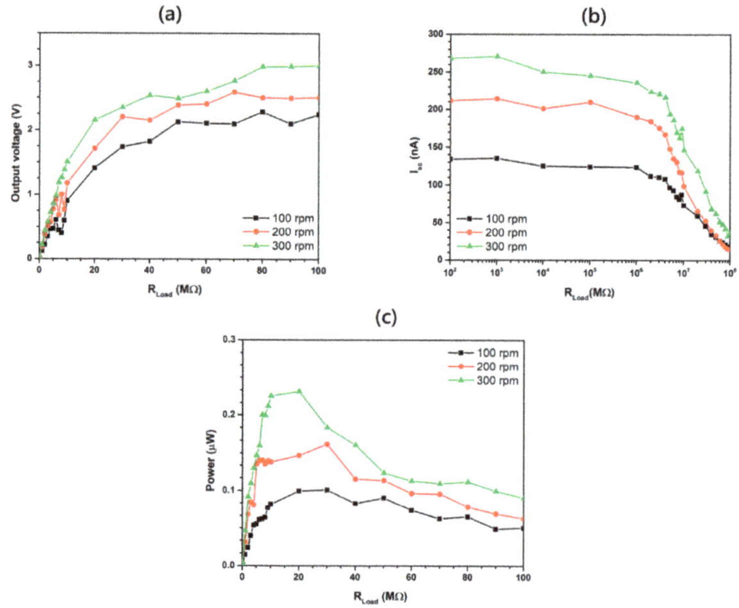

Figure 6. (**a**) Output voltage, (**b**) output current and (**c**) output power of the T1 triboelectric pair of the RL-TENG as a function of the external load for rotational velocities ranging from 100 to 300 rpm.

3.3. Connectivity of the Triboelectric Pairs

Since the RL-TENG consists of three triboelectric pairs, a question that has been raised is how to connect them in order to obtain the maximum power from the device. One obvious approach is to connect each triboelectric pair with a bridge rectifier and then connect the outputs together either in series or in parallel. Such a solution has been proven to be an efficient way for scavenging the total amount of generated energy [17]. However, this means that we need to use three separate bridge rectifiers, increasing the complexity of the electronic components required for device operation. To avoid that, we have implemented an alternative approach. We have investigated the influence of the total stored energy for two different cams: (a) During the rotation of the first cam (disk A, Figure 7a), the tribogenerators are moving in phase as can be seen from Figure 7b, where we notice that the triboelectric signal of T1 and T3 are almost identical. (b) During the rotation of the second cam (disk B, Figure 7c), the tribogenerators are out of phase, and only one triboelectric pair is set in motion during a specific time interval, as can be seen in Figure 7d, where we notice that T1 is active when T3 is inactive. As a metric for the comparison between the two methods, we use the energy stored in a capacitor using a single bridge rectifier during charging.

Figure 8 shows the energy stored at the capacitor as a function of time for triboelectric pairs T1 and T3 for two rotational velocities, i.e., 60 rpm (Figure 8a) and 300 rpm T3 (Figure 8b), when disk B is used. In the same figures, we see the energy stored when the triboelectric pairs are connected in parallel prior to the rectifier. We see that the energy of the combined triboelectric pairs is higher compared to each individual pair. Assuming that the total energy that should be stored in the capacitor is the sum of the energies that are stored in each triboelectric pair, from the analysis of the data, we see that for low rotational velocities, the stored energy is 67% of the total energy. However, as the rotational velocity increases, the amount of stored energy is reduced reaching 56% at 300 rpm. This energy loss is probably due to the random signal generated from the triboelectric pairs due to vibrations caused during device operation. These signals interfere in a destructive way, cancelling out each other and reducing the amount of stored energy. Similar results were obtained when disk A was used.

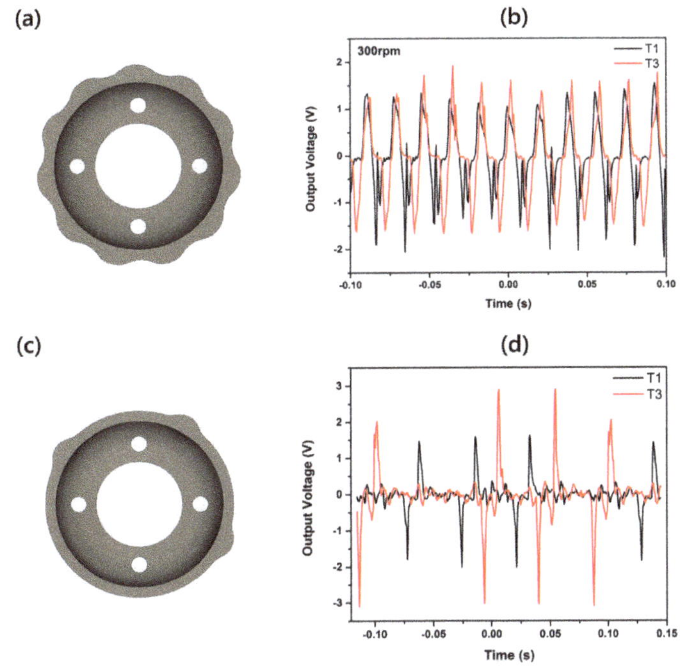

Figure 7. (**a**) Disk A for in-phase operation of triboelectric pairs, (**b**) output voltage of T1 and T3 for disk A, (**c**) disk B for out-of-phase operation of triboelectric pairs and (**d**) output voltage of T1 and T3 for disk B.

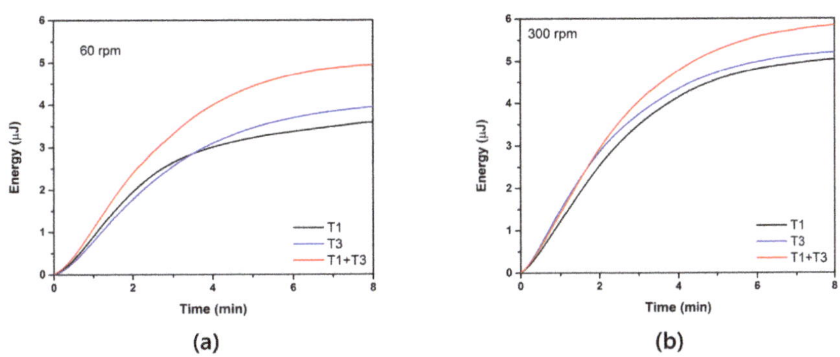

Figure 8. Energy stored in the capacitor as a function of time after charging for rotational velocities (**a**) 60 rpm and (**b**) 300 rpm.

3.4. Optimization of Performance Characteristics

From the analysis of the results, we see that the electrical performance of the RL-TENG is inferior to what we would anticipate compared to the results that we obtained for the Kapton/ZnO layers in previous experiments. When the same triboelectric surfaces of area 2×2 cm^2 were evaluated [18], the output voltage reached about 20 V, the short circuit current 12 μA, the open circuit voltage 105 V and the maximum output power 350 μW, which is much higher compared to the results obtained for a single triboelectric pair. To some extent, this is expected since the surface area of the triboelectric areas T1, T2 and T3 is reduced compared to the previous ones by 2.4 times (1.67 cm^2 compared to 4 cm^2).

There are also other parameters that could influence the triboelectric performance that are related to the operational principle of the triboelectric generator. Being a transient phenomenon, the triboelectric signal will depend strongly on the relative velocity between the two surfaces. Taghavi et al. [22] showed that the relative velocity of the triboelectric surfaces affects the instantaneous signal as well as the average power of TENGs. They also provided evidence that the separation phase of the two triboelectric surfaces could also impact the electrical performance of the tribogenerator. Another parameter that affects the triboelectric signal is the contact force that develops between the two triboelectric surfaces. Vasandani et al. [23] studied theoretically and experimentally the influence of the contact force of the triboelectric signal TENG using a PDMS film with microdome patterns. As a reference electrode, a PET film was used. The triboelectric signal was measured against a PET reference surface. They demonstrated that, by changing the contact force from 4.4 N to 5.1 N, the triboelectric signal increased by 100%.

To verify the influence of the above-mentioned parameters (surface area and transition time/force) on the triboelectric signal, we have fabricated a part of a more robust prototype tribogenerator. In this case, the surface area of the prototype was the same as that of the test triboelectric generators (Figure 9a). In addition, we have enhanced the mechanical stability of the structure and used a stronger return spring. Typical results are shown in Figure 9b. We see that the maximum voltage per tap is 12 V for a rotational velocity of 350 rpm. Moreover, the maximum power per tap is 10 times more compared to that of the triboelectric pairs used in the RL-TENG. Our results indicate that, with a more robust structure and with larger triboelectric contact areas, significantly enhanced performance can be obtained for the RL-TENG.

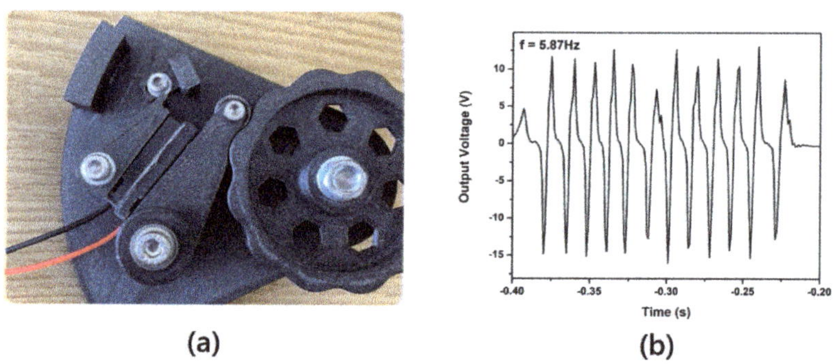

Figure 9. (**a**) Part of a more robust triboelectric generator and (**b**) output voltage as a function of time at 350 rpm.

3.5. Application for Harvesting Wind Energy

The RL-TENG was adapted to be a wind-harvesting device as shown in Figure 10a. The wind-harvesting device is composed of (a) a case that encloses the RL-TENG for mechanical support, providing a way for the vertical shaft to spin freely using deep-groove ball bearings, (b) the vertical shaft where the cam is mounted using an aluminum adapter and (c) three cups attached to the vertical shaft. As the wind blows, the cups create rotational torque, spinning the vertical shaft and as consequence the RL-TENG, as shown in Figure 10b (Video S1). For measuring the air velocity, a hand-held anemometer was utilized (UT363BT, UNI-T®, Dongguan, China) that provided real-time and data logging capabilities via Bluetooth®, displayed at a smartphone app (iENV2.0, UNI-T®).

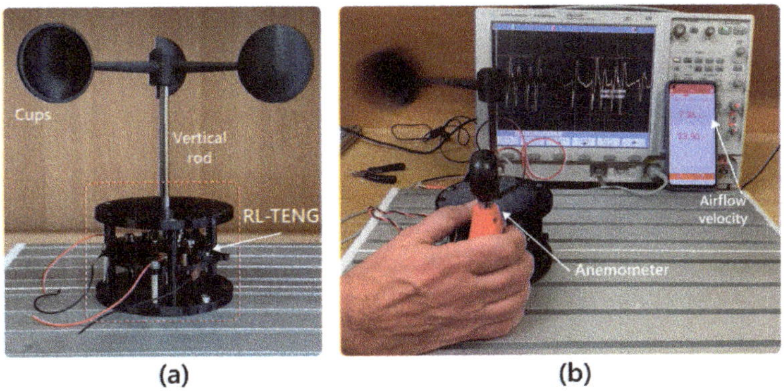

Figure 10. (**a**) RL-TENG attached to the wind-harvesting device and (**b**) setup for the wind-harvesting experiments.

Figure 11 shows the power generated by each triboelectric pair as a function of the wind velocity. We see that the different triboelectric pairs do not have the same performance. This is normal since the different pairs were prepared, to some extent, manually and not with a standardized manufacturing and assembly process. Moreover, the rotational speed of the device as a function of the wind velocity is presented in Figure 11b. We see that there is a very good match between the triboelectric pairs for which the measurements were not performed simultaneously.

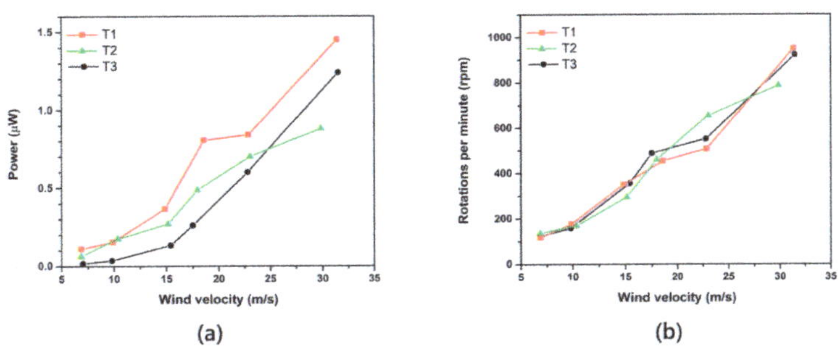

Figure 11. (**a**) Maximum power delivered by each triboelectric pair as a function of the wind velocity. (**b**) Rotational velocity of the RL-TENG as a function of wind velocity.

3.6. Advantages of the Proposed RL-TENG

Compared to the already available rotational tribogenerators, RL-TENG presents several advantages including the following:

1. Since RL-TENG operates in contact-separation mode, material wear is minimized compared to conventional rotational tribogenerators where continuous contact (in sliding mode) is required at high rotational velocities. It is thus expected that the operational time of the proposed RL-TENG will be significantly higher compared to that of rotational devices.
2. Since there is no sliding between the two triboelectric surfaces, no high temperatures are developed because of friction.
3. Due to its design, several material combinations can be easily used as triboelectric materials. By selecting a material combination that is further away in the triboelectric series, the triboelectric signal can be significantly increased.
4. It can be scaled up by increasing its size both radially as well as vertically to increase the surface area and thus improve the tribogenerator's performance. Another way to

increase the surface would be to implement origami-inspired structures which have been demonstrated to significantly increase triboelectric performance [24,25].

5. Another way to enhance the output performance of the RL-TENG is to use hybrid energy-harvesting approaches, such as combining piezoelectric, triboelectric and electromagnetic generators in a single device [25]. The advantage of the proposed design is that it can be easily adjusted to exploit all performance improvements that have been demonstrated for contact-separation mode tribogenerators [26].
6. It can be assembled in a modular way.
7. It can be fabricated using simple mechanical parts with reduced cost, increased lifetime and simple maintenance.

4. Conclusions

In this study, we have presented a rotational-based triboelectric generator based on a novel design. The proposed RL-TENG consists of three triboelectric pairs that operate in contact-separation mode, providing an alternative power generation mechanism in contrast to conventional rotational tribogenerators, mitigating the inherent problems found in common rotating tribogenerators such as the development of high temperature and increased material wear due to friction. The experimental results present the output characteristics of the RL-TENG, showing a clear dependence of open-circuit voltage, short-circuit current and output power for a wide range of rational speeds, with a maximum power of 0.23 μW at 20 MΩ and 300 rpm. The effect of connectivity between the triboelectric pairs was also explored showing that out-of-phase operation (disk B) does not require separate rectification for every triboelectric pair, reducing complexity and offering increased energy storage by connecting the pairs in parallel. Finally, a wind-harvesting device was developed, which presents the potential of the proposed RL-TENG in harvesting applications utilizing pure rotational motion and is able to provide power up to 1000 rpm. In conclusion, the proposed design provides flexibility, can be easily adapted to a variety of harvesting applications, exhibits ease of maintenance, has an ability to operate on a wide range of rotational velocities with reduced wear and can be optimized for the specific application by tailoring the relative contact velocity, contact force and contact surface between the triboelectric surfaces.

Supplementary Materials: The following supporting information can be downloaded at: https://www.mdpi.com/article/10.3390/app14062396/s1, Video S1: RL-TENG wind experiments.

Author Contributions: Conceptualization, A.B., A.S. and C.T.; methodology, A.B. and A.S.; validation, A.B. and C.T.; writing—original draft preparation, A.B.; writing—review and editing, all authors; supervision, C.T.; funding acquisition, C.T. All authors have read and agreed to the published version of the manuscript.

Funding: This research has been co-financed by the European Union and Greek national funds through the Operational Program Competitiveness, Entrepreneurship and Innovation, under the call RESEARCH-CREATE-INNOVATE (EFOS, project code: T2EΔK-00350).

Data Availability Statement: The raw data supporting the conclusions of this article will be made available by the corresponding authors on request.

Acknowledgments: The author acknowledge E. Hinari for the preparation of ZnO-based triboelectric surfaces.

Conflicts of Interest: The authors declare no conflicts of interest.

References

1. Choi, D.; Lee, Y.; Lin, Z.H.; Cho, S.; Kim, M.; Ao, C.K.; Soh, S.; Sohn, C.; Jeong, C.K.; Lee, J.; et al. Recent Advances in Triboelectric Nanogenerators: From Technological Progress to Commercial Applications. *ACS Nano* **2023**, *17*, 11087–11219. [CrossRef] [PubMed]
2. Yu, Y.; Li, H.; Zhao, D.; Gao, Q.; Li, X.; Wang, J.; Wang, Z.L.; Cheng, T. Material's Selection Rules for High Performance Triboelectric Nanogenerators. *Mater. Today* **2023**, *64*, 61–71. [CrossRef]

3. Delgado-Alvarado, E.; Elvira-Hernández, E.A.; Hernández-Hernández, J.; Huerta-Chua, J.; Vázquez-Leal, H.; Martínez-Castillo, J.; García-Ramírez, P.J.; Herrera-May, A.L. Recent Progress of Nanogenerators for Green Energy Harvesting: Performance, Applications, and Challenges. *Nanomaterials* **2022**, *12*, 2549. [CrossRef] [PubMed]
4. Zhang, R.; Örtegren, J.; Hummelgård, M.; Olsen, M.; Andersson, H.; Olin, H. A Review of the Advances in Composites/Nanocomposites for Triboelectric Nanogenerators. *Nanotechnology* **2022**, *33*, 212003. [CrossRef] [PubMed]
5. Chao, S.; Ouyang, H.; Jiang, D.; Fan, Y.; Li, Z. Triboelectric Nanogenerator Based on Degradable Materials. *EcoMat* **2021**, *3*, e12072. [CrossRef]
6. Lin, L.; Wang, S.; Xie, Y.; Jing, Q.; Niu, S.; Hu, Y.; Wang, Z.L. Segmentally Structured Disk Triboelectric Nanogenerator for Harvesting Rotational Mechanical Energy. *Nano Lett.* **2013**, *13*, 2916–2923. [CrossRef] [PubMed]
7. Lin, L.; Wang, S.; Niu, S.; Liu, C.; Xie, Y.; Wang, Z.L. Noncontact Free-Rotating Disk Triboelectric Nanogenerator as a Sustainable Energy Harvester and Self-Powered Mechanical Sensor. *ACS Appl. Mater. Interfaces* **2014**, *6*, 3031–3038. [CrossRef]
8. Zhang, C.; Zhou, T.; Tang, W.; Han, C.; Zhang, L.; Wang, Z.L. Rotating-Disk-Based Direct-Current Triboelectric Nanogenerator. *Adv. Energy Mater.* **2014**, *4*, 1301798. [CrossRef]
9. Bai, P.; Zhu, G.; Liu, Y.; Chen, J.; Jing, Q.; Yang, W.; Ma, J.; Zhang, G.; Wang, Z.L. Cylindrical Rotating Triboelectric Nanogenerator. *ACS Nano* **2013**, *7*, 6361–6366. [CrossRef]
10. Zhang, N.; Qin, C.; Feng, T.; Li, J.; Yang, Z.; Sun, X.; Liang, E.; Mao, Y.; Wang, X. Non-Contact Cylindrical Rotating Triboelectric Nanogenerator for Harvesting Kinetic Energy from Hydraulics. *Nano Res.* **2020**, *13*, 1903–1907. [CrossRef]
11. Qu, Z.; Huang, M.; Dai, R.; An, Y.; Chen, C.; Nie, G.; Wang, X.; Zhang, Y.; Yin, W. Using Non-Contact Eccentric Nanogenerator to Collect Energy Continuously under Periodic Vibration. *Nano Energy* **2021**, *87*, 106159. [CrossRef]
12. Yang, H.; Liu, W.; Xi, Y.; Lai, M.; Guo, H.; Liu, G.; Wang, M.; Li, T.; Ji, X.; Li, X. Rolling Friction Contact-Separation Mode Hybrid Triboelectric Nanogenerator for Mechanical Energy Harvesting and Self-Powered Multifunctional Sensors. *Nano Energy* **2018**, *47*, 539–546. [CrossRef]
13. Dong, X.; Liu, Z.; Yang, P.; Chen, X. Harvesting Wind Energy Based on Triboelectric Nanogenerators. *Nanoenergy Adv.* **2022**, *2*, 245–270. [CrossRef]
14. Park, S.; Ryu, H.; Park, S.; Hong, H.; Jung, H.Y.; Park, J.J. Rotating Triboelectric Generator Using Sliding Contact and Noncontact from 1D Fiber Friction. *Nano Energy* **2017**, *33*, 184–194. [CrossRef]
15. Xin, Y.; Du, T.; Liu, C.; Hu, Z.; Sun, P.; Xu, M. A Ring-Type Triboelectric Nanogenerator for Rotational Mechanical Energy Harvesting and Self-Powered Rotational Speed Sensing. *Micromachines* **2022**, *13*, 556. [CrossRef] [PubMed]
16. Segkos, A.; Tsamis, C. Rotating Triboelectric Nanogenerators for Energy Harvesting and Their Applications. *Nanoenergy Adv.* **2023**, *3*, 170–219. [CrossRef]
17. Mo, J.; Zhang, C.; Lu, Y.; Liu, Y.; Zhang, N.; Wang, S.; Nie, S. Radial Piston Triboelectric Nanogenerator-Enhanced Cellulose Fiber Air Filter for Self-Powered Particulate Matter Removal. *Nano Energy* **2020**, *78*, 105357. [CrossRef]
18. Hinari, E.; Bardakas, A.; Tsamis, C. Enhancement of the performance of ZnO-based triboelectric generators by nitrogen doping. *Energy Harvest. Syst.* **2024**. *submitted*.
19. Jayasvasti, S.; Thainiramit, P.; Yingyong, P.; Isarakorn, D. Technique for Measuring Power across High Resistive Load of Triboelectric Energy Harvester. *Micromachines* **2021**, *12*, 766. [CrossRef]
20. Niu, S.; Wang, S.; Lin, L.; Liu, Y.; Zhou, Y.S.; Hu, Y.; Wang, Z.L. Theoretical Study of Contact-Mode Triboelectric Nanogenerators as an Effective Power Source. *Energy Environ. Sci.* **2013**, *6*, 3576–3583. [CrossRef]
21. Dharmasena, R.D.I.G.; Jayawardena, K.D.G.I.; Mills, C.A.; Dorey, R.A.; Silva, S.R.P. A Unified Theoretical Model for Triboelectric Nanogenerators. *Nano Energy* **2018**, *48*, 391–400. [CrossRef]
22. Taghavi, M.; Beccai, L. A Contact-Key Triboelectric Nanogenerator: Theoretical and Experimental Study on Motion Speed Influence. *Nano Energy* **2015**, *18*, 283–292. [CrossRef]
23. Vasandani, P.; Mao, Z.H.; Jia, W.; Sun, M. Relationship between Triboelectric Charge and Contact Force for Two Triboelectric Layers. *J. Electrostat.* **2017**, *90*, 147–152. [CrossRef]
24. Hu, G.; Zhao, C.; Yang, Y.; Li, X.; Liang, J. Triboelectric Energy Harvesting Using an Origami-Inspired Structure. *Appl. Energy* **2022**, *306*, 118037. [CrossRef]
25. Pongampai, S.; Pakawanit, P.; Charoonsuk, T.; Hajra, S.; Kim, H.J.; Vittayakorn, N. Design and Optimization of Miura-Origami-Inspired Structure for High-Performance Self-Charging Hybrid Nanogenerator. *J. Sci. Adv. Mater. Devices* **2023**, *8*, 100618. [CrossRef]
26. Xie, L.; Zhai, N.; Liu, Y.; Wen, Z.; Sun, X. Hybrid Triboelectric Nanogenerators: From Energy Complementation to Integration. *Research* **2021**, *2021*, 9143762. [CrossRef]

applied sciences

Article

Design of a Mass Air Flow Sensor Employing Additive Manufacturing and Standard Airfoil Geometry

Dimitrios-Nikolaos Pagonis [1,*], Vasiliki Benaki [1], Grigoris Kaltsas [2] and Antonios Pagonis [3]

1 Naval Architecture Department, University of West Attica, 122 43 Athens, Greece; mscna1812@uniwa.gr
2 Electrical & Electronic Engineering Department, University of West Attica, 122 43 Athens, Greece; gkaltsas@uniwa.gr
3 Wärtsilä Greece S.A., 183 44 Moschato, Greece; antonis.pagonis@wartsila.com
* Correspondence: D.N.Pagonis@uniwa.gr

Abstract: This work concerns the design, fabrication, and preliminary characterization of a novel sensor for determining the air intake of low and medium power internal combustion engines employed at various applications in the marine industry. The novelty of the presented sensor focuses on the fabrication process, which is based on additive manufacturing combined with PCB technology, and the design of the sensing elements housing geometry, which is derived through suitable CFD simulations and is based on standard airfoil geometry. The proposed process enables low-cost, fast fabrication, effective thermal isolation, and facile electrical interconnection to the corresponding circuitry of the sensor. For initial characterization purposes, the prototype device was integrated into a DIESEL engine testbed while a commercially available mass air flow sensor was employed as a reference; the proper functionality of the developed prototype has been validated. Key features of the proposed device are low-cost, fast on-site manufacturing of the device, robustness, and simplicity, suggesting numerous potential applications in marine engineering.

Keywords: engine air intake sensor; additive manufacturing; PCB technology; hot-wire; MAF sensor; DIESEL engine; internal combustion engine; NACA airfoils

Citation: Pagonis, D.-N.; Benaki, V.; Kaltsas, G.; Pagonis, A. Design of a Mass Air Flow Sensor Employing Additive Manufacturing and Standard Airfoil Geometry. *Appl. Sci.* **2021**, *11*, 11579. https://doi.org/10.3390/app112411579

Academic Editor: Gabriele Di Blasio

Received: 4 November 2021
Accepted: 1 December 2021
Published: 7 December 2021

Publisher's Note: MDPI stays neutral with regard to jurisdictional claims in published maps and institutional affiliations.

1. Introduction

Marine industry employs a wide range of internal combustion (IC) engines for various applications; while their role as the prime mover of a vessel is undoubtedly the most important, it is broadly accepted that IC engines have a fundamental role in ship operations in general, making their proper operation crucial; for example, a typical Ro-Ro vessel has four medium power 4-stroke engines installed onboard, which are required for the operation of the corresponding generator-sets [1]; a possible failure even only at one of the installed IC engines will immediately result in a safety compromise, according to the corresponding safety regulations [2], with severe consequences.

Mass Air Flow (MAF) sensors are measuring devices integrated to IC engines. Their role is to measure the air intake during the operation of the engine in order to consequently efficiently adjust the amount of fuel supplied during combustion. The majority of the MAF sensors currently employed are thermal flow microsensors, since, during the last two decades, microsystem technologies have greatly contributed to the reduction of the cost for this specific type of sensor [3,4]. Printed Circuit Board (PCB) technology is a new sensor fabrication technology that combines printed circuit and MEMS technology and addresses some drawbacks of the standard microelectronic fabrication process [5]. Briefly, the sensing elements of a sensor developed with PCB technology are inherently connected to the macroworld, overcoming the necessity for performing wire bonding (a delicate process for the creation of the necessary electrical pathways between microscale sensing elements and the corresponding macroworld circuitry), with immediate benefits in device complexity, mechanical reliability, process time, and cost [5,6].

Airfoils find many applications in the industry, including, but not limited to, aeronautics, naval, and automotive engineering. An airfoil is a streamlined body, capable of generating significantly more lift than drag, which provides much smoother flow than non-streamlined bodies, keeping airflow attached to the surface rather than breaking free and causing turbulence. Numerous airfoils can be designed for different applications [7]. The most renown are the NACA series developed by the National Advisory Committee for Aeronautics airfoils during the late 1920s and into the 1930s [8,9].

In this paper, we present a novel air intake sensor, which is based on additive manufacturing and PCB technology. In more detail, the sensor's sensing element was fabricated, utilizing only standard PCB fabrication techniques and commercially available sensing components, while 3D printing technology was employed for fabricating an appropriate air flow, shaping housing. The housing's geometry was derived through appropriate Computational Fluid Dynamic (CFD) simulations-in order to investigate the effect of the main design parameters, employing SimScale Online CAE Simulation Platform [10]. The contribution of additive manufacturing is of paramount importance, as it significantly minimizes the design constrains during manufacturing [11], allowing implementation of the derived topology as per the optimum deduced from CFD simulations.

As a general remark, the obtained optimal geometry efficiently addresses the measuring challenges present on the intake flow of an IC engine; it is originated in a flow-bypass topology studied in previous work [12] combined with standard airfoil geometry; the airfoil geometry implementation drastically improves the bypass flow profile compared to a previous study [12]. It is the authors' belief that no research work is currently available focusing on the employment of standard airfoil geometries in sensor housing design through appropriate CFD simulations, nor on employing additive manufacturing for delivering the optimal housing; we should note that its final geometry is quite challenging but it results in a lightweight and robust housing that can be manufactured in a relevant short time, even on-site, employing a commercially available 3D printing device. This contributes to reduced manufacturing cost and space saving, especially onboard a vessel, while at the same time, promotes the general objective of minimizing the financial and environmental impact of transportation [11,13].

2. Materials and Methods

The required air flow during the operation of an Internal Combustion (IC) engine can vary depending on its operating conditions (i.e., the load applied); a typical air flow intake for a low power industrial engine is 0 to 615 kg/h [14], which is equal to about 0 to 8806 SLPM (25 °C at 1 bar), a noticeably large range. In reference [15], the authors successfully demonstrated a flow sensor based on PCB low-cost technology that was able to measure the air intake of a working IC engine up to 180 kg/h under the hot-wire principle of operation for constant current (CC) and constant resistance (CR) mode.

However, saturation in the sensor's signal was noticed for flows above the specific value of air intake, limiting the measuring range of the device. Furthermore, during IC engine operation, the intake flow was not constant but pulsating, due to the engine's principle of operation [16]; this effect was not taken into consideration in the specific PCB-based sensing device.

In reference [12], an engine air intake sensor based on PCB and 3D printing technology was presented by the authors. Briefly, the sensor's sensing element was fabricated utilizing standard PCB fabrication techniques and commercially available components, while 3D printing technology was employed for fabricating an appropriate air flow shaping housing with an integrated flow-bypass pathway as a first attempt to address the measuring challenges present on the intake flow (Figure 1). The functionality of the device has been well demonstrated for a range of 0 to 260 kg/h (0 to 3723 SLPM), which is increased by 44% compared to the increase initially obtained in [15]. However, we should note that for flow rate values approximately above 260 kg/h (3723 SLPM), saturation in the sensor's signal was still present. A series of appropriate CFD simulations concerning the developed flow profile inside the bypass pathway revealed a significant circulating flow at the end of the pathway, near the sensing element's mounting location; furthermore, the sudden expansion at the inlet of the employed bypass design can lead to boundary layer separation with a reattachment point that can be situated near the sensing element area as the flow rate increases [17]. As mentioned by the authors, both of these geometry-induced effects restricted the device's measuring range.

(a)

(b)

Figure 1. Airfoils: (**a**) streamlines over a typical airfoil [9]; (**b**) NACA 0006-6121 family of air foils [8].

The current work focuses on further optimizing the engine air intake sensor presented in [12] and drastically increasing its measuring range. In more detail, an optimal bypass pathway design, employing inlet and outlet designs, which are based on airfoil geometries (i.e., National Advisory Committee for Aeronautics airfoils (NACA) [18], Joukowski airfoils [19]), was derived through suitable CFD simulations. As we present in the later sections, by appropriately shaping the flow profile at both inlet and outlet ports of the bypass, boundary layer separation and circulating flow were minimized significantly, expanding the sensor's measuring range.

A. Initial Housing Design Concepts Investigated

The initial housing design (Figure 2) was re-designed regarding both the inlet and outlet of the bypass section geometries, along with the necessary orifice for appropriately decreasing the bypass inflow and the outflow topology for constraining the back flow induced during IC engine operation. Three test design concepts (i.e., test designs 1, 2, and 3) were investigated through CFD simulations. In more detail, a series of simulations were performed, assuming a maximum engine air intake of 1000 kg/h (14076 SLPM), which corresponds to an air velocity of 29.87 m/s at the main inflow pathway of the housing, taking into consideration its cross-section area, equal to 78.54 cm^2. For the specific value of air velocity entering the bypass' inlet, with a much smaller cross-section area of 3.14 cm^2, the corresponding Reynolds number for the flow inside the Bypass pathway can be easily estimated at 38,248 pointing to turbulent flow.

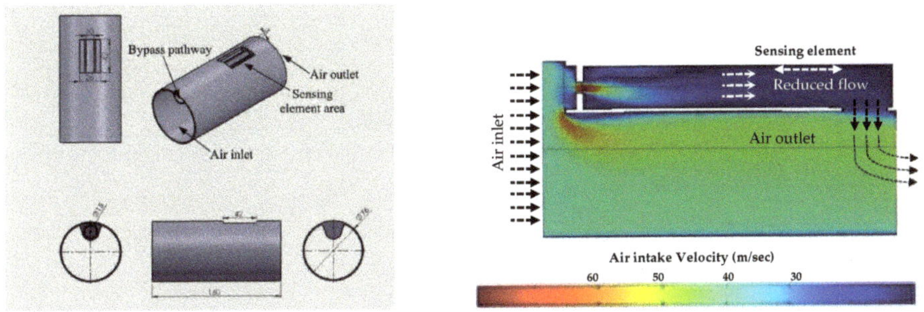

(**a**) (**b**)

Figure 2. (**a**) Schematic of the housing geometry; (**b**) corresponding simulation results for the developed flow [12].

Considering the above remark, the turbulence model k-omega (k−ω) SST [20] was employed for all the simulations performed, since the developed flow profile inside the Bypass region should be determined as accurate as possible both in the near-wall region (where the sensing element is situated) and in the fully turbulent region, far from the wall. Each geometry and the corresponding characteristic simulation results are presented in Figures 3–5, while the main simulation results and the overall outcome for each test design are summarized in Table 1. Note that a brief presentation follows for test designs 1 and 2, while a more thorough description is presented in the next section for the third test design.

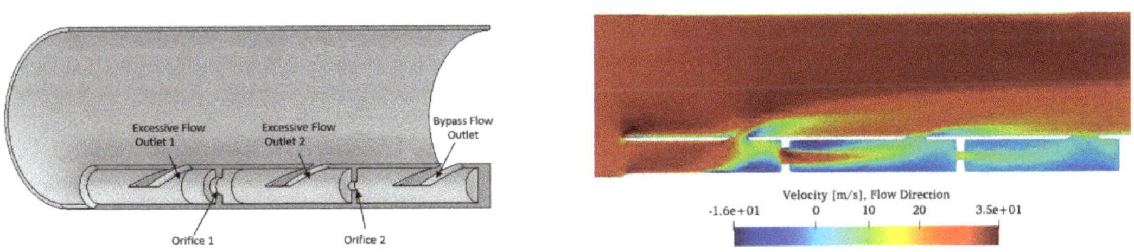

(**a**) (**b**)

Figure 3. (**a**) Test design concept (1) 3D, mid-section; (**b**) rest design concept (1) simulation result.

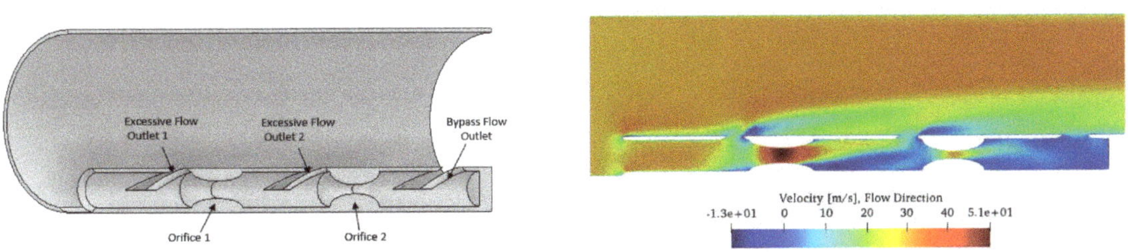

(**a**) (**b**)

Figure 4. (**a**) Test design concept (2) 3D, mid-section; (**b**) test design concept (2) simulation result.

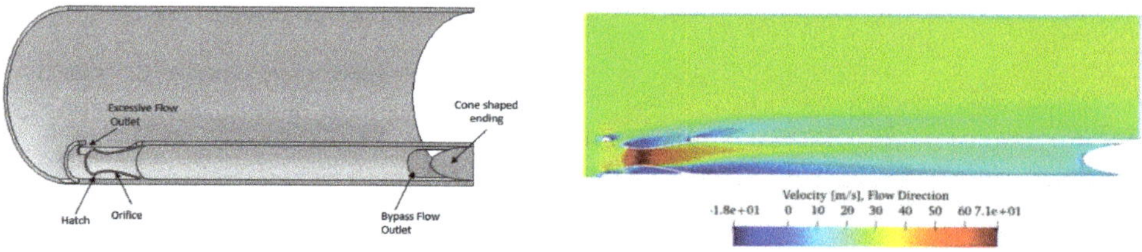

(a) (b)

Figure 5. (**a**) Test design concept (3) 3D, mid-section; (**b**) test design concept (3) simulation result.

Table 1. Test design concepts.

Design Concept	Main Simulation Results	Overall Outcome
Test Design concept (1) and (2)	Significant velocity reduction	Rejected
	Unsteady velocity profile across the entire bypass pathway	
	High circulating flow present	
Test Design concept (3)	Significant velocity reduction	Accepted
	Stable velocity profile obtained	
	Minimized circulating flow	

As we can see in Figure 3, the first test design was based on a topology similar to the one employed at reference [12]; the basic difference is that the reduction of the incoming bypass flow was performed in two consecutive steps instead of employing a single orifice with a drastically reduced diameter (compared to the bypass tube), which lead to a sudden increase of inflow velocity locally. In addition, in order to minimize flow circulation throughout the bypass, an outlet was added before each orifice for the diversion of excessive flow outside the bypass pathway. The topology of the second test design was similar (Figure 4), decreasing the incoming flow in two steps; the main difference in this case was the orifice topology. In more detail, the concept of a "curved orifice" was introduced (imitating the pipe eccentric reducers design/shape [21]) to allow the orifice diameter to be gradually reduced and then re-expanded, aiming for minimum flow circulation around the orifice area.

Summarizing the obtained CFD simulation results for test designs (1) and (2), we noticed that a high circulating flow was present (dark-blue areas at Figures 3b and 4b), while the velocity varied significantly throughout the bypass pathway.

In the third test design (Figure 5), considering their important contribution to the automobile industry [22], airfoil geometries (NACA and Joukowski) were employed for the basis of the inlet orifice's geometry to obtain a stable velocity profile near the sensing area with a significant reduction of speed. Thus, several standard airfoil types were tested through appropriate CFD simulations based on the literature [22–24] and the subject geometry constrains.

We should point out that, apart from employing an airfoil-based geometry for the inlet orifice, a "hatch-orifice pair" topology was introduced at the inflow, as presented in Figure 6. In more detail, instead of adding an outlet prior to the orifice, as in the previous two test designs, a circumferential pathway was placed around the orifice for the diversion of the excessive flow. Furthermore, a cone-shaped ending was also employed at the rear-end of the bypass in order to subside the circulating flow near the outflow region; note that the flow outlet was perpendicular to the bypass pathway, as shown in Figure 5a. The above alterations led to substantial minimization of flow circulation throughout the bypass, as shown in Figure 5b.

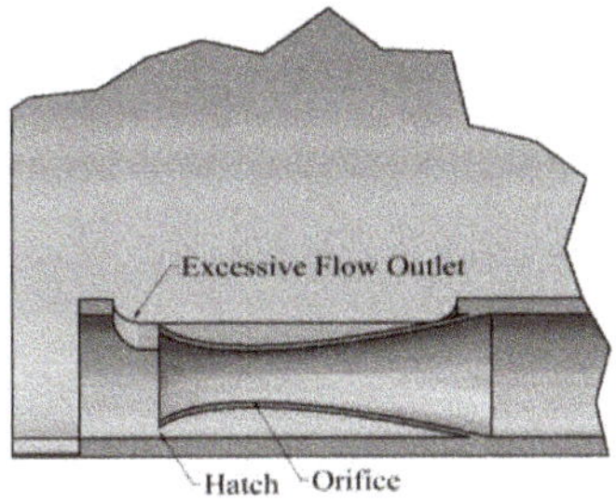

Figure 6. Detailed view of Figure 5a; (Hatch and Orifice, cross section).

B. Determining the Optimum Geometry of the Sensor's Housing

Having determined the most appropriate bypass topology, we proceeded with the necessary study for finalizing the sensor's complete housing geometry. A maximum engine air intake of 1000 kg/h (14076 SLPM) was assumed, exceeding by over 50% the maximum air intake for a typical low power industrial IC engine [14]. We should note that, in order to expand the measuring range of the final sensor as much as possible, the employed housing design featured two bypass sections, each with a separate sensing element. That is, one bypass ("High Velocity Bypass") with an optimized geometry for a moderate incoming flow range (0–350 kg/h) and a second bypass ("Low Velocity Bypass") for a considerably higher flow range (350–1000 kg/h); note that the names of each bypass refer to the magnitude of the developed velocity inside the pathway for the same incoming flow and not to the magnitude of the incoming flow itself. In addition, a third sensing element was mounted directly on the main inflow pathway of the housing for reference. The final topology is presented in Figure 7.

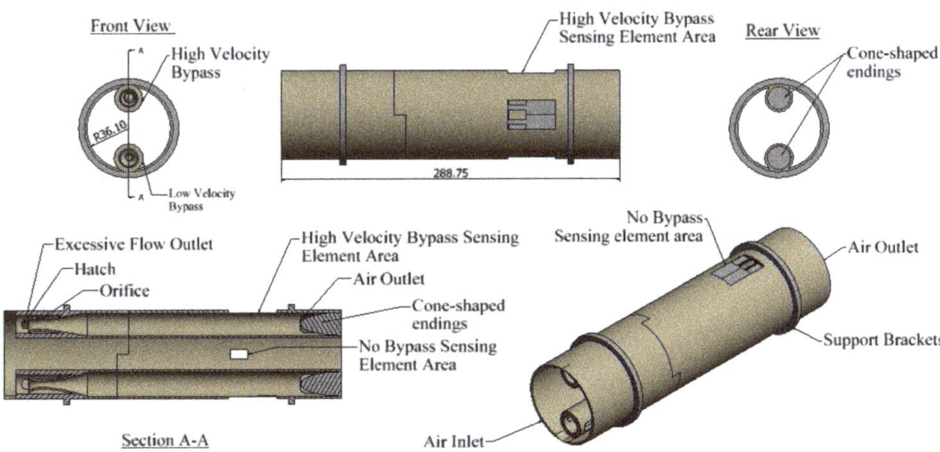

Figure 7. Design geometry schematic (dimensions in mm).

As in the previous study steps, the exact geometry of the overall design (i.e., orifice diameters, bypass diameters, outlet areas, etc.) was obtained through a series of CFD simulations. For the maximum engine air intake (1000 kg/h or 14076 SLPM), an inflow velocity of 50.82 m/s was obtained, considering the main area at the housing's inlet. For the specific velocity, present at each of the bypass inlets (with a much smaller cross-section area of 3.14 cm^2), the corresponding Reynolds number was estimated at 65,064, pointing to turbulent flow for both bypass pathways.

Considering the above remark, the turbulence model k-omega (k−ω) SST was employed, as in the previous series of simulations performed. Characteristic simulation results (for fully developed flows, corresponding to intakes of 1000 kg/h and 500 kg/h) are presented at Figure 8 for the deduced optimum geometry.

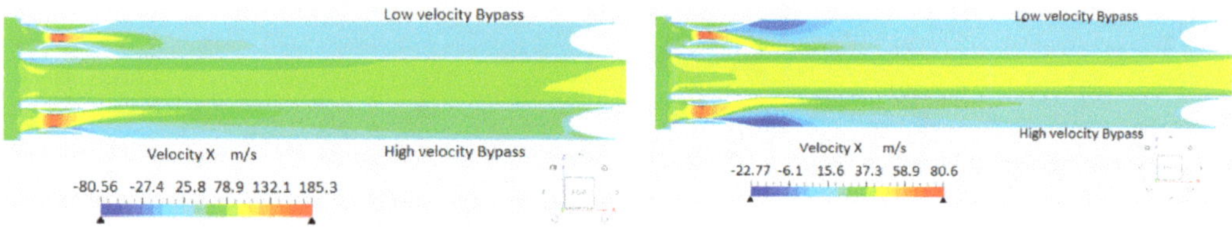

(a) (b)

Figure 8. Design geometry simulation results. (**a**) 1000 kg/h (14,076 SLPM); (**b**) 500 kg/h (7038 SLPM).

As we can see in the specific figure, a drastic drop of over 50% for the Low Velocity bypass and 80% for the High Velocity Bypass of the initial value for the air velocity entering the housing inlet was predicted, resulting in reduced air velocity at the sensing element, as presented in Table 2; an approximately three-fold increase was noticed at the orifice areas, as expected.

Table 2. Design geometry simulation results, velocity reduction at sensing element area.

Bypass Design	Flow Entering the Housing Inlet		Velocity Magnitude at the Housing Inlet_ V_{in} (m/s)	Estimated Velocity Magnitude at Sensing Element Area_ V_s (m/s)	Velocity Decrease (%)	Decrease Factor (V_s/V_{in})
	(kg/h)	(SLPM approx.)				
High Velocity Bypass	1000	14318	56	27	51	0.48
	615	8806	34	17	50	0.5
	350	5011	16.7	8.3	50	0.5
Low Velocity Bypass	1000	14318	56	10	82	0.18
	615	8806	34	6.4	81	0.19
	350	5011	16.7	3.3	80	0.20

Having finalized the geometry details, an appropriate model was built, employing additive manufacturing (i.e., 3D printing technology). In more detail, Fused Deposition Modelling (FDM) [25] was employed through a typical commercial 3D printer (Ultimaker S5), while the building material selected was standard PLA in combination with supportive material PVA; the constructed housing is presented at Figure 9.

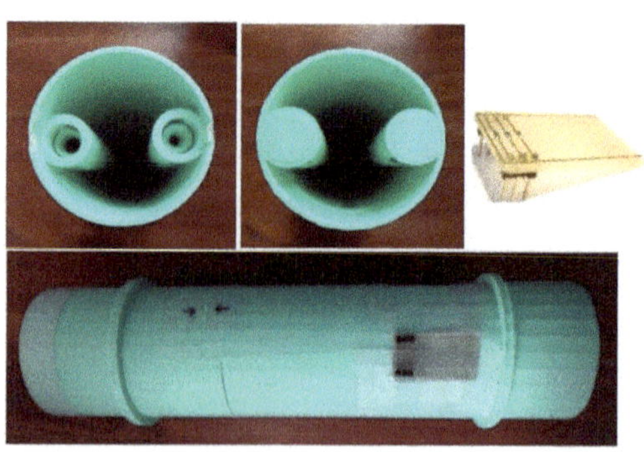

Figure 9. 3D printed sensor housing fabricated with PLA material in combination with PVA supportive material; the PT100 RTDs-integrated on PCB substrate are also presented (upper right image).

C. Sensing Element Employed

The sensing element of the device was fabricated, employing PCB technology (more details can be found in [15]); the specific process combines limited process steps, low cost, and increased robustness. Laconically, three commercially available PT100 Resistance Temperature Detectors (RTDs) were mounted on a typical PCB substrate after the necessary copper patterning for forming the corresponding Cu tracks; as a next step, appropriate vias were formed in order to connect the RTDs to the contact pins at the backside of the PCB. The resulting sensing element is presented in Figure 9 (upper right image).

D. Experimental Setup for the Initial Characterization of the Developed Sensor

An initial characterization of the developed sensor was performed on a DIESEL engine testbed. In more detail, the sensor was integrated into a typical four-cylinder low-power industrial engine (IVECO®, N45 MST series), with rated power of 125 HP at 2200 rpm, coupled with an Eddy Current dynamometer (DYNOmite®), with maximum absorbed power 400 HP at 5000 rpm. As can be seen in Figure 10, the device under characterization was mounted in a series with the integrated Mass Air Flow (MAF) reference sensor of the testbed (BOSCH®, HFM5 series), which records the engine air intake. Both the engine speed and the mechanical load applied by the dynamometer were monitored continuously and controlled accordingly by the testbed's control station.

(a) (b)

Figure 10. (**a**) Engine testbed experimental setup; (**b**) the developed sensor mounted in series with the integrated MAF of the testbed.

3. Results

For the purposes of a preliminary electrical characterization of the developed sensor, hot-wire principle of operation on constant current (CC) mode was selected. In more detail, a constant current of 40 mA was supplied to one of the RTDs on the sensing element, while the corresponding voltage decrease was monitored for different operating points of the engine. The power supplied, taking into consideration the initial resistance of the RTD element, was 238 mW. The air intake varied from 0 to 485 kg/h (0 to 6944 SLPM) by running the engine at appropriate operating points of speed and mechanical load applied; the maximum tested inflow corresponded to a mechanical load of 78.8%, which was not further increased due to safety reasons related to the testbed.

We should note that the estimated pressure drop (ΔP) between the inlet and outlet of the housing, employing the optimum topology with two bypasses pathways, was approximately 1000 Pa, according to the performed CFD simulations for a maximum air intake of 615 kg/h, while the maximum allowed air intake restriction in the testbed's IC engine is 6500 Pa [14], significantly higher than the estimated pressure drop. Furthermore, during all the performed sequential runs of the engine, no alteration was noticed at any of the engine performance parameters monitored, which could be induced by the presence of the developed device under characterization. Thus, the presence of the bypasses on the airflow did not influence in any way the performance of the testbed's IC engine.

For each operating point, three sets of measurements were performed, with respect to the three different configurations available; that is, for the three locations on which the RTD element was mounted (main inflow pathway, High Velocity Bypass, and Low Velocity Bypass). The recorded sensor's output is presented in Figures 11a, 12a and 13a respectively; note that approximation curves were fitted in the specific experimental data, employing simple polynomial functions with a high accurate fit ($R^2 = 0.997$ or above). A comparative graph for the three sensor configurations is presented at Figure 14. According to the IEC 60751:2008 standard (international standard for industrial PT resistance sensors), the sensing RTD's resistance variation, with respect to temperature at a range of 0 to 155 °C, is given using the following equation [26,27]:

$$R(T) = R_0 \times (1 + A \cdot T + B \cdot T^2) \tag{1}$$

where:

R_0 is the nominal resistance value at 0 °C
T is the operating temperature in °C
A is an appropriate coefficient according to IEC 60751 ($A = 3.9083 \cdot 10^{-3}$ °C^{-1})
B is an appropriate coefficient according to IEC 60751 ($B = -5.775 \cdot 10^{-7}$ °C^{-2})

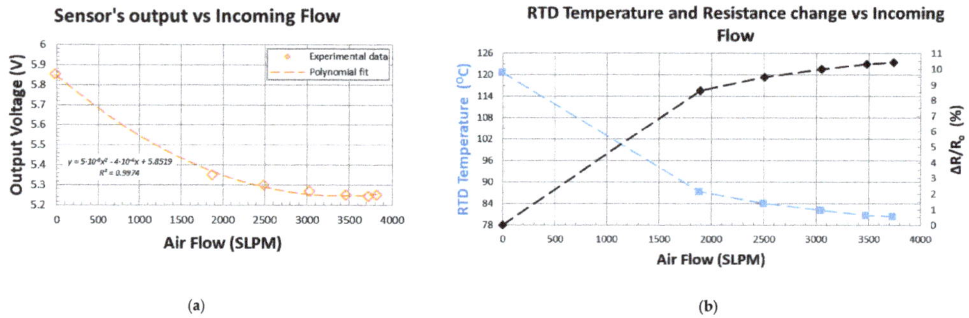

Figure 11. RTD on main inflow pathway. (**a**) Sensor's output vs. incoming flow; (**b**) RTD temperature and resistance change vs. incoming flow.

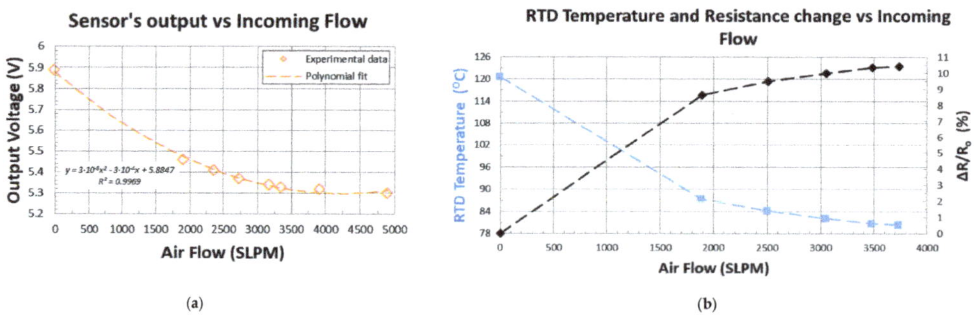

Figure 12. RTD on High Velocity Bypass. (**a**) Sensor's output vs. incoming flow; (**b**) RTD temperature and resistance change vs. incoming flow.

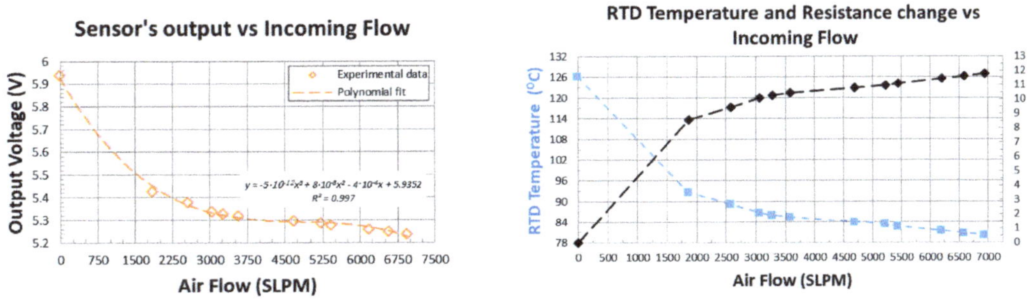

(a) (b)

Figure 13. RTD on Low Velocity Bypass. (**a**) Sensor's output vs. incoming flow; (**b**) RTD temperature and resistance change vs. incoming flow.

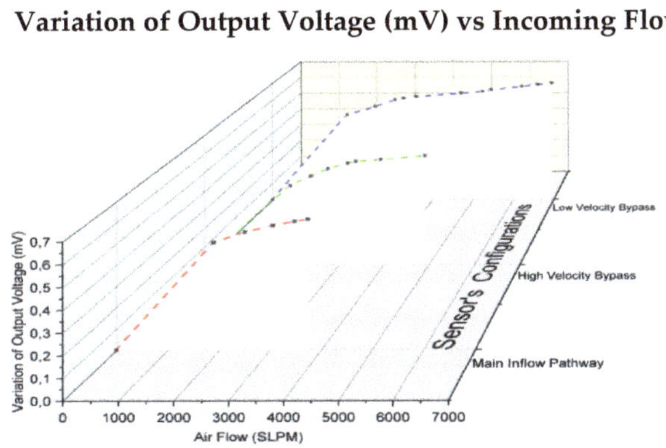

Figure 14. Output voltage variation vs. incoming flow for the three sensor configurations (i.e., for the three locations on which the sensing RTD element was mounted).

Since the resistance value of the sensing element was recorded during characterization, we can easily calculate its temperature for each point of incoming flow, employing the above equation. The resulting RTD's temperature decrease, together with the monitored resistance change, are presented at Figures 11b, 12b and 13b.

As we can easily observe from the above figures, a significant variation in the sensor's output signal was recorded in all three cases, with respect to the incoming flow, validating the functionality of the fabricated sensor. Furthermore, a drastic extension of the measuring range can be noticed when employing either bypass, compared to mounting the sensing element directly on the main inflow pathway, as expected.

In more detail, saturation in the sensor's signal manifests itself for flow rate values above 3000 SLPM (210 kg/h) and 4000 SLPM (280 kg/h), approximately, when no bypass or when the High Velocity Bypass was employed, respectively (Figures 11a and 12a), while no saturation was present in the case of the Low Velocity Bypass, even for the maximum engine air intake (Figure 13a). Similarly, the sensing element's temperature was drastically reduced by 40 °C, approximately, when no bypass or when the High Velocity Bypass was employed (Figures 11b and 12b), while a 46 °C drop was present in the case of Low Velocity Bypass (Figure 13b).

4. Discussion

As a general remark, based on the previously presented experimental findings, the adaptation of standard airfoil geometry to the housing design leads undoubtedly to an extensive increase in the measurement range of the final sensing device. Taking into consideration the findings of previous relative research work [12], but also the simulations presented in this paper, this increase is attributed mainly to the drastic reduction of the circulating flow inside the bypass pathway, especially near the sensing area, by appropriately shaping the flow profile at both inlet and outlet ports of the bypass pathway.

Furthermore, we should note that the obtained experimental results are in agreement with those theoretically predicted by the performed simulations, since the recorded measurement range for both bypass topologies was significantly larger compared to the obtained range when the RTD element was placed directly on the housing 's main inlet. In more detail, an increase of 33.3% and over 130% was measured for the High Velocity Bypass and Low Velocity Bypass topologies, respectively. Regarding the numerically estimated reduction in velocity over the sensing element for each topology (see previous section, Table 2), we should consider the following based on our findings.

The convective heat flux of air over a surface can be expressed by the following equation:

$$q = h_{c\_air} \cdot \Delta T \tag{2}$$

where:

q is the rate of heat flux perpendicular to a surface (W/m^2)
h_{c_air} is the convective heat transfer coefficient of air (W/m^2·K)
ΔT is the temperature difference between the air and surface (K).

The convective heat transfer coefficient of air (h_{c_air}) can be estimated though by the following empirical equation [28]:

$$h_{c\_air} = 12.12 - 1.16\,v + 11.6\,v^{1/2} \tag{3}$$

where v is the relative speed between the surface and air (m/s).

Equation (3) is graphically presented at Figure 15 for velocity values from 0 to 17 m/s; we note that the value of the convective heat transfer coefficient (h_{c_air}) increases negligibly for velocities exceeding 16 m/s, while for velocities exceeding 14 m/s, the increase is less than 2% (i.e., from 39.3 to 40 W/m^2·K). Thus, regardless of the topology employed in the developed sensor, even if we consider the Constant Temperature (CT) mode of operation, where the temperature difference (ΔT) between the incoming air and the sensing element is maintained constant, saturation of the sensor's signal should be expected at an air velocity of 14 m/s, approximately.

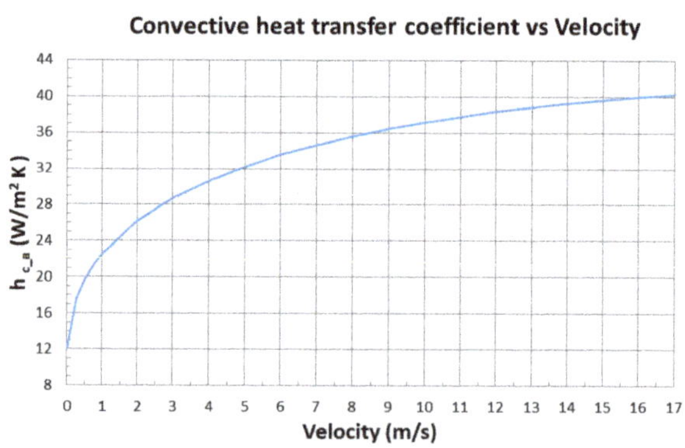

Figure 15. Convective heat transfer coefficient of air over a surface vs. velocity.

Additionally, when the sensor is operating at Constant Current (CC) mode, as in the performed characterization, the sensing element's temperature declines considerably as the incoming flow (i.e., velocity) increases. Note that the maximum temperature decrease deduced from the experimental data presented was from 40 °C to 46 °C (Figures 11b, 12b and 13b), which corresponds to a decrease of 31.6% to 36.5% of the initial temperature. Therefore, we can safely conclude that saturation of the sensor's signal is expected at a significantly lower velocity of 14 m/s (which theoretically applies considering CT mode of operation) for all three topologies.

Considering the velocity decrease factor, which is deduced from the performed CFD simulations for each bypass topology (see Table 2), we can estimate the air velocity over the sensing element for any of the specific inflow values tested. As already mentioned in the results section, saturation at the sensor's output signal was recorded for a flow rate of 4000 SLPM (280 kg/h), approximately, for the High Velocity Bypass topology employed. The specific flow rate for this particular housing inlet area corresponds to an incoming velocity of 15.2 m/s; thus, the reduced velocity inside the bypass is estimated at 7.6 m/s (considering an approximate 0.5 decrease factor), which is significantly lower than 14 m/s and in accordance with the previously performed analysis, concerning the saturation of the sensor's signal.

Similarly, in the case of the Low Velocity Bypass pathway for the maximum applied inflow of 6944 SLPM (485 kg/h), the estimated incoming velocity is 26.3 m/s, while the reduced velocity inside the bypass is estimated at 5 m/s, considering an approximate 0.19 decrease factor according to the CFD simulations performed. Therefore, no saturation at the sensor's signal is expected for the specific bypass topology, even for the maximum air intake (615 kg/h), since the estimated value is substantially below 14 m/s; note that this is supported by the obtained experimental results. As a reference, in the case of the sensing element directly mounted on the main inflow pathway of the housing, saturation occurred at an incoming velocity of 11.4 m/s, which was closer to the value of 14 m/s in comparison to the other topologies, as expected.

We should note at this point that the exact values of air velocity above the sensing element cannot be directly compared for the three different topologies without incorporating a non-negligible error, since even a modest difference regarding the location of the element inside each bypass pathway, with respect to the one assumed for the simulated flow profile, can result in a reasonable difference between the velocity estimated and experimentally measured; apparently, presuming that two or three elements are situated concurrently at the exact same locations, with regard to the assumed positions at the corresponding simulations, can enhance this error. Furthermore, the obtained results from the simulation study, regarding the optimization of each topology, can potentially incorporate, to some extent, different modelling errors (different models/meshing topology, etc.).

Regarding the foremost measurement errors present in the specific study, we should take into consideration the errors introduced by the sensing element (RTD) employed and the reference MAF sensor integrated in the diesel engine testbed. In more detail, considering 126 °C as the maximum RTD temperature, the introduced error by the sensing element (IEC 60751:2008, Class F0.6) [26,27] was equal to ±1.86 °C, less than 4.6% of the measured temperature decrease for all three topologies; for constant current mode of operation with 40 mA supplied to the element, the above figure corresponds to a maximum variation of the sensor's output of ±0.074 V. With respect to the integrated MAF sensor, the absolute value of the measurement deviation was less than or equal to 3% [29]; thus, for the engine air intake range employed, a maximum error of ±14.55 kg/h concerning the reference MAF sensor should be considered.

At this point, we should mention that, in the context of the specific preliminary electrical characterization of the developed sensor, no reference temperature sensing element was available in order to compensate for the expected output variation, which was induced due to ambient temperature variation. Therefore, a reasonable fluctuation of intake air temperature is expected to lead to a drift at the device's characteristic curve. However, in order to quantify the above statement, further electrical characterization of the sensor should be performed at the engine testbed at different air temperatures, a procedure that is practically not feasible for adequate temperature fluctuation and with no additive value for a preliminary investigation of the developed device's functionality, employing the CI mode of operation. That being said, the above analysis, based on the experimental findings, strongly supports the CFD study results, concerning the optimum design geometry of the housing, while the adaptation of standard airfoil geometry to the housing design leads undoubtedly to a substantial increase in the measurement range of the final sensing device.

5. Conclusions

A thermal air flow sensor for use as a Mass Air Flow (MAF) measuring device for small to medium size IC engines was designed and fabricated successfully, employing additive manufacturing technology combined with PCB technology; the proper functionality of the developed prototype was validated by performing a preliminary electrical characterization.

The design of the sensor's housing was derived from an appropriate Computational Fluid Dynamic (CFD) study; three basic design concepts, which employ different approaches for the design of an internal air flow bypass pathway, were investigated. As a result, the optimum design determined is based on a "Hatch and Orifice" topology for the bypass air inlet, while well-established airfoil geometries in the automobile industry (NACA and Joukowski) are the basis for the challenging air flow profile shaping over the sensing element, minimizing the developed circulating air flow. An appropriate set of simulations were additionally performed in order to further optimize the selected concept's geometry, with regard to an assumed maximum engine air intake of 1000 kg/h (14,076 SLPM), which exceeds by over 50% the maximum air intake for a typical low power Industrial IC engine [14]. Subsequently, the final housing design features two bypass sections, each with a separate sensing element: one bypass ("High Velocity Bypass") with an optimized geometry for a moderate incoming flow range (0–350 kg/h) and a second bypass ("Low Velocity Bypass") for a considerably higher flow range (350–1000 kg/h).

Additive manufacturing was successfully employed for the fabrication of the challenging housing design by employing a commercially available desktop 3D printer and standard building materials (PLA/PVA), while PCB technology was selected for the sensing element for increased robustness and limited process steps.

The developed device was integrated into a typical four-cylinder low power engine on a DIESEL engine testbed, in series with the integrated Mass Air Flow (MAF) reference sensor. The air intake was varied from 0 to 485 kg/h (6944 SLPM) by running the testbed at appropriate operating points of speed and mechanical load applied, while the hot-wire principle of operation on constant current (CC) mode was selected.

A significant variation in the sensor's output signal was recorded for both pathways, with respect to the incoming flow, validating the functionality of the fabricated sensor. Furthermore, a drastic extension of the measuring range was noticed when employing either bypass compared to mounting the sensing element directly on the main inflow pathway. Saturation in the sensor's signal manifested itself for flow rate values above 280 kg/h (4000 SLPM), approximately, when High Velocity Bypass topology was employed, while no saturation was present in the case of the Low Velocity Bypass, even for the maximum engine air intake of 485 kg/h (6944 SLPM). The sensing element's temperature was drastically reduced by 40 °C or 46 °C depending on the topology employed.

Based on the findings of previous relative research work [12], but also on the current experimental findings, the adaptation of standard airfoil geometry to the housing design, in combination with the "Hatch and Orifice" topology for the bypass air inlet, lead undoubtedly to an extensive increase in the measurement range of the sensing device, mainly attributed to the drastic reduction of the circulating flow inside the bypass pathway. Furthermore, the experimental results strongly supported the CFD study performed with regard to the optimum design geometry of the housing.

Key features of the developed device are low-cost, fast on-site manufacturing, robustness, and simple fabrication procedure, suggesting numerous potential applications in marine engineering and in industrial applications in general. Future work includes the characterization of the sensor at increased air intake, employing a different experimental setup, together with appropriate characterization (including reproducibility evaluation), both on Constant Current (CC) and Constant Temperature (CT) modes of operation, under the hot-wire principle and calorimetric principle of operation.

Author Contributions: Conceptualization, D.-N.P. and V.B.; methodology, D.-N.P., V.B., G.K., and A.P.; investigation, A.P.; validation, G.K. and A.P.; resources, D.-N.P. and G.K.; data curation, A.P.; writing—original draft preparation, D.-N.P. and V.B.; writing—review and editing, D.-N.P.; visualization, V.B.; supervision, D.-N.P. All authors have read and agreed to the published version of the manuscript.

Funding: This research received no external funding.

Data Availability Statement: The data presented in this study are available on request from the corresponding author.

Conflicts of Interest: The authors declare no conflict of interest.

References

1. Livanos, G.A.; Theotokatos, G.; Pagonis, D.-N. Techno-economic investigation of alternative propulsion plants for Ferries and RoRo ships. *Energy Convers. Manag.* **2014**, *79*, 640–651. [CrossRef]
2. International Maritime Organization (IMO). SOLAS Chapter II-1, Regulation 26.3. Available online: https://www.imo.org/en/About/Conventions/Pages/International-Convention-for-the-Safety-of-Life-at-Sea-(SOLAS),-1974.aspx (accessed on 2 December 2021).
3. Sazhin, O. Novel mass air flow meter for automobile industry based on thermal flow microsensor. *Flow Meas. Instrum.* **2014**, *30*, 60–65. [CrossRef]
4. Marek, J.; Illing, M. Microsystems for the automotive industry. In Proceedings of the International Electron Devices Meeting 2000, San Francisco, CA, USA, 10–13 December 2000; pp. 3–8.
5. Kaltsas, G.; Petropoulos, A.; Tsougeni, K.; Pagonis, D.N.; Speliotis, T.; Gogolides, E.; Nassiopoulou, A.G. A novel microfab-rication technology on organic substrates—Application to a thermal flow sensor. *J. Phys. Conf. Ser.* **2007**, *92*, 012046. [CrossRef]
6. Petropoulos, A.; Pagonis, D.N.; Kaltsas, G. Flexible PCB-MEMS Flow Sensor. *Procedia Eng.* **2012**, *47*, 236–239. [CrossRef]
7. UIUC Applied Aerodynamics Group, Department of Aerospace Engineering. Available online: https://m-selig.ae.illinois.edu/ads/coord_database.html (accessed on 19 September 2021).
8. NACA Airfoil Series. Available online: https://www.nasa.gov/image-feature/langley/100/naca-airfoils (accessed on 20 September 2021).
9. Definition of Streamlines. Available online: https://www.grc.nasa.gov/www/k-12/airplane/stream.html (accessed on 20 September 2021).
10. SimScale Online CAE Simulation Platform. Available online: http://www.simscale.com (accessed on 14 April 2021).
11. Salmi, M.; Partanen, J.; Tuomi, J.; Chekurov, S.; Björkstrand, R.; Huotilainen, E.; Kukko, K.; Kretzschmar, N.; Akmal, J.; Jalava, K.; et al. *Digital Spare Parts*; Aalto University: Espoo, Finland; VTT Technical Research Centre of Finland Ltd.: Espoo, Finland, 2018; ISBN 978-952-60-3746-2. Available online: http://urn.fi/URN:ISBN:978-952-60-3746-2 (accessed on 22 November 2021).
12. Pagonis, D.N.; Kaltsas, G.; Koutsis, T.; Pagonis, A. A Novel Engine Air Intake Sensor Based on 3D Printing and PCB Technology. In Proceedings of the IEEE Sensors Conference 2021, Virtual, 31 October–4 November 2021.
13. Wärtsilä Corp. Next in the Print Queue: Critical Engine Parts. Available online: https://www.wartsila.com/insights/article/next-in-the-print-queue-critical-engine-parts (accessed on 22 November 2021).
14. IVECO N45 MST Technical Manual. Available online: https://www.fptindustrial.com/global/en/engines/off-road/construction/n45 (accessed on 14 April 2021).
15. Pagonis, D.N.; Moschos, A.; Kaltsas, G. A PCB Based Engine Air Intake Sensor—Application to a Typical Low Power Engine. *Procedia Eng.* **2016**, *168*, 59–62. [CrossRef]
16. Tabatabaei, H.; Boroomand, M.; Taeibi-Rahni, M. Experimental and Numerical Study of the Effect of Pulsating Flow on the Turbocharger Turbine Performance Parameters. In *SAE Technical Paper Series*; SAE International: Warrendale, PA, USA, 2013.
17. Darmawan, S. Reynolds number effects on swirling flows intensity and reattachment point over a backward-facing step geometry using STD k-ε turbulence model. *IOP Conf. Ser. Mater. Sci. Eng.* **2020**, *852*, 012073. [CrossRef]
18. Jain, R.; Sheikh, M.Y.; Singh, D.; Tripathi, M. Numerical Analysis of Joukowski (T = 12%) Airfoil by k-e Turbulence Model at High Reynolds Number. In *Proceedings in Adaptation, Learning and Optimization*; Springer: Berlin/Heidelberg, Germany, 2020; pp. 320–329. ISBN 978-3-030-44757-1.

19. Modi, V.; Deshpande, V. A Joukowski airfoil with momentum injection. In *Atmospheric Flight Mechanics Conference;* American Institute of Aeronautics and Astronautics (AIAA): Reston, VA, USA, 2000.
20. Versteeg, H.K.; Malalasekera, W. *An Introduction to Computational Fluid Dynamics: The Finite Volume Method*, 2nd ed.; Prentice Hall: London, UK, 2007; pp. 40–98.
21. Mahaffey, R.M. Review and Optimisation of Pump Suction Reducer Selection. Master's Dissertation, University of Pretoria, Pretoria, South Africa, September 2014.
22. Pal, S.; Kabir, S.M.H.; Talukder, M.M.M. Aerodynamic Analysis of a Concept Car Model. In Proceedings of the International Conference on Mechanical Engineering and Renewable Energy 2015 (ICMERE2015), Chittagong, Bangladesh, 26–29 November 2015.
23. Mohebbi, F.; Sellier, M. On the Kutta Condition in Potential Flow over Airfoil. *J. Aerodyn.* **2014**, *2014*, 676912. [CrossRef]
24. Airfoil Tools. Available online: http://www.airfoiltools.com (accessed on 19 September 2021).
25. Shahrubudin, N.; Lee, T.; Ramlan, R. An Overview on 3D Printing Technology: Technological, Materials, and Applications. *Procedia Manuf.* **2019**, *35*, 1286–1296. [CrossRef]
26. Vishay Intertechnology, Inc. *Temperature Dependent Platinum Thin Film Chip Resistor (RTD) Datasheet;* PTS Series; Vishay Intertechnology, Inc.: Malvern, PA, USA, 2019.
27. IEC 60751:2008 Standard, International Electrotechnical Commission. 17 July 2008. Available online: https://webstore.iec.ch/publication/3400 (accessed on 23 November 2021).
28. Khabari, A.; Zenouzi, M.; O'Connor, T.; Rodas, A. Natural and Forced Convective Heat Transfer Analysis of Nanostructured Surface. In Proceedings of the World Congress on Engineering 2014, WCE 2014, London, UK, 2–4 July 2014; Volume I.
29. Bosch HFM 5 Technical Datasheet. Available online: https://www.bosch-ibusiness.com/media/images/products/sensors/xx_pdfs_1/sensors_i-business.pdf (accessed on 25 November 2021).

applied sciences

MDPI

Article

A Case Study of Floating Offshore Super-Long Steel Pipeline Combing with Field Monitoring

Jin Yu [1,*], Chonghong Ren [1], Yanyan Cai [1], Jian Chen [1], Yuanqing Wang [2] and Weiyun Chen [3]

[1] Fujian Research Center for Tunneling and Urban Underground Space Engineering, College of Civil Engineering, Huaqiao University, Xiamen 361021, China; chren173@163.com (C.R.); yycai@hqu.edu.cn (Y.C.); CJ12011201@163.com (J.C.)
[2] Xiamen Water Group Co. Ltd., Xiamen 361008, China; 13806084166@139.com
[3] Institute of Geotechnical Engineering, College of Civil Engineering, Nanjing Tech University, Nanjing 210009, China; cwy@njtech.edu.cn
* Correspondence: bugyu0717@163.com

Abstract: How to control deformation and avoid resonance is the key to ensuring the safety of the super-long pipeline when it is floating in the sea. Based on the deformation warning value of pipeline prototype composite material obtained from laboratory tests, the raw water pipeline project in Tong'an Xiamen adopts wireless communication equipment to transmit data, supplemented by aerial photography technology to monitor and feedback the strain and vibration during the dynamic construction of long-distance pipeline floating transportation. Combined with dynamic construction, this monitoring method avoids excessive deformation and resonance of the steel pipeline during floating transportation, and prevents the destruction of the anticorrosive coating. The airtightness test after completion shows that the whole pipeline meets the acceptance requirements. The monitoring results show that the strain at the bent position of the pipeline is large in the process of floating transportation, and the jacking speed and position of the tugboats have an important influence on the deformation of the pipeline. The same type of project should focus on these aspects and timely feedback monitoring data. At the same time, the study also provides detailed strain, modal analysis and effective monitoring technology for the safety of offshore steel pipeline floating transportation.

Keywords: marine floating transport; field monitoring; dynamic construction; steel pipeline strain; vibration mode

Citation: Yu, J.; Ren, C.; Cai, Y.; Chen, J.; Wang, Y.; Chen, W. A Case Study of Floating Offshore Super-Long Steel Pipeline Combing with Field Monitoring. *Appl. Sci.* **2021**, *11*, 10186. https://doi.org/10.3390/app112110186

Academic Editor: Dimitrios-Nikolaos Pagonis

Received: 11 September 2021
Accepted: 27 October 2021
Published: 30 October 2021

1. Introduction

The construction of the trans-sea pipeline is an important part of the municipal pipeline project. Due to the complexity of the sea environment, the construction of the pipeline floating on the sea is difficult. The existing floating transportation method is mostly a single pipeline segment floating by segment, while the overall floating transportation technology of super-long pipeline is to float the whole pipeline to the sinking position after individual pipelines make up the whole. Ye et al. [1] studied the mechanical properties of the whole sinking process of super long asymmetric pipe. Compared with the pipe-laying vessel method [2] and pipeline jacking method [3], this construction method has less navigation influence, shorter construction period, less influence from geological conditions and low construction cost, it can complete the laying of underwater long-distance water conveyance pipeline with high efficiency and high quality. However, floating transportation on the sea is greatly affected by wind, wave and flow, so real-time monitoring feedback and dynamic adjustment are the keys to ensuring the safety of floating transportation of steel pipelines.

The Marine environment is complex. There are three main methods to study ocean engineering: theoretical research [4–6], numerical simulation [7–9], indoor modal and on-site test [10–13]. Compared with other methods, field monitoring can obtain the raw data of Marine structures and track the effective information in real-time, which provides essential support for offshore installations and reduce the risk of accidents [14,15]. At present, the monitoring methods of offshore pipelines mainly include fiber Bragg grating, acoustic emission and wireless sensor testing. Brower et al. [16] used optical fiber to monitor submarine pipelines as early as 2004. Ren et al. [17] apply Distributed Optical Fiber Sensor (DOFS), including strain and temperature sensors, to the health monitoring of the Bohai CB271 oil production platform. The experimental investigations on a method for distributed detection of lateral buckling in subsea pipelines with Brillouin fiber optic sensor were conducted by Feng et al. [18]. Xu et al. [19] developed a new Fiber Bragg Grating (FBG)-based bundle-structure riser stress monitoring sensor which could meet the requirements of riser safety monitoring in offshore oil fields. Fiber Bragg gratings are easy to move and paste, and the installation process has less impact on the pipeline surface. However, it also has some disadvantages, such as chirps and creeps. Jin and Eydgahi [20] proposed monitoring of distributed pipeline systems by acoustic sensor networks. Based on the Underwater Acoustic Sensor Networks, Wang et al. [21] proposed an energy-efficient data transmission scheme called EGRC (Energy-efficiency Grid Routing based on 3D Cubes). Mahmutoglu and Turk [22] proposed a novel passive acoustic-based system that could locate the leak holes and determine their size remotely. The acoustic sensor is portable and easy to install, but it is sensitive to environmental noise and prone to false positives. The Independent Remote Monitoring System (IRMS) of Mexico deepwater floating is fully independent of the primary Integrated Marine Monitoring Systems (IMMS) system in terms of sensors [23]. An energy-efficient cooperative scheme for wireless sensor network (WSN) nodes used in long-distance water pipeline monitoring systems was proposed by Seddiq et al. [24]. Huang and Nagarajaiah [25] developed new structural health monitoring (SHM) techniques, in which, the degree of damage was determined by local monitoring using a robotic crawler with magnetic flux leakage (MFL) sensor. It could monitor damage due to vibration fatigue. Production Systems Unmanned aerial vehicles (UAVs) technology prototypes to monitor pipelines were introduced by Gómez and Green [26]. The unmanned aerial vehicles technology is suitable for the overall monitoring of long and large pipelines, but it is not mature yet. Furthermore, besides the monitoring method, the mechanical and deformation characteristics of the pipeline body are also worthy to attend. Many efforts focused on the material characteristics and deformation mechanism of the pipe body. Wang Y et al. [27] investigate the transverse impact response for ultra-lightweight cement composite (ULCC) filled pipe-in-pipe structures through a parametric study using both a validated finite element procedure and a validated theoretical model. ESPÍN-LAGOS et al. [28] have determined through mechanical tests, metallographic analysis, and morphological studies that the degree of discoloration of the HAZ is directly affected by the atmospheric oxygen content during purging, and it has been proven to have a good electrical resistance in welded joints. Kiss I et al. [29] measured the characteristics of stainless steel during large plastic deformation through a hot torsion test, and obtained the best heating temperature range for the studied steel deformation.

Offshore pipeline monitoring is a challenging task. The main concerns are pipeline stress and strain, pipeline leakage and vibration. Relevant scholars have conducted a large number of studies on the offshore pipeline. They focused on a single fixed pipeline joint and ignored the deformation of the pipeline itself. For ultra-long pipelines, there will inevitably be an error in the calculation of force exerted on them by these methods. Guo et al. [30] analyzed stress states of pipelines upon floating by using the Finite Element Calculation method. Chen et al. [31] presented a probabilistic methodology for monitoring the condition of offshore pipelines and predicting the reliability considering stress observation and structural deterioration. There were many ways to monitor pipeline corrosion and leakage. Arumugamr et al. [32] described the application of the finite element method (FEM) and the development of equations to predict the failure pressure of single corrosion affected pipelines subjected to internal pressure and axial compressive stress. Aniskin A et al. [33] introduced the application of the boundary element method in the calculation of a closed cylindrical shell of stepped thickness, and compared it with two numerical methods: boundary and finite element. Negative pressure wave [34] and dynamic modelling [35,36] methods were used to monitor pipeline leaks. Almazyad et al. [37] presented a scalable design and simulation of a water pipeline leakage monitoring system using Radio Frequency Identification and Wireless Sensor Network technology. The study on the vibration of the pipeline is mostly about the vibration characteristics of the submarine pipeline during the period of use, while the study on the vibration of the floating pipeline is less. Jin and Shao [38] presented numerical simulations and experimental verification of a vibration-based damage detection technique. Goldsmith et al. [39] suggested that fatigue could be calculated based on the modal shape of response and amplitudes from the monitoring data. Yang et al. [40] investigated the transverse vortex-induced vibrations (VIVs) of a submarine pipeline near an erodible sandy seabed under the influence of ocean currents by a series of experiments. Considering the function of inertia and dampness, under the optimal span, a frequency response analysis was carried out. Zhang et al. [41] proved that resonance did not happen between the pipeline and wave. One marine riser fatigue acoustic telemetry scheme was proposed by Li et al. [42] to monitor deep and ultra-deepwater risers' vibration fatigue. Sollund et al. [43] studied the dynamic response of multiple ocean pipelines. A semi-analytical method for the multi-spans offshore pipeline was developed by them, which provided a fast and accurate alternative to finite element modeling of potentially interacting spans. Also, Cook et al. [44] proposed 'Threshold' monitoring by comparing measured data at alert levels to determine whether key performance indicators have reached a critical or near-critical level.

There is a lack of on-site data for the super-long steel pipeline floating project, and it is also difficult. Determination of deformation warning value of steel pipeline and 3PE composite material, monitoring instrument accuracy, waterproofing and long-distance signal attenuation, the monitoring scheme and the dynamic construction method which can effectively reduce the deformation and resonance of the pipeline, these are problems urgent to solve and design. Based on the project in Tongan Xiamen, this paper attempted to carry out on-site monitoring and real-time feedback of long-distance offshore steel pipeline floating transportation, and studies the monitoring scheme, dynamic construction method and deformation characteristics of offshore pipe floating transportation.

2. Project Characteristics

2.1. Project Profiles

Raw water pipeline project locates in Tongan Xiamen (Pile No. C2 + 187.77~C3 + 565.91). The length of pipeline 1 is 595 m in the straight pipeline and 120m in the oblique pipeline section, totally 715 m; the length of pipeline 2 is 370 m in the straight pipeline section and 290 m in the oblique pipeline, totally 660 m. The whole pipeline is 1375 m long, 1600 mm in diameter, about 1087 t in weight, 20 mm in the wall thickness, and coated with 3PE anticorrosive coating on the surface. The whole pipeline is similar to the inverted rainbow pipeline structure. Both of the two pipelines are relatively long, so it is difficult to arrange the position of tugboats and towing force during floating transportation. The reasonable arrangement of tugboats can make the pipelines bear even force in the floating transportation process and avoid the local deformation of the pipelines. The floating pipeline section operates in a cross-flow direction from the downstream direction, with a turning angle of about 90°, and the detailed floating routes of the two pipelines are illustrated in Figure 1. The vertical distance of the floating pipeline is 800 m and the route is 1.15 km. Due to the influence of sea surface width and the Tong'an bridge, it is difficult to float the whole pipeline at a time. Therefore, the pipeline 1 was floated to the sinking position firstly, then the pipeline 2 was floated, the two pipelines were welded at the sinking position in the sea. Finally, the whole pipeline was sunk at once. The floating transportation process is shown in Figure 2.

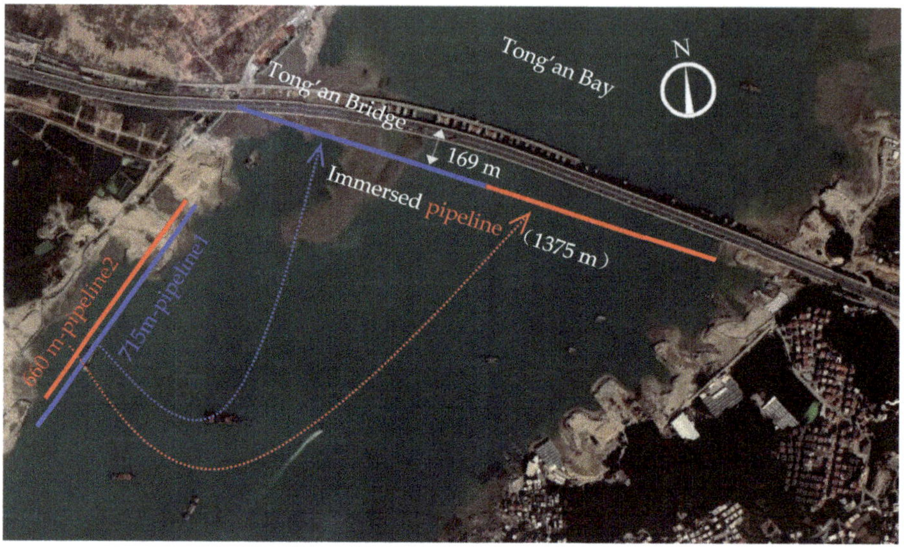

Figure 1. Satellite image of immersed pipeline segment (taken from Google map).

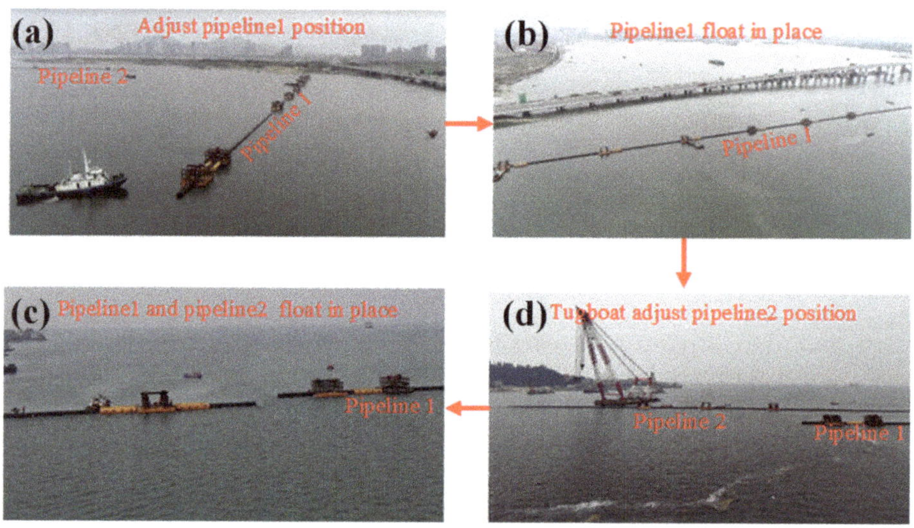

Figure 2. Aerial photo of floating transportation on site.

2.2. *Key Engineering Problems*

2.2.1. The Particularity of Offshore Construction

Offshore construction tools and monitoring have certain particularities. The wind, wave and flow on the sea are unpredictable and non-artificial. The change of weather has a great impact on the construction environment on the sea, and the change of the flow at the estuary is more drastic. Long and large pipelines are susceptible to wind, wave and flow during floating transportation. If the pipeline body deforms too much under the action of external forces, the damage is irreversible, and the pipeline quality and later service will have very adverse consequences. If the raw water pipeline is damaged, it is difficult to repair. Akram A et al. [45] providing insights on the use of thermosetting liner for the repair of offshore pipelines exposed to corrosion and leakage. The seawater erodes the pipeline body and pollutes the inside of the pipeline and the airtightness of the pipeline cannot meet the requirements, resulting in a huge loss of social benefits and economic benefits. The flow of liquids from the pipes into the ocean may also cause pollution and affect its sustainable development. The impact of natural elements was quantified using the ICI index (Impact Quantification Index) proposed by López et al. [46], which summarizes different sources of variation in a single value.

Secondly, offshore construction operations are not as convenient and flexible as those on land. If there is no reliable construction monitoring, operators cannot make timely and correct responses, which is likely to cause irreparable damage to engineering structures. In addition, metal objects are prone to decay after immersion in seawater, so offshore construction also requires effective waterproofing and erosion protection of precision instruments and structures.

2.2.2. The Particularity of Materials

The deformation test of steel-3PE composite material shows that the steel pipeline and 3PE would be separated from each other after large deformation. The indoor test material was the Q345 steel plate as same as the field steel pipeline material. The length, width and thickness of the steel plate were 40 cm, 20 cm and 2 cm respectively. Its surface was coated with 3PE anticorrosive and was tested on the universal material testing machine. During the testing process, the steel plate was on the top and the 3PE anticorrosive coating was on the bottom. The strain gauge was attached to the position corresponding to the steel plate and the 3PE anticorrosive coating (as shown in Figure 3), and the force was applied on the middle line of the steel plate.

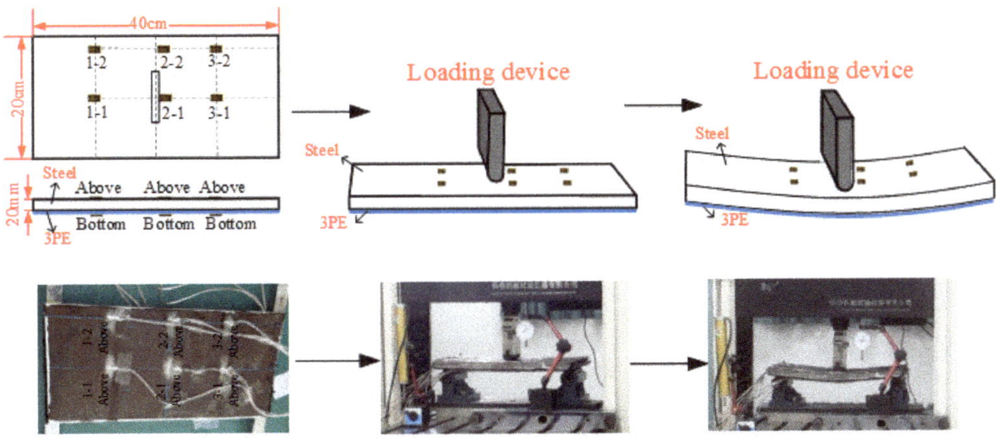

Figure 3. Laboratory test drawing of steel-3PE.

The steel-3pe deformation curve is shown in Figure 4. It can be seen that the strain at the centerline of the steel plate is the largest. When the load is about 30 kN, the center point of the steel plate (the measuring point 2-1) reaches the elastic limit state, and there is a significant difference between the steel plate and 3PE strain at the same position. The maximum strain of the corresponding steel plate is 437 με. If the load on the composite continues to increase, the steel plate and the 3PE material will be separated from each other. Based on the test situation and considering that the compressive strength of general welds is similar to that of the base metal, the monitoring tester suggests that the safety range of the monitoring and warning strain value could be 260 με. As soon as the monitoring strain approaches or reaches this value, the monitoring tester should immediately send out early warning signals and location information to the operators on the construction site.

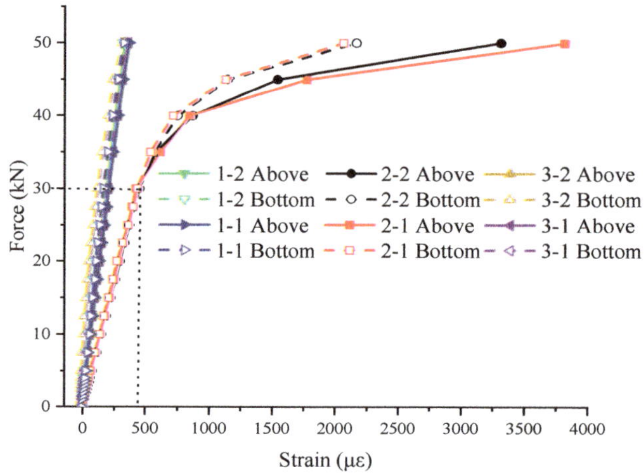

Figure 4. Deformation curve of steel-3PE.

3. Monitoring Instruments and Monitoring Programs

3.1. Monitoring Instruments

In the Marine environment, waterproofing and data transmission of monitoring equipment are the first problems to be solved. The measures adopted in this monitoring are as follows: after the strain gauge was pasted on the pipeline, it was further fixed with AB epoxy resin glue, and then sealed and coated with 703 waterproof glue, the wireless anti-interference communication system was also used to solve the problem of the interference and long-distance signal attenuation.

Figure 5 shows the specific monitoring equipment and data transmission process. The advanced isolation system of DH3819 has a strong anti-interference ability. Furthermore, it can automatically complete some functions such as the setting of parameters and the collection of data. This system has eight measurement points for each acquisition module. The maximum acquisition rate of each measurement point is 5 Hz, the obtainable strain range is $\pm$19,999 με, and the measurement error is less than 0.5% $\pm$ 3 με. The adopted communication mode is the wireless network communication; ZigBee is lightweight and portable, which can be used for wireless communication extension to reduce the field workload and improve the field test safety; DH5902N can be synchronized with GPS and can be tested and monitored in extreme environments such as strong vibration, high/low temperature and high humidity. The strain gauge was connected with the DH3819 static strain acquisition instrument through a waterproof cable-tied and fixed on the pipeline. 4 to 32 channels are provided for the system of DH3819, and all the channels can synchronously collect data, and the maximum frequency of each channel is 256 kHz. ZigBee was used to extend the wireless transmission of data. Finally, the data was transmitted to the computer through USB. The acceleration sensor was connected with the DH5902N mode acquisition instrument via a waterproof cable, and the computer received data through WiFi. The floating transportation dynamic construction could be completed through real-time monitoring and data analysis and the real-time feedback of aerial photography technology.

A lifting test was carried out before floating to ensure the reliability of testing equipment and the accuracy of steel pipe monitoring data. The test method is shown in Figure 6. The mid-point of the steel pipeline was carried out by a hoist on-site with the spacing of the buoyancy tank was 325 m. The strain at the mid-point was monitored with the results shown in Table 1. The pipeline was regarded as a statically determined structure with two hinged ends. The theoretical strain value was calculated according to the lifting height, and compared with the measured strain value.

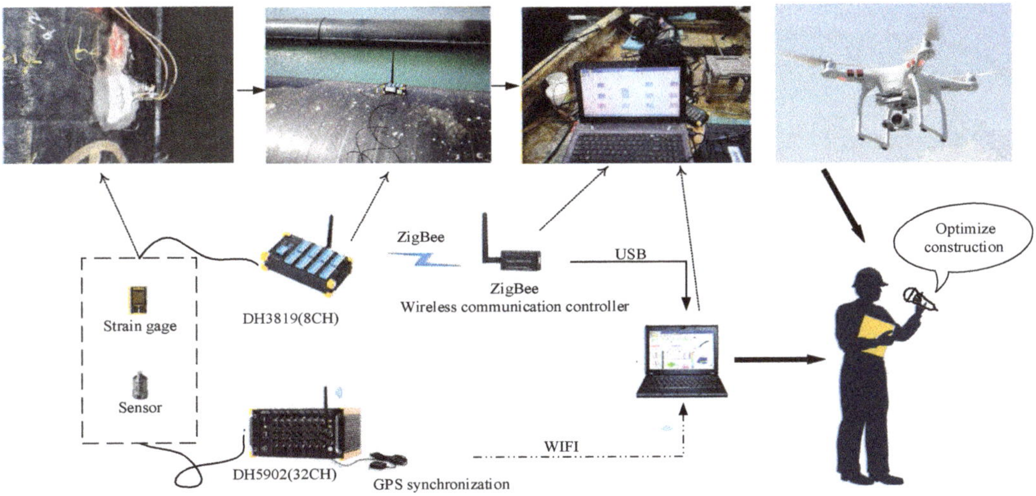

Figure 5. Flow chart of field monitoring data transmission.

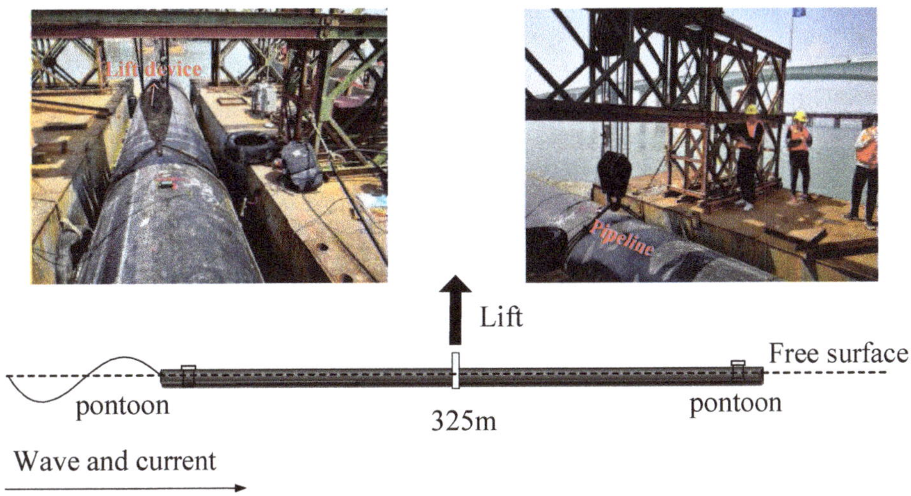

Figure 6. Lifting test on site.

Table 1. Lifting test results.

Case	Lift Height f (cm)	Theoretical Strain $\varepsilon=12df/l^2$ ($\mu\varepsilon$)	Measured Strain ε ($\mu\varepsilon$)	Deviation
1	7.8	14.18	14.76	4.09%
2	9.5	17.27	17.8	3.07%
3	11.2	20.35	18.91	7.08%
4	12.9	23.45	21.55	8.10%
5	14.6	26.54	24.66	7.08%

3.2. *Monitoring Programs*

3.2.1. Arranging the Number of Tugboats

The water flow force on the steel pipeline was estimated to determine whether the number of floating tugboats was sufficient. The calculation method of water flow force was referred to in the literature of Chen and Ren [47], and the calculation formula of the maximum water flow force of the pipeline Section 1 is as follows:

$$F_D = 1/2 C_D \rho u^2 S \tag{1}$$

where: F_D is water flow force, N; C_D is the drag coefficient, which is 1.2; ρ is the density of water, which is 1000 kg/m^3 according to the test data. u is the water flow velocity, and 0.35 m/s is taken according to the test data. S is the area facing the flow (the submergence depth is 0.4 D, D is the diameter, m), m^2.

It is generally believed that 1 KW = 1.36 horsepower, and the drag force per 1 KW is 199.92 N. Therefore, one 700-horsepower tugboat and three 120-horsepower tugboats were equipped in the floating transport process of pipeline Section 1, and the maximum towing force was 155,812.2 n > 33,633.6 n, which met the requirements of floating transport towing force.

During site construction, the tugboats should be arranged uniformly with the pontoons as far as possible, and rotated from the downstream to the transverse direction with a larger radius of curvature. Under the condition of overcoming buoyancy resistance, the tugboats kept the same power pushing on the buoyancy tank to avoid excessive deformation and 3PE destruction caused by uneven thrust. Each tugboat was equipped with a walkie-talkie and 1~2 tugboats were reserved to respond to abnormal situations timely.

3.2.2. Field Monitoring Arrangement

As shown in Figure 7 that Considering that there may be a problem of stress concentration near the bending point, strain monitors were performed near the bending section of the pipeline Sections 1 and 2 with a larger bending angle. At the same time, select a strain monitoring section near the middle of the straight pipeline section and the inclined pipeline section. Because the inclined pipeline section is longer in pipeline Section 2, two strain monitoring sections are selected here, one at the midpoint of the entire inclined pipeline, and one between the two pontoons with a larger distance. Three measuring points were arranged in each section. In the case that the water flow was perpendicular to the straight pipeline section, taking Section 1 of pipeline 1 as an example, the measuring point 1-1 belongs to the facing water, 1-2 was the top, and 1-3 was the backwater. As the bending Angle of pipeline Section 2 was only 1°, and the cable had been uncoiled to meet the construction needs before floating, the pipeline section had been flipped by 90°. Taking Section 4 as an example, the measuring point 4-1 was the top, 4-2 was the backwater, and 4-3 was at the bottom. In general, the number of measurement points had little influence on the modal parameters of natural frequency. Four modal monitoring parts were arranged in Section 2, and two measuring points were arranged on each monitoring part. Taking Section 1 as an example, measurement points 1-1 were the top and 1-2 were the backwater.

During the process of pipeline floating, the pilot boat moved synchronously with the pipeline to obtain monitoring data. The test boat of pipeline Section 1 was equipped with DH3819, which was used to receive 9 strain signals in 3 sections of the inclined pipeline section. DH3819 and DH5902N were arranged on the test boat of pipeline Section 2 to receive 12 strain signals in 4 sections of the inclined section and 8 vibration acceleration signals in 4 sections of the straight section.

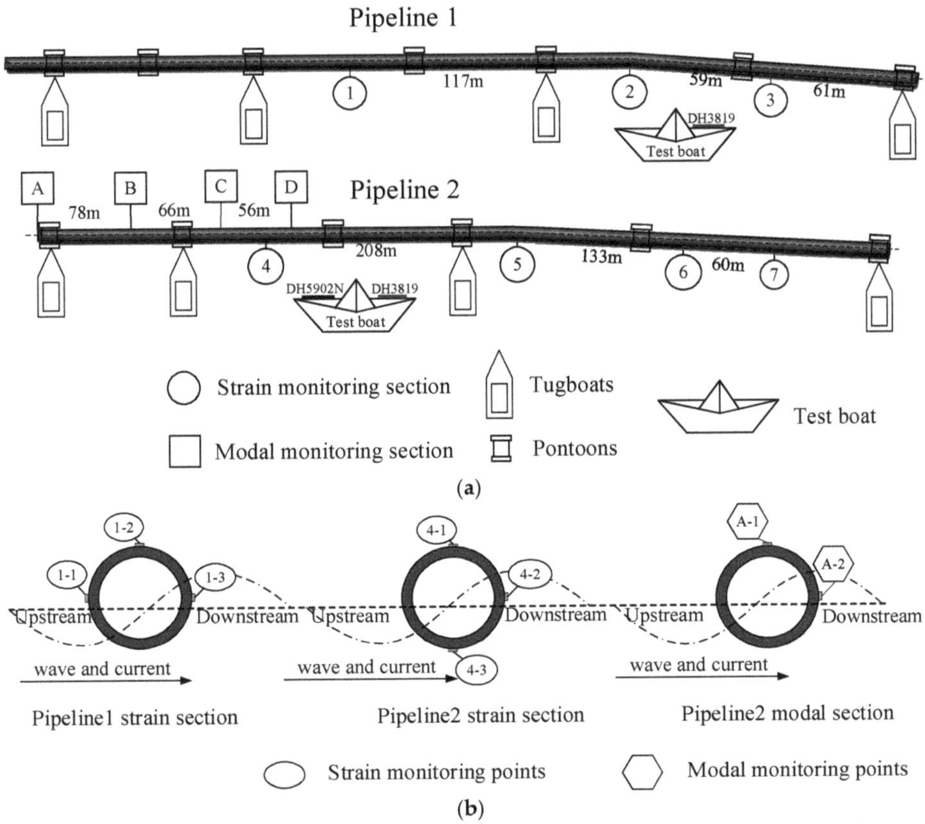

Figure 7. Arrangement of strain and modal measuring points. (**a**) Section arrangement of pipeline. (**b**) Point arrangement of section.

4. Analysis of Monitoring Results

4.1. Strain Monitoring Results

The pipeline body followed the direction of the flow before the floating movement began. After the floating movement started, the pipeline body was rotated counterclockwise at a speed of about 2°/min. It rotated to the direction perpendicular to the flow at 45 min, and to the designated position at 110 min, with the strain value tending to be stable.

Figure 8 shows the strain during the floating movement process of pipeline Section 1. The maximum tension strain at the point 1-1 facing the water surface is 99.1 με, and the maximum pressure strain is −122.3 με; at the top measuring point 1-2, the maximum tensile strain is 55.4 με, and the maximum pressure strain is −48.6 με; the maximum tensile strain is 150.9 με and the maximum pressure strain is −52.4 με at the backwater detection point 1-3. The measuring point 2-1 was located at the corner outside the surface of the water. It used a unidirectional strain gauge and the maximum strain was along the axis of the straight pipeline section. The maximum tension strain is 57.9 με, and the maximum pressure strain is −100.4 με. The measuring point 2-2 was at the top, the main stress direction is clear, and the right-angle strain flower is adopted, 0° along the axis direction and 90° vertical pipeline axis direction. The maximum tensile strain of 0° is 44.5 με, and the maximum pressure strain is −78.4 με. The maximum tensile strain of 90° is 87.7 με, and the maximum pressure strain is −60.2 με. Measuring points 2-3 at the angle of the backwater, the direction of principal stress was unknown, so the tri-directional strain flower is adopted. The maximum tensile strain of 0° is 177.9 με, and the maximum pressure strain is −99.9 με. The maximum tensile strain of 45° is 60.4 με, and the maximum pressure strain is −24.5 με. The maximum tensile strain of 90° is 44.7 με, and the maximum pressure strain is −135.7 με. The maximum tensile strain at the facing water surface measuring point 3-1 is 66.9 με, and the maximum pressure strain is −130.6 με. The maximum tension strain at the top measuring point 3-2 is 14.5 με, and the maximum pressure strain is −68.2 με. The maximum tension strain at the backwater detection point 3-3 is 85.3 με, and the maximum pressure strain is −53.1 με. The cables of measuring points 3-1 and 3-2 were pulled during floating transportation, and the strain curve shows a downward trend.

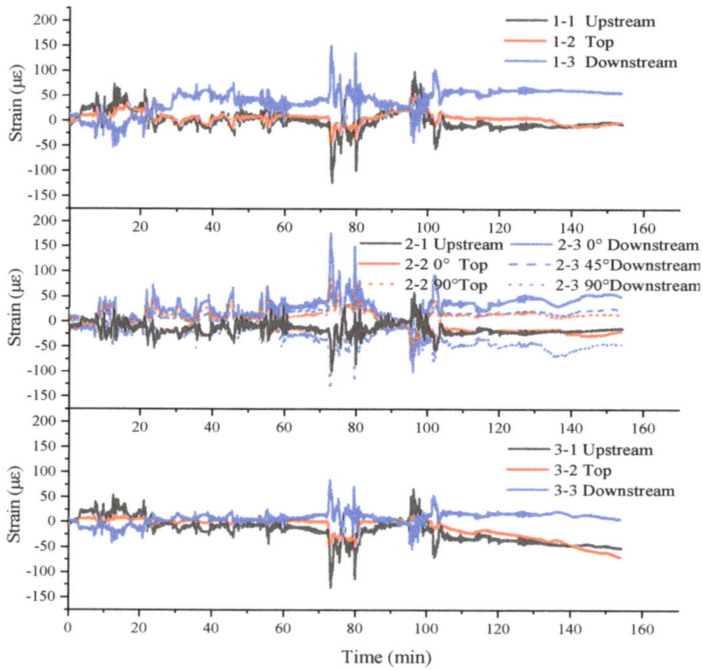

Figure 8. Pipeline Section 1 floating strain.

In the early stage of floating transport of pipeline segment 1, the pipeline rotated from downstream to cross-flow, and the middle tugboat had a large jacking force to correct it, so there was a short tension stage on the upstream and a short compression stage on the water-carrying side. At the time of floating transport of 72 min, a strong wind suddenly blew on the sea surface, causing an abrupt increase in the strain of the pipeline. In general, the strain in the floating transportation of pipeline Section 1 does not exceed the warning value. The steel pipeline is pressed against the surface of the water or pulled away from the surface of the water, and the strain curve is anti-symmetric. The strain of the windward side and the dorsal side of pipeline Section 1 is significantly greater than that of the top. The strain of the windward side and the dorsal side is the most dramatic under the influence of sea waves and flows, while the strain of the top is relatively small, and the strain of the measurement point 1 at the bend is greater than that of the measurement point 2 and 3 at the flat section.

The strain in the floating transportation of pipeline segment 2 is shown in Figure 9. The maximum tension strain at the top of measuring point 4-1 is 0.2 με, and the maximum pressure strain is −167.8 με; The maximum tension strain at the backwater detection point 4-2 is 278.4 με, and the maximum pressure strain is −151.4 με; the maximum tension strain at the measuring point 4-3 is 26.6 με, and the maximum pressure strain is 14.8 με. The maximum tension strain at the backwater detection point 5-2 is 337.8 με, and the maximum pressure strain is −123.3 με. The maximum tension strain at the top measuring point 6-1 is 29.9 με, and the maximum pressure strain is −53.6 με; the maximum tension strain at the backwater detection point 6-2 is 150.9 με, and the maximum pressure strain is -167.5 με; the measuring point 6-3 at the bottom, the maximum tensile strain is 46.0 με, and no pressure strain is observed. The maximum tension strain at the top measuring point 7-1 is 25.0 με, and the maximum pressure strain is −99.0 με; the maximum tension strain at the backwater detection point 7-2 is 89.0 με, and the maximum pressure strain is −18.6 με; the maximum tension strain at the measuring point 7-3 at the bottom is 21.4 με, and the maximum pressure strain is −5.0 με.

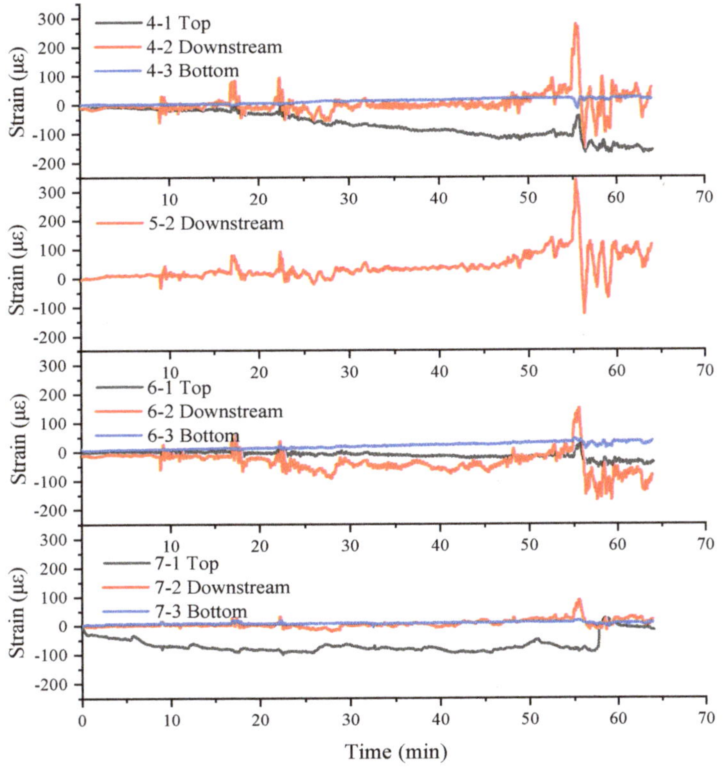

Figure 9. Pipeline Section 2 floating strain.

In the early stage of buoyancy movement of pipeline segment 2, the strain value of each measuring point is small, with the maximum strain less than 100 με. When the buoyancy movement reaches 48 min, a gale of six sprang up off the sea. In general, the strain in the floating transportation of pipeline Section 2 does not exceed the warning value. The steel pipeline is under pressure at the top, under tension at the bottom, and tension at the surface of the back. The strain change on the waterside of the pipeline is the most drastic, followed by the top and the bottom. The strain monitoring curves of test sections 4-1 show good consistency.

4.2. Dynamic Regulation

By real-time monitoring and timely warning and feedback to the construction institution, the construction institution can dynamically adjust the construction operation according to the monitoring feedback situation, avoid the pipeline damage caused by the deformation of steel pipeline and 3PE composite material, and ensure the safety of pipeline floating transportation.

At the beginning of floating transportation of pipeline Section 1, the wind speed was 3 m/s, the flow velocity was 0.35 m/s, and the wave height was 0.15 m, the axis of the pipeline was along the direction of water flow. The strain of the pipeline obtained by on-site monitoring is small and the change amplitude tends to be gentle. When floating at 72 min, a level 6 gale suddenly appears at sea, the wind speed was 6.5 m/s, the flow velocity was 0.55 m/s, and the wave height was 0.45 m. At this time, the angle between the pipeline axis and the flow direction was about 90°. The strain of the pipeline increases and the construction institution notifies the tugboats to increase the towing capacity. However, it is found that the strain suddenly increases and quickly approaches the warning value through monitoring. The monitoring personnel immediately reports the situation to the construction institution. Considering that the situation is caused by too much tugboat thrust, the operator at the construction site should be informed to reduce the tugboats thrust slowly, and always pay attention to the monitoring data, the strain data to be monitored returns to the normal level, that is, stop reducing the thrust of the tugboats. As shown in Figure 10, real-time monitoring and construction cooperate to dynamically adjust the construction operation (Section 2 at the bend is the maximum main strain).

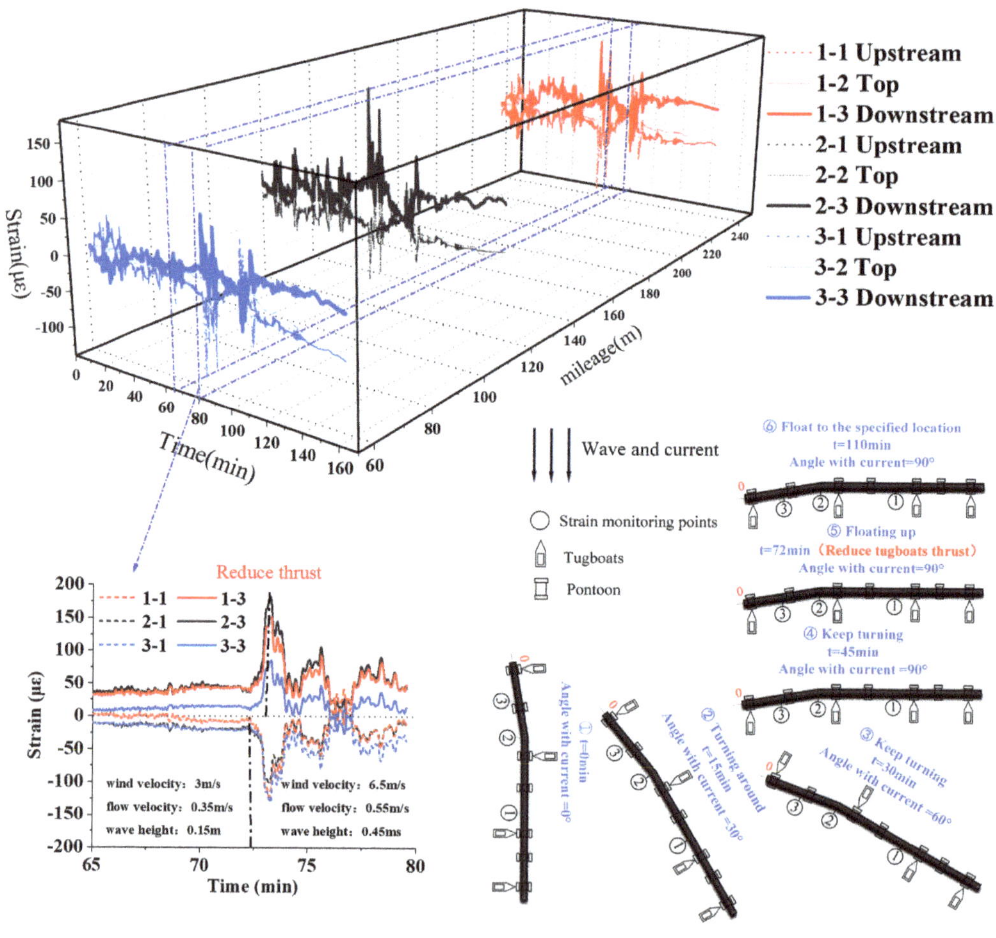

Figure 10. Dynamic floating construction of pipeline Section 1.

When the floating movement of pipeline Section 2 begins, the wind speed was 3 m/s, the flow velocity was 0.35 m/s, and the wave height was 0.15 m, the axis of the pipeline was along the direction of water flow. At 48 min of floating transportation, a strong wind suddenly appeared at sea, which was opposite to the floating transportation direction of the pipeline section. The wind speed was up to 6.5 m/s, the wave was up to 0.45 m, and the flow velocity was 0.55 m/s. At this time, the angle between the pipeline axis and the flow direction was about 90°. Dynamic construction is shown in Figure 11. Through monitoring, it is found that the strain at the water surface measurement point on the back of the pipeline continues to increase when it reaches 200 με, which is mainly caused by the increase in wind speed and flow velocity. The monitoring personnel immediately gave warning to the construction institution and reports the position of the maximum strain, informing the construction site operators to send an additional tugboat near Section 6 near the midpoint of the inclined pipeline section and continuously adjusting the position of the additional tugboat according to the monitoring data. Although the maximum strain of measuring point 5-2 at 55 min was 337.8 με, exceeding the alarm value, it was still within the safety range. After the additional tugboat starts to work, the overall strain of the pipeline gradually reduced and then becomes stable. Finally, the pipeline Section 2 was transported to the designated position safely.

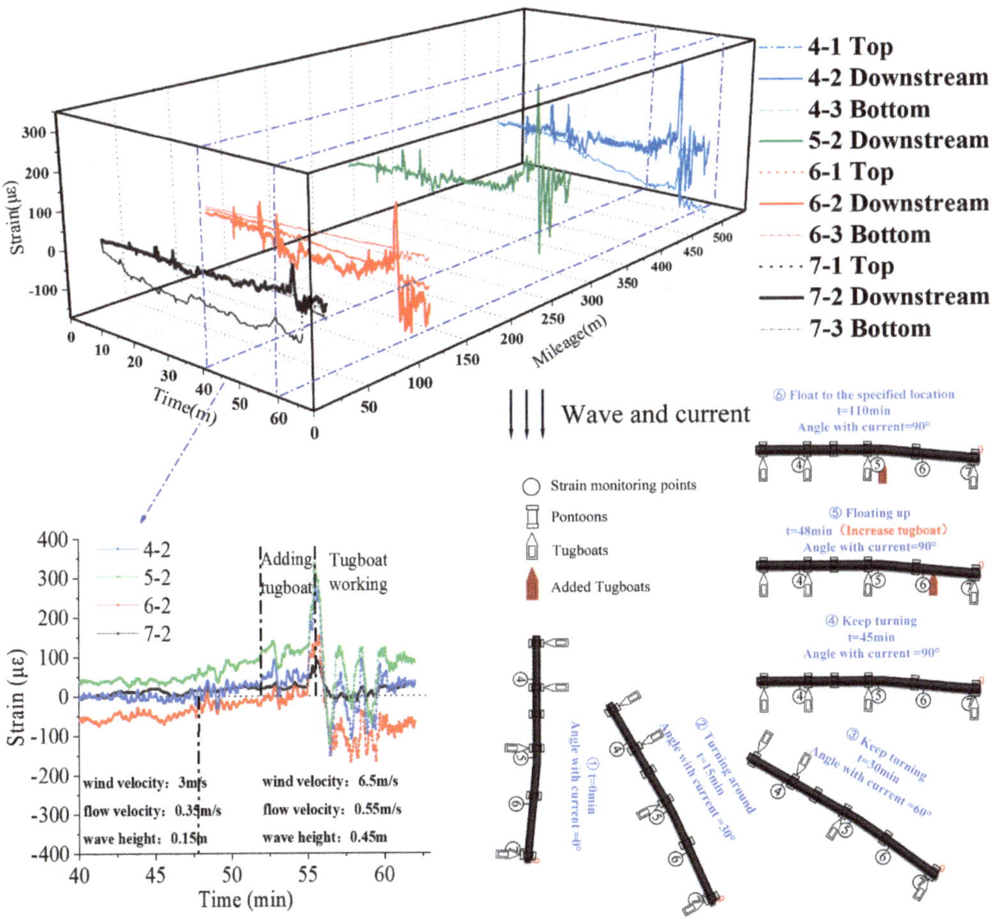

Figure 11. Dynamic floating construction of pipeline Section 2.

4.3. Modal Monitoring Results

The pipeline modal test adopted the environmental excitation method. The response signals of all the measurement points arranged on the pipeline were collected at once, and then the response signals were analyzed in the frequency spectrum to identify the modal parameters of the pipeline. The spectrum analysis adopted the power spectrum method to identify the system modal parameters by using the self-power spectrum of the pipeline response output and the mutual power spectrum amplitude of the reference point response input. The high-frequency acquisition could make the collected signal closer to the original signal and reduce the influence of industrial noise. As the floating time was as long as 3 h, the modal test of the floating process of pipeline segment 2 was divided into two stages, and the parameter balance was carried out respectively.

As shown in Figure 12 that A-1, B-1, B-2, C-2, D-1, D-2 are measured in the pipeline modal test. In the first stage of the test, at about 55 min, the time-history curve of acceleration increases significantly, which is the same as the time of strain surge. Since the sailing speed of the additional tugboat was not synchronized with that of other tugboats and the tugboats had been adjusting their positions, the movement of the extra tugboats had a temporary impact on the steel pipeline and the waves and flows near the pipeline. Therefore, there are some big fluctuations in the acceleration time-history curve between 55–93 min, among which the acceleration time-history curve at 62 min and 92 min has a large increase, but the duration is short.

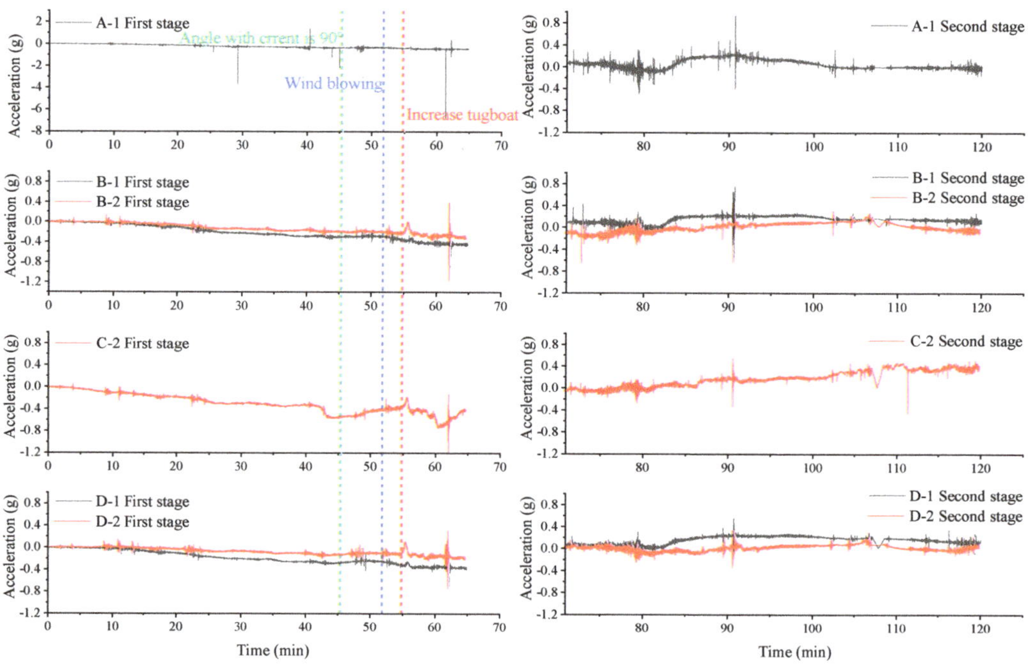

Figure 12. Data diagram of floating vibration of pipeline segment 2.

As shown in Figure 13 that according to the spectrum analysis, from the beginning of the floating movement to the wind rise, the horizontal and vertical first-order vibration frequency is 16.602 Hz, the frequency changes with a small increase to 17.578 Hz after adding a tugboat. At the beginning of floating transportation, the axis of the pipeline was parallel to the direction of the water flow, and the amplitude was the smallest. When the pipeline angle was transferred to 90°, the amplitude increased slightly, and the amplitude reached the maximum value after the wind rose. Then the tugboat was added, and the amplitude of the pipeline decreased significantly while the frequency increased.

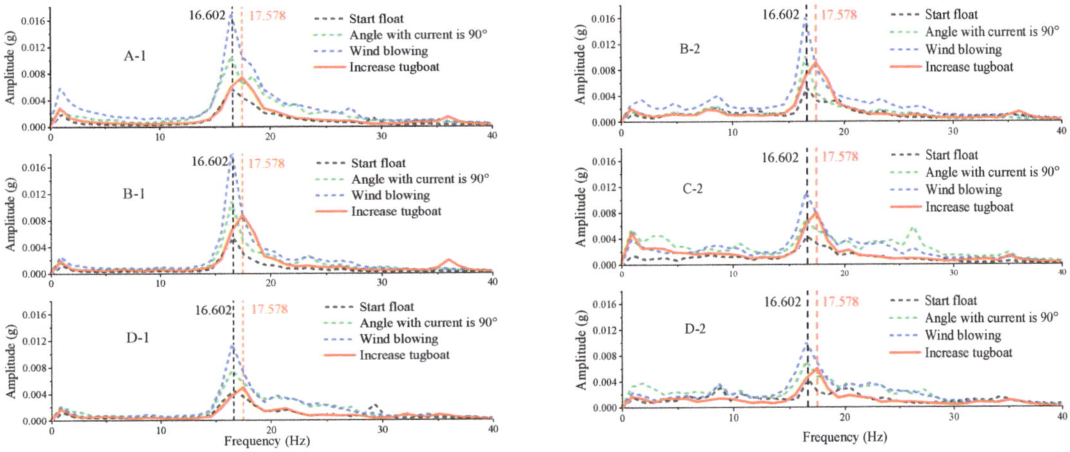

Figure 13. Frequency spectrum of pipeline segment 2 floating transport.

During the whole process of floating transportation, on-site real-time monitoring and dynamic construction reduced the impact of abrupt weather on the vibration of the pipeline, and the monitoring results show that there is no resonance phenomenon with excessive amplitude. All these further verify the rationality of the dynamic construction scheme based on the timely feedback of monitoring data.

5. Discussion

1. During the floating transportation of pipelines 1 and 2, the strain of the upstream and the downstream was obviously greater than that of the top, which was because that the upstream and the downstream were directly affected by water flow and waves, while the top part was less affected. In addition, the strain at the bent point of the pipeline was also larger than that of other parts. The bend point was not only affected by the vertical action of water flow and waves, but also affected by the component force transmitted by the bend section.
2. Floating posture and wind speed will affect the amplitude of the pipeline, but not the vibration frequency of the pipeline. However, increasing the number of tugboats will increase the vibration frequency of the pipeline and decrease the amplitude. For the same vibration system, the amplitude may change, but the vibration frequency is unchanged. Floating posture and wind speed do not change the weight and stiffness of the pipeline structure (i.e., vibration system), but increasing or decreasing the number of tugboats will change the weight and stiffness of the whole structure composed of tugboats and steel pipelines.
3. The strain at the bending point of the pipeline must be paid special attention to during the dynamic construction of the floating pipeline. The construction institution shall reasonably arrange the number of tugboats according to the real-time monitoring data, pay attention to controlling the speed of tugboats and the size of the towing force, and prevent the damage of anticorrosive coating and the vibration damage of pipeline due to excessive bending strain.
4. Through the reasonable arrangement of tugboats and buoyancy tanks for the super-long pipeline, the dynamic construction of the long-distance floating process can be realized with on-site monitoring. Ensure that the deformation and vibration of the

steel pipeline during floating transport are within a safe range. The obtained data and conclusions can provide a useful reference for similar pipeline floating construction.
5. The limitation of the discrete sensor is that it can only measure a finite section of the structure and cannot cover the whole length. If conditions permit, construction institutions or researchers can use distributed optical fiber sensors to realize non-destructive monitoring of a full kilometer distance, with more accurate event identification and positioning capabilities and more accurate and timely early warning.

6. Conclusions

Considering that the floating process of super-long steel pipeline at sea is a nonlinear geometric large deformation problem and is very sensitive to the load. This paper conducts field monitoring and dynamic construction of the super-long steel pipeline offshore floating project to feedback the deformation state of the pipeline to the construction institution, enabling construction institutions to adjust construction operations. Strictly control the deformation of the pipeline to ensure the safety of the extra-long steel pipeline floating, and obtain valuably measured and analyzed data, which provides a reference for similar projects. The specific conclusions are as follows:

- In the process of floating transportation, the pipeline was under the stress state of the upstream and the downstream, and the strain of the upstream and the downstream was obviously greater than that of the top of the steel pipeline. The strain at the bend was larger than that at the straight section. The long-distance pipeline could also maintain good consistency in its stress state while floating. Therefore, in the process of construction and monitoring, special attention should be paid to the upstream, downstream and bend of the floating steel pipeline, to keep the force on the pipeline uniform.
- If other structures (such as tugboats) were not added, the vibration frequency of the pipeline would remain basically unchanged. The amplitude would increase with the increase of wind, wave and water flow, and decrease obviously with the increase of the number of tugboats. The number, position and speed of tugboats both affected the pipeline, so the tugboats must be adjusted reasonably in the construction operation.
- The offshore floating transport pipeline should be subjected to on-site strain and modal monitoring at the same time. The monitoring personnel should feedback the data to the construction institution in real-time, giving a warning when necessary, and use the monitoring data to guide the construction institution to adjust the construction plan reasonably and dynamically. To ensure the strain and anticorrosion coating of the steel pipeline are within a safe range during the floating transportation and prevent the vibration damage of the pipeline. If it is possible, distributed optical fiber sensors can also be used for full-length and overall non-destructive monitoring of the kilometer pipeline, its event recognition and positioning capability are more accurate and early warning will be more accurate and timely.

Author Contributions: Conceptualization, J.Y.; Data curation, C.R., J.C. and W.C.; Formal analysis, Y.C.; Funding acquisition, J.Y.; Investigation, J.Y. and Y.W.; Methodology, C.R.; Supervision, Y.W.; Validation, Y.W.; Visualization, Y.C.; Writing—original draft, J.Y.; Writing—review & editing, C.R. All authors have read and agreed to the published version of the manuscript.

Funding: This work was supported in part by the National Natural Science Foundation of China under Grant 51874144, Grant 51978292, and Grant 42077254, in part by the Scientific and Technology Plan Projects in Xiamen, China under Grant 3502Z20193040, in part by the Subsidized Project for Postgraduates' Innovative Fund in Scientific Research of Huaqiao University under Grant 18014086003 and Grant 18013086018.

Informed Consent Statement: Informed consent was obtained from all subjects involved in the study.

Data Availability Statement: The data presented in this study are available on request from the corresponding author.

Acknowledgments: This work was supported by the National Natural Science Foundation of China (51874144, 5197829 and 42077254); Scientific and Technology Plan Projects in Xiamen, China (3502Z20193040); and Subsidized Project for Postgraduates' Innovative Fund in Scientific Research of Huaqiao University (18014086003 and 18013086018).

Conflicts of Interest: The authors have declared that we have no financial and personal relationships with other people or organizations that can inappropriately influence our work.

References

1. Ye, J.; Zhou, X.; Yu, J.; Fu, D.; Wang, L. Experimental Study on Integral Sinking of Ultralong Asymmetric Steel Pipeline. *J. Coast. Res.* **2020**, *106*, 672. [CrossRef]
2. Cowan, R.M.; Andris, R. Total pipelaying system dynamics. In Proceedings of the Offshore Technology Conference, Houston, TX, USA, 2–5 May 1977; Document IDOTC-2914-MS. [CrossRef]
3. Nomura, Y.; Hoshina, H.; Shiomi, H.; Umezu, T. Pipe Jacking Method for Long Curve Construction. *J. Constr. Eng. Manag.* **1985**, *111*, 138–148. [CrossRef]
4. Guo, A.; Fang, Q.; Li, H. Analytical solution of hurricane wave forces acting on submerged bridge decks. *Ocean Eng.* **2015**, *108*, 519–528. [CrossRef]
5. Wei, G.; Lu, S.-J.; Wang, Z.; Huang, X. A Theoretical Model for the Circumferential Strain of Immersed Tunnel Elements under Tidal Load. *Geotech. Geol. Eng.* **2017**, *36*, 1633–1645. [CrossRef]
6. Ning, D.; Chen, L.; Lin, H.; Zou, Q.; Teng, B. Interaction mechanisms among waves, currents and a submerged plate. *Appl. Ocean Res.* **2019**, *91*, 101911. [CrossRef]
7. Hu, Z.Z.; Causon, D.M.; Mingham, C.G.; Qian, L. Numerical simulation of floating bodies in extreme free surface waves. *Nat. Hazards Earth Syst. Sci.* **2011**, *11*, 519–527. [CrossRef]
8. Zhao, X.; Ye, Z.; Fu, Y.; Cao, F. A CIP-based numerical simulation of freak wave impact on a floating body. *Ocean Eng.* **2014**, *87*, 50–63. [CrossRef]
9. Ong, M.C.; Kamath, A.; Bihs, H.; Afzal, M.S. Numerical simulation of free-surface waves past two semi-submerged horizontal circular cylinders in tandem. *Mar. Struct.* **2017**, *52*, 1–14. [CrossRef]
10. Guo, A.; Fang, Q.; Bai, X.; Li, H. Hydrodynamic Experiment of the Wave Force Acting on the Superstructures of Coastal Bridges. *J. Bridge Eng.* **2015**, *20*, 04015012. [CrossRef]
11. Banks, M.; Abdussamie, N. The response of a semisubmersible model under focused wave groups: Experimental investigation. *J. Ocean Eng. Sci.* **2017**, *2*, 161–171. [CrossRef]
12. Chen, H.; Li, X.; Yan, W.; Chen, S.; Zhang, X. Shaking table test of immersed tunnel considering the geological condition. *Eng. Geol.* **2017**, *227*, 93–107. [CrossRef]
13. Fang, Q.; Guo, A. Analytical and Experimental Study of Focused Wave Action on a Partially Immersed Box. *Math. Probl. Eng.* **2019**, *2019*, 9850302. [CrossRef]
14. Du, Y.; Wu, W.; Wang, Y.; Yue, Q. Prototype data analysis on LH11-1 semi submersible platform in South China Sea. In Proceedings of the 33rd Internationa Conference on Ocean, Offshore and Arctic Engineering, Volume 1B: Offshore Technology, San Francisco, CA, USA, 8–13 June 2014. [CrossRef]
15. Wang, P.; Tian, X.; Peng, T.; Luo, Y. A review of the state-of-the-art developments in the field monitoring of offshore structures. *Ocean Eng.* **2018**, *147*, 148–164. [CrossRef]
16. Brower, D.V.; Prescott, N. Real time subsea monitoring and control smart field solutions. In Proceedings of the Subsea Rio Conference, Rio, Brazil, 3 June 2004.
17. Ren, L.; Li, H.-N.; Zhou, J.; Li, D.-S.; Sun, L. Health monitoring system for offshore platform with fiber Bragg grating sensors. *Opt. Eng.* **2006**, *45*, 084401. [CrossRef]
18. Feng, X.; Wu, W.; Li, X.; Zhang, X.; Zhou, J. Experimental investigations on detecting lateral buckling for subsea pipelines with distributed fiber optic sensors. *Smart Struct. Syst.* **2015**, *15*, 245–258. [CrossRef]
19. Xu, J.; Yang, D.; Qin, C.; Jiang, Y.; Sheng, L.; Jia, X.; Bai, Y.; Shen, X.; Wang, H.; Deng, X.; et al. Study and Test of a New Bundle-Structure Riser Stress Monitoring Sensor Based on FBG. *Sensors* **2015**, *15*, 29648–29660. [CrossRef]
20. Jin, Y.; Eydgahi, A. Monitoring of distributed pipeline systems by wireless sensor networks. In Proceedings of the 2008 IAJC-IJME International Conference, Banff, AB, Canada, 17–19 November 2008.
21. Wang, K.; Gao, H.; Xu, X.; Jiang, J.; Yue, D. An Energy-Efficient Reliable Data Transmission Scheme for Complex Environmental Monitoring in Underwater Acoustic Sensor Networks. *IEEE Sens. J.* **2016**, *16*, 4051–4062. [CrossRef]
22. Mahmutoglu, Y.; Turk, K. A passive acoustic based system to locate leak hole in underwater natural gas pipelines. *Digit. Signal Process.* **2018**, *76*, 59–65. [CrossRef]
23. von Aschwege, J.; Jassal, K.; Barker, R. An independent remote monitoring system for gulf of mexico deepwater floating production systems. In Proceedings of the Offshore Technology Conference, Houston, TX, USA, 30 April–3 May 2007. [CrossRef]
24. Seddiq, Y.M.; Alotaibi, A.M.; Al-Nasheri, A.Y.; Almazyad, A.S.; BenSaleh, M.S.; Qasim, S.M.; Ahmed, Y. Evaluation of energy-efficient cooperative scheme for wireless sensor nodes used in long distance water pipeline monitoring systems. In Proceedings of the 2013 Fifth International Conference on Computational Intelligence, Communication Systems and Networks, Washington, DC, USA, 5–7 June 2013. [CrossRef]

25. Huang, C.; Nagarajaiah, S. Robotic MFL sensor and vibration based deepwater riser monitoring system. In Proceedings of the Offshore Technology Conference, Houston, TX, USA, 30 April–3 May 2012. [CrossRef]
26. Gómez, C.; Green, D.R. Small unmanned airborne systems to support oil and gas pipeline monitoring and mapping. *Arab. J. Geosci.* **2017**, *10*, 202. [CrossRef]
27. Wang, Y.; Qian, X.; Liew, J.R.; Zhang, M.-H. A numerical and theoretical investigation on composite pipe-in-pipe structure under impact. *Steel Compos. Struct.* **2016**, *22*, 1085–1114. [CrossRef]
28. Lagos, S.M.E.; Fiallos, D.C.C.; Toasa, J.P.G.; Salinas, C.P.; Zumbana, J.P. Influencia del contenido de oxígeno atmosférico en el proceso de purgado con argón en la soldadura TIG de tubería de acero inoxidable AISI 304. *DYNA* **2019**, *86*, 355–361. [CrossRef]
29. Kiss, I.; Alexa, V. Study on Deformation Behavior of Non–Hardenable Austenitic Stainless Steel (Grade X5CrNi18–10) by Hot Torsion Tests. *Teh. Glas.* **2020**, *14*, 396–402. [CrossRef]
30. Wenpeng, G.; Yuqing, L.; Chao, L. Stress analysis and mitigation measures for floating pipeline. In Proceedings of the 2nd International Conference on Advances in Energy Resources and Environment Engineering, Guangzhou, China, 30–31 December 2016. [CrossRef]
31. Chen, L.; Arzaghi, E.; Abaei, M.M.; Garaniya, V.; Abbassi, R. Condition monitoring of subsea pipelines considering stress observation and structural deterioration. *J. Loss Prev. Process. Ind.* **2018**, *51*, 178–185. [CrossRef]
32. Arumugam, T.; Karuppanan, S.; Ovinis, M. Finite element analyses of corroded pipeline with single defect subjected to internal pressure and axial compressive stress. *Mar. Struct.* **2020**, *72*, 102746. [CrossRef]
33. Aniskin, A.; Lymarenko, A.M. Calculation of displacements and stresses in cylindrical shells by the boundary elements method. *Teh. Glas.* **2018**, *12*, 196–203. [CrossRef]
34. Chen, Q.; Shen, G.; Jiang, J.; Diao, X.; Wang, Z.; Ni, L.; Dou, Z. Effect of rubber washers on leak location for assembled pressurized liquid pipeline based on negative pressure wave method. *Process. Saf. Environ. Prot.* **2018**, *119*, 181–190. [CrossRef]
35. Yang, Z.; Fan, S.; Xiong, T. Simulation and numerical calculation on pipeline leakage process. In Proceedings of the 2010 2nd International Symposium on Information Engineering and Electronic Commerce, Ternopil, Ukraine, 23–25 July 2010. [CrossRef]
36. Xinhong, L.; Guoming, C.; Renren, Z.; Hongwei, Z.; Jianmin, F. Simulation and assessment of underwater gas release and dispersion from subsea gas pipelines leak. *Process. Saf. Environ. Prot.* **2018**, *119*, 46–57. [CrossRef]
37. Almazyad, A.S.; Seddiq, Y.M.; Alotaibi, A.M.; Al-Nasheri, A.Y.; BenSaleh, M.S.; Obeid, A.M.; Qasim, S.M. A Proposed Scalable Design and Simulation of Wireless Sensor Network-Based Long-Distance Water Pipeline Leakage Monitoring System. *Sensors* **2014**, *14*, 3557–3577. [CrossRef]
38. Jin, W.-L.; Shao, J.-W. A practical algorithms on health monitoring of submarine pipeline and Its application. In Proceedings of the 25th Internationa Conference on Ocean, Offshore and Arctic Engineering, Hamburg, Germany, 4–9 June 2006. [CrossRef]
39. Goldsmith, B.; Foyt, E.; Hariharan, M. The role of offshore monitoring in an effective deepwater riser integrity management program. In Proceedings of the International Conference on Offshore Mechanics and Arctic Engineering-OMAE, San Diego, CA, USA, 10–15 June 2007. [CrossRef]
40. Yang, B.; Gao, F.-P.; Jeng, D.-S.; Wu, Y.-X. Experimental study of vortex-induced vibrations of a pipeline near an erodible sandy seabed. *Ocean. Eng.* **2008**, *35*, 301–309. [CrossRef]
41. Zhang, S.; Wang, Y.; Liu, L.; Zhang, Q. Dynamic analysis of a steel floating pipeline on sea. *Procedia Eng.* **2011**, *12*, 60–65. [CrossRef]
42. Li, B.; Wang, H.; Shen, X.; Yan, Y.; Yang, F.; Hua, F. Deep-water riser fatigue monitoring systems based on acoustic telemetry. *J. Ocean. Univ. China* **2014**, *13*, 951–956. [CrossRef]
43. Sollund, H.A.; Vedeld, K.; Hellesland, J.; Fyrileiv, O. Dynamic response of multi-span offshore pipelines. *Mar. Struct.* **2014**, *39*, 174–197. [CrossRef]
44. Cook, H.; Dopjera, D.; Thethi, R.; Williams, L. Riser integrity management for deepwater developments. In Proceedings of the Offshore Technology Conference, Houston, TX, USA, 1–4 May 2006. [CrossRef]
45. Akram, A.; Mustaffa, Z.; Albarody, T.M.B. Burst capacity of pipe under corrosion defects and repaired with thermosetting liner. *Steel Compos. Struct.* **2020**, *35*, 171–186. [CrossRef]
46. López, J.P.A.; Jaramillo-Justinico, A.; Salazar-Uribe, J.C.; Mazo-Lopera, M.A.; Rodríguez-Cortés, F.J. Una metodología para análisis de impactos ambientales en regiones de offshore. Ejemplo de caso territorios marinos profundos frente a la costa del departamento de Córdoba Colombia. *DYNA* **2020**, *87*, 222–231. [CrossRef]
47. Ren, Y.X.; Chen, H.X. *Basis of Computational Fluid Mechanics*; Tsinghua University Press: Beijing, China, 2006. (In Chinese)

applied sciences

MDPI

Article

Temperature Compensated Wide-Range Micro Pressure Sensor with Polyimide Anticorrosive Coating for Harsh Environment Applications

Mengru Jiao [1], Minghao Wang [1,2,*], Ye Fan [1], Bangbang Guo [1], Bowen Ji [3], Yuhua Cheng [1,2,*] and Gaofeng Wang [1,*]

1 MOE Engineering Research Center of Smart Microsensors and Microsystems, School of Electronics & Information, Hangzhou Dianzi University, Hangzhou 310018, China; jmr@hdu.edu.cn (M.J.); 16041701@hdu.edu.cn (Y.F.); bbguo@hdu.edu.cn (B.G.)
2 Wenzhou Institute of Hangzhou Dianzi University, Wenzhou 325038, China
3 The Unmanned System Research Institute, Northwestern Polytechnical University, Xi'an 710060, China; bwji@nwpu.edu.cn
* Correspondence: mhwang@hdu.edu.cn (M.W.); chengyh@hdu.edu.cn (Y.C.); gaofeng@hdu.edu.cn (G.W.)

Abstract: In this work, a MEMS piezoresistive micro pressure sensor (1.5 × 1.5 × 0.82 mm) is designed and fabricated with SOI-based micromachining technology and assembled using anodic bonding technology. In order to optimize the linearity and sensitivity over a wide effective pressure range (0–5 MPa) and temperature range (25–125 °C), the diaphragm thickness and the insulation of piezoresistors are precisely controlled by an optimized micromachining process. The consistency of the four piezoresistors is greatly improved by optimizing the structure of the ohmic contact pads. Furthermore, the probability of piezoresistive breakdown during anodic bonding is greatly reduced by conducting the top and bottom silicon of the SOI. At room temperature, the pressure sensor with 40 μm diaphragm demonstrates reliable linearity (0.48% F.S.) and sensitivity (33.04 mV/MPa) over a wide pressure range of 0–5.0 MPa. In addition, a polyimide protection layer is fabricated on the top surface of the sensor to prevent it from corrosion by a moist marine environment. To overcome the linearity drift due to temperature variation in practice, a digital temperature compensation system is developed for the pressure sensor, which shows a maximum error of 0.43% F.S. in a temperature range of 25–125 °C.

Keywords: temperature compensation; wide-range; pressure sensor; polyimide anticorrosive coating; harsh environment

Citation: Jiao, M.; Wang, M.; Fan, Y.; Guo, B.; Ji, B.; Cheng, Y.; Wang, G. Temperature Compensated Wide-Range Micro Pressure Sensor with Polyimide Anticorrosive Coating for Harsh Environment Applications. *Appl. Sci.* **2021**, *11*, 9012. https://doi.org/10.3390/app11199012

Academic Editor: Dimitrios-Nikolaos Pagonis

Received: 18 July 2021
Accepted: 24 September 2021
Published: 28 September 2021

1. Introduction

The piezoresistive effect of a semiconductor was first discovered in the 1950s, which is the theoretical basis of most pressure sensors [1–3]. In addition to piezoresistive sensors, several different transduction mechanisms have been applied to convert pressures or strains to signals, such as inductance, capacitance, piezoelectricity, and resonance, which can be detected on circuitry [4,5]. Among these designs, piezoresistive sensors are generally accepted due to their simple construction and low energy consumption [6]. The micro pressure sensors have been greatly developed and played a pivotal role in the automobile industry, aerospace industry, and medical equipment with the promotion of MEMS technology [7–11]. In marine engineering, pressure sensors are commonly used in applications involving high pressure up to several megapascal (MPa), high temperature up to several hundred degrees Celsius, and moist external conditions. However, some intrinsic pitfalls of the silicon piezoresistive pressure sensor, such as inadequate performance in harsh environment, e.g., high pressure or high temperature, and output drift due to temperature variation, have limited its application in marine settings.

In order to improve the performance of the micro piezoresistive pressure sensor, numerous studies have focused on optimization of structure, fabrication, and encapsulation.

Dong et al. [12] introduced a monolithic composite MEMS pressure sensor with an effective range of 450 kPa. Nag et al. [13] modified the structure by integrating rod beams in a silicon diaphragm. A pressure range of 0–689.5 kPa was achieved. Sheeparamatti et al. [14] fabricated two micro pressure sensors based on polysilicon on insulator (PolySOI) and amorphous silicon on insulator (a-SOI). These two sensors provided a pressure range of 0–1 MPa. Alcheikh et al. [15] presented the fabrication and electromechanical characterization of silicon-based 3D force micro sensors with piezoresistive gauges. The high sensitivity, the high linearity, and the low hysteresis of the sensors submitted to a normal force are obtained. Sujit et al. [16] incorporated FEM (finite element method) to give suggestions on the selection of parameters for the piezoresistors, such as shape and the position on the sensitive diaphragm.

To prevent micro sensors from being damaged by external environment, most pressure sensors are encapsulated by silicone rubber. However, a large deformation of silicone rubber under a high pressure could lead to a fracture of bonding wires. Previous published work [17,18] has presented an anticorrosive PDMS (polydimethylsiloxane) coating to prevent the conductive layer of a strain sensor from being damaged by corrosive external environment. In addition, Won et al. [19] spun polyimide film onto the electrode of the sensor to protect it from the humid environment. However, hardly any previous research has focused on the performance and protection of the micro sensors under a harsh environment, such as high pressure, high temperature, or moist corrosive electrolyte environment, which will lead to a shortening of the micro sensor life. Therefore, our work aims at resolving the contradiction between the output performance and stability with a miniature size in harsh environment.

Previous research has established two approaches to treat nonlinearity temperature drift. The first approach is to use an analog compensation circuit, such as the external thermistor. Aryafar et al. [20] presented a passive technique for temperature compensation of a silicon pressure sensor based on extra polysilicon resistors with negative temperature coefficient of resistivity. The second approach is to use a digital compensation system with the combination of electronic circuits and specific software. However, the first approach has intrinsic pitfalls such as poor reliability, low accuracy, and high design difficulty [21]. Concerning the second approach, neural network has recently been applied. Yang et al. proposed an artificial neural network (ANN) [22] to establish a temperature shift compensated model. Zhou et al. developed an extreme learning machine (ELM) to calibrate the pressure shift [23]. Neural networks, however, require a long training time and a large amount of calculation, which is a barrier for marine engineering applications. Therefore, our work proposes a low-cost pressure measurement system with temperature compensation based on embedded technology and rational polynomial fitting to reduce the cost of temperature compensation and increase the flexibility of the compensation system.

In this work, a MEMS piezoresistive pressure sensor with ideal performance over a wide effective range by optimizing the micromachining process is designed, simulated, fabricated, and tested. Firstly, the optimization greatly enhances the performance and the stability of the pressure sensor under harsh environment such as high temperature and pressure. Secondly, a novel polyimide anticorrosive layer is fabricated by polyimide film on the top of the micro sensor to prevent it from corrosion by an electrolyte such as seawater. Finally, to compensate the temperature drift, a digital low-cost temperature compensation system is built for this micro pressure sensor.

2. Design and Fabrication

2.1. Design and Fabrication of the Micro Sensor

In general, a piezoresistive pressure sensor has a sensing diaphragm [24–27]. For piezoresistive sensors, the sensitive diaphragms above the vacuum cavities are pressured and then generate stress. The variation of piezoresistors caused by the applied pressure is transferred into voltage due to the piezoresistive effect [28,29]. Therefore, the performance of a piezoresistive sensor depends largely on the pressure sensitive diaphragm. Firstly, the dimensions of the pressure sensor's diaphragm were designed based on previous published work [12]. Then, the thickness of the diaphragm was determined by theoretical calculation from Expression (1), where P_{burst} is five times the maximum applied pressure, $\sigma_{fracture}$ is the maximum stress of the diaphragm, A is the area of the diaphragm, h is its thickness, and v is the Poisson's ratio. The calculated minimum thickness is 26.87 μm. Therefore, the simulation was done in COMSOL Multiphysics FEA (finite element analysis) software for a thickness of more than 30 μm. Based on this, the pressure sensor models have an n-type silicon diaphragm with the dimensions 900 × 450 μm and thicknesses of 40 and 50 μm. The von Mises stress induced in the diaphragm with various thicknesses were determined and evaluated with the analytic explanation for a face-loaded pressure of 5 MPa. The simulation results obtained from the FEA tool are shown in Figure 1a–d. Along the centerline of the diaphragm, the tensile stress at the diaphragm edge reaches its maximum and the compressive stress in the center of the diaphragm reaches its maximum. This resulted from the silicon diaphragm contracting at the center and stretching at the edges when the pressure was applied uniformly from the top. Therefore, the piezoresistors should be located in these places to increase the sensitivity. By comparing the simulation results of the pressure sensor models with different diaphragm thicknesses, it can be found that the surface stress of the pressure sensor with 40-μm-thick membrane is larger, indicating that the thinner membrane thickness exhibits the higher sensitivity.

$$P_{burst} = \frac{3.4}{1-v^2} \times \frac{\sigma_{fracture} \times h^2}{A} \tag{1}$$

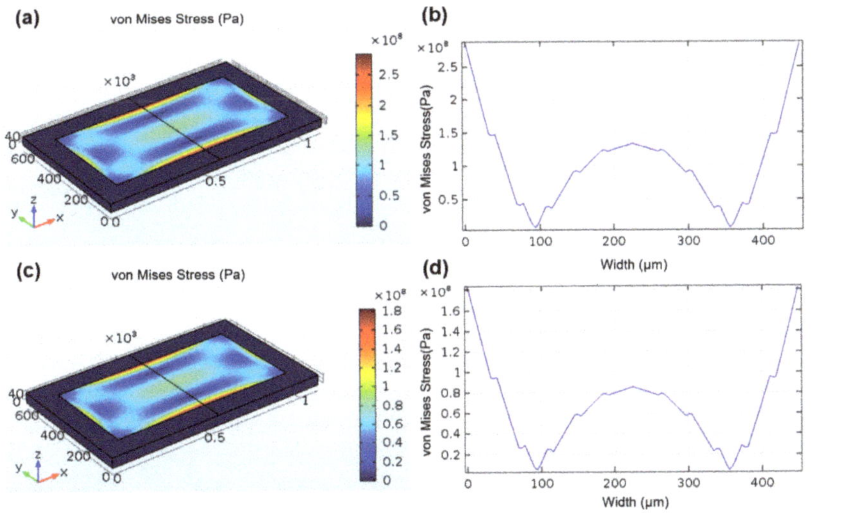

Figure 1. Von Mises stress distribution under finite element analysis: (**a**,**c**) stress distribution under 5 MPa of face-loaded pressure; (**b**,**d**) transverse stress distribution along the centerline.

The schematic diagram of the proposed silicon piezoresistive pressure sensor is shown in Figure 2. The pressure sensor has a rectangular silicon diaphragm, which is fabricated on the device layer of a SOI wafer [30,31]. By using buried-oxygen-layer etching stop technology [32], the diaphragm thickness of the pressure sensor can be precisely controlled. After one-mask deep reactive ion etching (DRIE), the pressure sensor's rectangular diaphragm is shaped. The thickness of the diaphragm is the same as that of the device layer. The piezoresistors are formed on the silicon diaphragm by boron ion implantation. Four piezoresistors are connected to a Wheatstone bridge structure by forming ohmic contact with aluminum wires. The contact area between the aluminum conductor and the heavily doped region of the piezoresistors has a great influence on the resistance value of the four piezoresistors in the Wheatstone bridge. Therefore, the consistency of ohmic contact area should be ensured by improving the accuracy of lithography and etching, and increasing the width of aluminum wire relative to that of piezoresistors. Furthermore, a layer of SiO_2 is deposited and patterned on the substrate in the plasma enhanced chemical vapor deposition (PECVD) system, as a passivation layer to improve the reliability of pressure sensors in harsh environments. Dimensions of the diaphragm and the piezoresistors are shown in Figure 2c. The silicon diaphragm has dimensions of 900 × 450 × 50/40 μm. Four boron-doped piezoresistors (250 × 20 μm) are arranged in parallel on the diaphragm along the <110> direction and connected with aluminum wire to form a Wheatstone bridge.

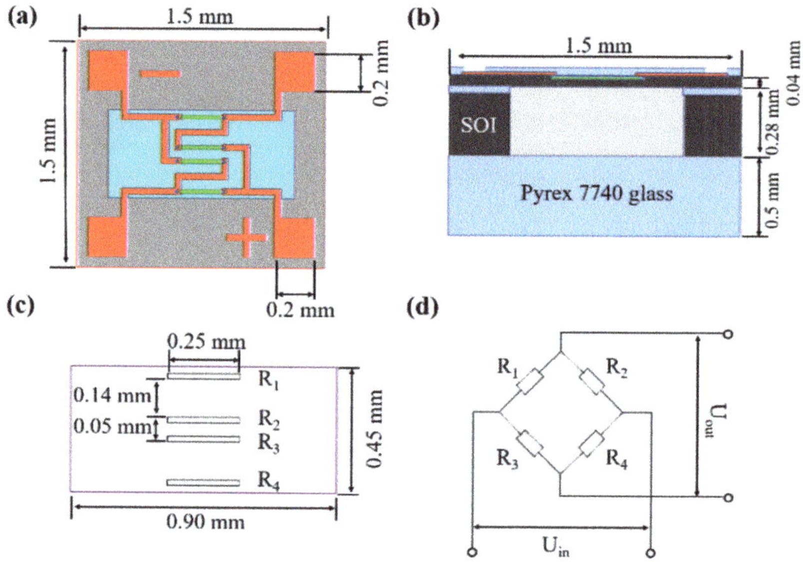

Figure 2. Schematic diagrams of the silicon piezoresistive pressure sensor: (**a**) top view and (**b**) cross-section view of the pressure sensor; (**c**) dimensional parameters; (**d**) Wheatstone full-bridge of the pressure sensor.

The silicon piezoresistive pressure sensor is fabricated with the micromachining technology and assembled using the anodic bonding technology. As illustrated in Figure 3, the substrate of the pressure sensor is an n-type (100) SOI wafer with a resistivity of 1–10 Ω·cm. The thicknesses of the silicon device layer, the buried oxide layer, and the handle layer of the SOI are 50/40, 0.5, and 280 μm, respectively. The main processes follow: Firstly, the device layer is patterned by photoresist and doped by boron implantation with a doping concentration of 4×10^{14} atoms/cm^2 to form piezoresistors (steps 1 and 2). The doped region is then covered by a polyimide film, exposing only the alignment marks. By using the deep reactive ion etching (DRIE) technique, the alignment marks are transferred to the silicon substrate (step 3). Secondly, the contact areas between the Al wires and the piezoresistors are exposed by patterned photoresist and doped by heavy boron implantation with a doping concentration of 2×10^{15} atoms/cm^2 to form the ohmic contact (step 4). Subsequently, a layer of 500 nm-thick silicon oxide is deposited in the PECVD system to form the bottom passivation layer, while the ohmic contact areas are exposed by RIE (steps 5 and 6). Then, a layer of 1 μm-thick Al is sputtering deposited and patterned by wet etching to form interconnects and four output pads of the Wheatstone bridge (step 7). After that, a layer of 500 nm-thick silicon oxide is deposited in the PECVD system to form the top passivation layer with the output pads exposed by RIE (steps 8 and 9).

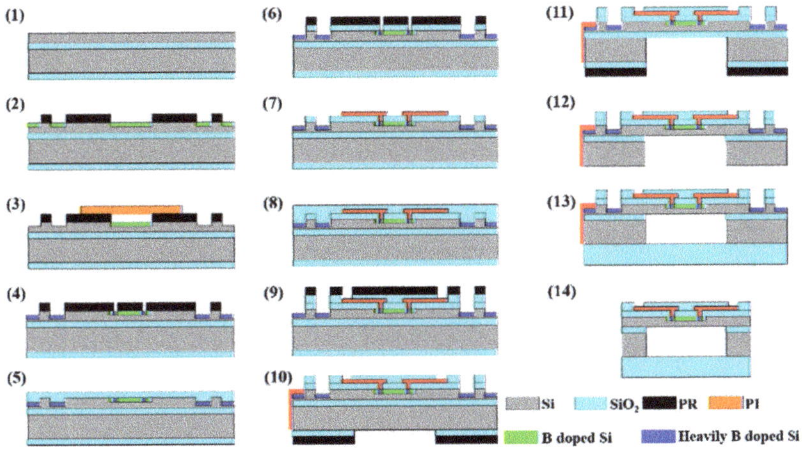

Figure 3. Fabrication process of the silicon piezoresistive pressure sensor.

In order to prevent the pizeoresistors from breakage by the voltage applied during anodic bonding, the passivation layers on the edge of the SOI are also etched to expose the device layer and another 1 μm-thick Al layer is sputtered and patterned to connect the device layer and handle layer. Next, the pressure sensor's sensing diaphragm is patterned by double-sided alignment lithography and etched by DRIE and RIE (steps 10, 11, and 12). After polishing by CMP, the handle layer of SOI is bonded to Pyrex7740 glass by anodic bonding with voltage of 1600 V, current of 20 mA, temperature of 300 °C, pressure of 1600 N, and time of 20 min (step 13). Finally, the sensor dies are released from the SOI wafer by using a dicing saw (step 14).

2.2. Fabrication of Polyimide(PI) Protection Layer

To protect the micro sensor from the corrosion of the external environment, such as the electrolytic salty seawater, a protective layer composed of a PI (polyimide) film was developed. Polyimide is a kind of organic polymer material with excellent comprehensive properties, such as better high temperature resistance than PDMS, excellent insulation properties, and stability under electrolyte condition. Meanwhile, polyimide can be well attached on the surface of the devices due to its ductility and flexibility. Therefore, a polyimide film was selected as the protective layer for the micro sensor and for protection of the bonding wires.

After bonding the sensor with packaged PCB by gold wire, polyimide solution of 15% concentration was placed into a vacuum chamber for 30 min to remove the air bubbles. Then, the solution was uniformly dripped onto the top surface of the micro pressure sensor. A high-speed rotary coating machine was used to make the polyimide solution form a uniform thin film. The rotation speed, which is an important factor for the thickness of the polyimide film, was set at 500 rpm for 10 s, then 3000 rpm for 30 s. After spinning, the micro sensor with polyimide film was placed in an oven for further solidification. The heating process for the polyimide film was 80 °C for 10 min, 120 °C for 30 min, 150 °C for 10 min, 180 °C for 10 min, and then 220 °C for 40 min. Finally, a thin polyimide film with a thickness of 10 μm was covered on the top surface of the micro pressure sensor as a protective layer to prevent corrosion by moist air and electrolyte droplets. The device with a 10 μm-thin PI protection film and the dice itself are shown in Figure 4.

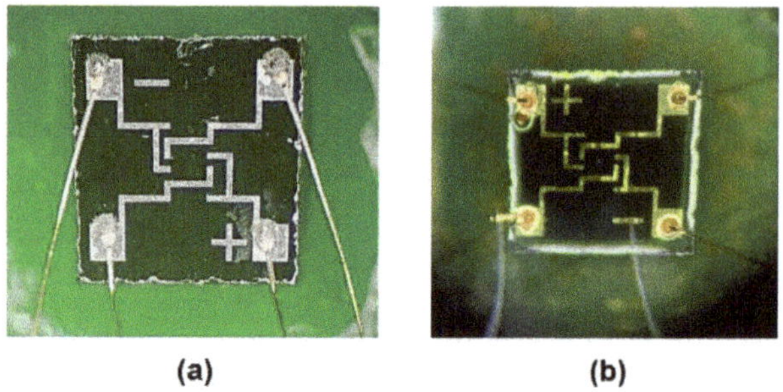

Figure 4. (**a**) Micro pressure sensor after bonding without a protection layer. (**b**) Micro pressure sensor after bonding with a 10 μm PI protection layer.

3. Results and Discussions

3.1. Output Characteristics under High Temperature

The output characteristic of the pressure sensor was measured by a digital pressure test system consisting of a high-precision piston pressure meter, a hot air gun, and a control system. A constant DC voltage of 3.3 V was applied as the incentive power of the Wheatstone bridge, and the output voltage was captured by a digital multimeter (VC8246B).

To test the output characteristics under different temperature conditions, a piezoresistive pressure sensor with a 50 μm sensing diaphragm was pressured from 0 to 3 MPa and the output voltage of the Wheatstone bridge was recorded for every 0.2 MPa. The temperature test points were set from 25 to 125 °C for every 10 °C. The V–P characteristic curve of the pressure sensor under different temperatures is depicted in Figure 5. As shown in Figure 5a, the output voltage exhibits an expected linear correlation in a wide effective pressure range from 0 to 3 MPa and a temperature range from 25 to 125 °C. Sensitivity of the pressure sensor is the ratio of response change to excitation pressure change, and the nonlinearity is the maximum deviation between the actual output of the device and the ideal output curve that changes linearly with the input pressure. Fitting the experimental data with a linear least-square technique, the sensitivity of the pressure sensor was 22.99 mV/MPa and the nonlinearity was 0.25% F.S. at 25 °C. Due to the application of SOI wafer, the pressure sensor showed a stable performance with the nonlinearity of 0.31% F.S. under 125 °C.

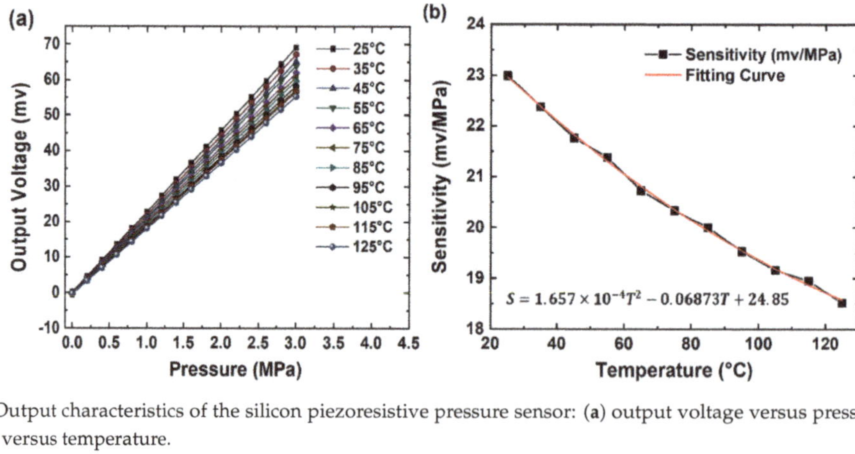

Figure 5. Output characteristics of the silicon piezoresistive pressure sensor: (**a**) output voltage versus pressure and (**b**) sensitivity versus temperature.

The ambient temperature would greatly influence the output characteristics of the pressure sensor. The curve of the sensitivity variation versus the temperature is shown in Figure 5b. To depict the relation between sensitivity and temperature more accurately, the experimental data is fitted by a two-degree polynomial. The results show that the increase of ambient temperature can still lead to a distinct decline of the sensitivity. This is because the buried oxide layer can only reduce the leakage current of the piezojunction in the handle layer, but not in the device layer of the SOI wafer.

3.2. Output Characteristics under High Pressure

In subsequent experiments of pressure measurement range, two devices with different diaphragm thicknesses, 40 and 50 μm, were tested. The V–P curves of the two devices are presented in Figure 6. As depicted, the device with the thinner diaphragm showed a higher sensitivity but with higher nonlinearity over a wide pressure range. Gratifyingly, the two pressure sensors were operated up to 5 MPa with an acceptable linearity. In order to evaluate the performance of the silicon pressure sensors, the main characteristics of tested devices are presented in Table 1, and the comparison among tested devices and the other pressure sensor is presented in Table 2. As can be seen, the increase of ambient temperature leads to the decrease of the full-scale output and sensitivity of the pressure sensor, and the increase of its nonlinearity. The micro sensor is designed for harsh applications; therefore, the thickness of the sensing diaphragm is increased to enhance the performance under high pressure. Note that the increase of diaphragm thickness leads to decreased full-scale output, nonlinearity, and sensitivity of the pressure sensor.

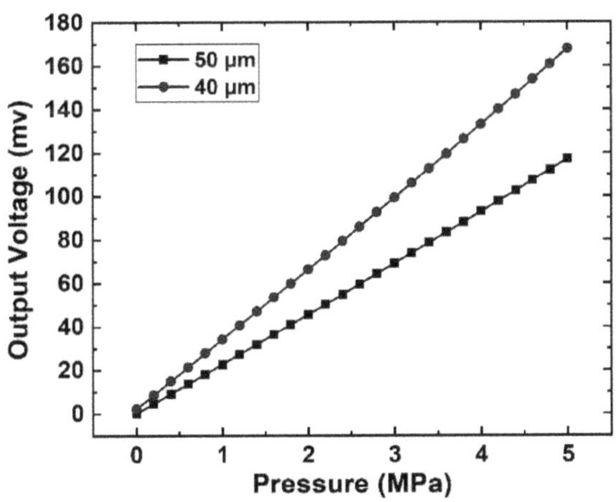

Figure 6. Output voltage of the sensor with different diaphragm thickness.

Table 1. Main characteristics of the tested devices under different test conditions.

Parameters	Units	Tested Devices			
Operation Temperature	°C	25	125	25	25
Thickness of Diaphragm	μm	50	50	40	50
Excitation Voltage	V	3.3	3.3	3.3	3.3
Pressure Range	MPa	0–3	0–3	0–5	0–5
Full-Scale Output	mv	69.28	55.23	167.87	117.1
Sensitivity	mv/MPa	22.99	18.52	33.04	23.55
Nonlinearity	F.S.%	0.25	0.31	0.48	0.41

Table 2. Comparison among our devices, a commercial device, and others described in literature.

Parameters	Units	Fabricated Devices		MSD700-ASO [1]	[14]	[31]
Thickness of Diaphragm	μm	50	40	–	40	1.2
Chip Size	mm	1.5 × 1.5 × 0.82		0.95 × 0.95 × 0.6	1.5 × 1.5	1.6 × 1.6 × 0.9
Excitation Voltage	V	3.3	3.3	5	10	3
Pressure Range	MPa	0–5	0–5	0–0.7	0–1	0–0.55
Full-Scale Output	mv	117.1	167.87	100	105	45.99
Sensitivity	mv/V/MPa	7.85	10.01	28.57	10.5	27.87
Nonlinearity	F.S.%	0.41	0.48	0.3	–	0.34

[1] Product code: MSD700-ASO, MEMSensing Microsystems Co., Ltd., Suzhou, China.

3.3. Output Characteristics under PI Protection

The output characteristics under high pressure range of the PI protected micro pressure sensor were tested. The device with a PI protection layer was pressured from 0 to 5 MPa. In order to compare the influence of the protective layer on the output characteristics, V–P curves of the devices with and without the top surface PI layer are presented in Figure 7.

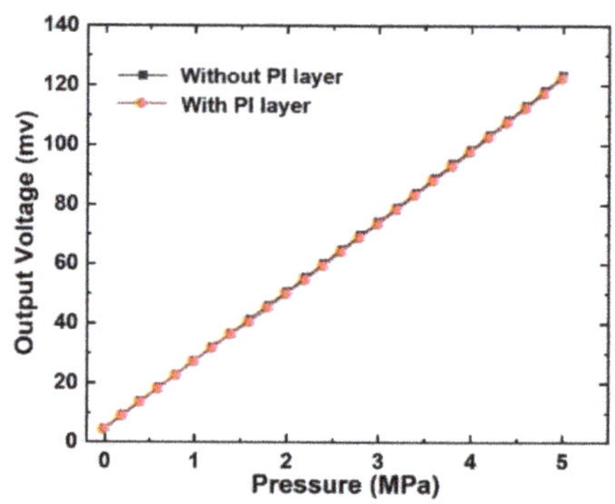

Figure 7. Output voltage of the devices with and without the PI protection layer.

The output curves show that the PI protection layer has almost no effect on the output characteristics of the micro sensor. A further experiment was designed to test the anticorrosive performance of the polyimide layer. Both devices, with and without a polyimide protection layer, were immersed in seawater and heated to 80 °C in an oven to accelerate the corrosion of the devices. The devices were removed from the seawater and dried before a resistance test every 30 min. Normally, the resistance of the Wheatstone bridge was between 1.7 and 2.0 kΩ; if the output resistance was not in the normal range, the device was considered out of operation. The experiment results are shown in Table 3.

Table 3. Resistance of the Wheatstone bridge.

Test Devices	Number	0 min	30 min	60 min	90 min	120 min
Devices without polyimide protective layer	1	normal	open circuit	–	–	–
	2	normal	open circuit	–	–	–
Devices with polyimide protective layer	3	normal	normal	normal	normal	normal
	4	normal	normal	normal	normal	normal

Results show that the devices with the polyimide protective layer could still work after 2 h immersion in seawater, while the devices without PI layer were out of operation after 30 min immersion. It can be inferred that the PI protective layer possesses an exceptional anticorrosive performance. Therefore, the PI protection layer has a broad prospect on the micro pressure sensor for harsh environment applications, such as gas pressure measurement on ships or other marine engineering applications.

In order to compare the stability of PI protection layer and traditional silicone rubber encapsulation under instantaneous violent pressure variation, a dynamic pressure test was designed. Tested pressure sensors were covered with a PI protection layer and silicone rubber, nonencapsulated sensors served as a control group. The start pressure was set at 3, 4, and 5 MPa. After each instantaneous decompression to normal atmospheric pressure, the output resistance was measured and compared with the control group. The process was repeated three times; if the output resistance was abnormal, the process was discontinued. Test results are presented in Table 4.

Table 4. Encapsulation stability tests.

Test Cycle	Protective Layer	Pressure	1	2	3
Cycle 1	Silicone rubber	3 MPa	Pass	Pass	Pass
		4 MPa	Pass	Pass	Abnormal
	Polyimide coat	3 MPa	Pass	Pass	Pass
		4 MPa	Pass	Pass	Pass
		5 MPa	Pass	Pass	Pass
Cycle 2	Silicone rubber	3 MPa	Pass	Pass	Pass
		4 MPa	Abnormal	–	–
	Polyimide coat	3 MPa	Pass	Pass	Pass
		4 MPa	Pass	Pass	Pass
		5 MPa	Pass	Pass	Pass

It can be concluded from the table that the polyimide coat exhibits better stability than traditional silicone rubber encapsulation. Due to less deformation under high pressure compared with silicone rubber, the PI protection layer leads to less potential for abnormal output resistance caused by a fracture of the bonding wires. In conclusion, the polyimide protection layer has reliable corrosion resistance and high stability under violent pressure variation. Thus, it has great potential for application on micro sensors designed for harsh environments and high-pressure applications.

3.4. Temperature Compensation System

In addition to the large temperature difference between day and night on the sea, the operating temperature of the instruments onboard also varies from moment to moment. To overcome the sensitivity drift due to the temperature variation, a practical and flexible pressure measurement system with temperature compensation is developed to calibrate the drift caused by the ambient temperature variation. The design purpose of the embedded digital compensation system is to simplify the complexity and reduce the cost to satisfy the requirement of marine engineering. Hardware of the compensation system consists of a commercial monolithic-integrated 24-bit analog–digital convertor and signal amplifier, a high-precision temperature sensor, power management system, and a low-cost MCU. The system diagram is shown in Figure 8.

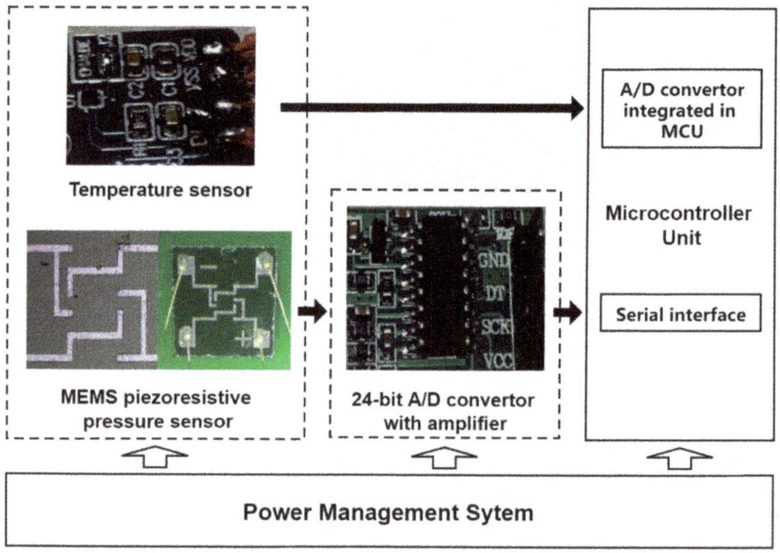

Figure 8. The diagram of the digital pressure measurement system with temperature compensation.

Generally, the output of the pressure sensor at a constant voltage can be expressed as follows:

$$V_{out} = S_T \times P + V_0 \quad (2)$$

where S_T is the sensitivity at T °C, P denotes the incentive pressure, and V_0 represents the zero-drift voltage due to fabrication error when no incentive pressure exists.

Firstly, the zero-drift voltage V_0 is compensated by average approximation. The compensation value is obtained preliminarily by averaging the zero-drift voltages at various temperatures. After compensation, the output voltage can be described as a proportional function:

$$V_Z = V_{out} - \overline{V_0} \approx S_T \times P \quad (3)$$

Secondly, the temperature-drift voltage is compensated by a sensitivity coefficient. After compensation, the output voltage can be described as follows:

$$V_{comp} = K \times V_Z \approx K \times S_T \times P \quad (4)$$

where K denotes the compensation coefficient of S_T The coefficient K can be calculated from:

$$K = \frac{S_{ref}}{S_T} = \| \ S_T = a_1 T^2 + a_2 T + a_3, S_{ref} = 22.99 \ \| \quad (5)$$

where a_1, a_2, and a_3 are the coefficients of the rational polynomial model, and S_{ref} is the sensitivity at 25 °C (which is defined as the reference). After the curve fitting of the sensitivity, the polynomial coefficients a_1, a_2, and a_3 were obtained: 1.657×10^{-4}, -0.06873, and 24.58, respectively.

By simultaneously collecting output data from the pressure sensor and the temperature sensor through the serial ports of the microcontroller, the voltage of the pressure sensor at different temperatures can be mapped to that at the reference temperature (i.e., 25 °C). Therefore, the output voltage of the measurement system becomes weakly correlated with the temperature. The output characteristics of the pressure measurement system before and after compensation are shown in Figure 9a,b. The results show that the temperature compensation system for the pressure sensor could successfully implement the sensitivity drift compensation. The maximum error of the output value after compensation was 0.43% F.S.

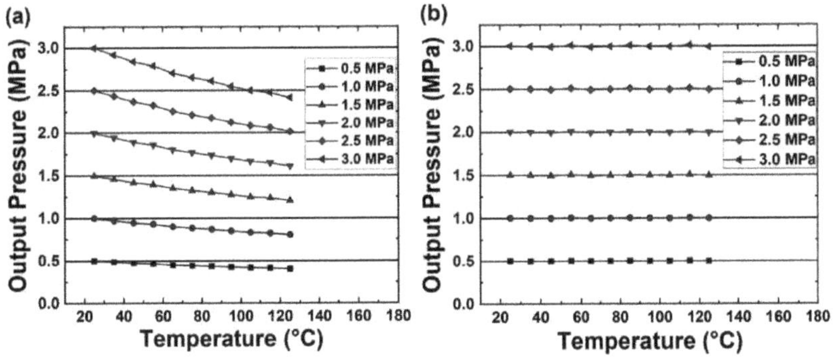

Figure 9. (**a**) Output without temperature compensation. (**b**) Output with temperature compensation.

4. Conclusions

Our work aimed to develop a novel solution for micro pressure sensors for harsh marine environment applications. A wide-range MEMS piezoresistive micro pressure sensor was presented, which was fabricated with the optimized bulk-micromachining technology and packaged using Si-glass bonding technology. The pressure sensors achieved reliable performance in a harsh environment such as high pressure and high temperature while using a miniature size. To manage the moist and liquid external environment for marine application, a flexible polyimide protection layer was fabricated on the top surface of the micro sensor, which showed reliable performance and stability of protection. Meanwhile, a flexible digital compensation system was developed to calibrate linearity drift in a temperature range, which showed potential for large-scale applications with low cost. However, the sensitivity of the sensor was sacrificed for measurement range and stability to some extent. In future work, we will keep focusing on structure and material optimization of piezoresistive pressure sensors to enhance the performance of pressure sensors.

Author Contributions: Conceptualization, M.W.; methodology, M.J. and M.W.; validation, M.J., B.G. and Y.F.; writing—original draft preparation, M.J.; writing—review and editing, M.W., Y.C., B.J. and G.W.; supervision, M.W.; project administration, G.W. All authors have read and agreed to the published version of the manuscript.

Funding: This work was supported in part by the National Key R&D Program of China (2018YFE0120000); National Natural Science Foundation of China (62104056,61771175); the Fundamental Research Funds for the Provincial Universities of Zhejiang (GK199900X001); Zhejiang Provincial Natural Science Foundation of China (LQ21F010010); Zhejiang Provincial Key Research & Development Project (2019C04003); Wenzhou Major Science and Technology Innovation Project (ZG2020016) China; Post-doctoral Science Foundation (2020TQ0246, 2021M692638); Shanghai Sailing Program (21YF1451000); and the Fundamental Research Funds for the Central Universities (31020200QD013).

Data Availability Statement: Not applicable.

Acknowledgments: The authors would like to thank the Institute of Flexible Electronics Technology of THU, Zhejiang. The authors are also grateful to Suzhou Institute of Nana-Tech and Nano-Bionics, Chinese Academy of Science.

Conflicts of Interest: The authors declare that they have no known competing financial interests or personal relationships.

References

1. Fiorillo, A.S.; Critello, C.D.; Pullano, A.S. Theory, technology and applications of piezoresistive sensors: A review. *Sens. Actuators A Phys.* **2018**, *281*, 156–175. [CrossRef]
2. Barlian, A.A.; Park, W.-T.; Mallon, J.R.; Rastegar, A.J.; Pruitt, B.L. Semiconductor piezoresistance for microsystems. *Proc. IEEE* **2009**, *97*, 513–552. [CrossRef] [PubMed]
3. Balavalad, K.B.; Sheeparamatti, B.G. Design, simulation & analysis of SOI based micro piezoresistive pressure sensor for high temperature applications. In Proceedings of the 2018 3rd IEEE International Conference on Recent Trends in Electronics, Information & Communication Technology (RTEICT), Bangalore, India, 18–19 May 2018; IEEE: Bangalore, India, 2018; pp. 2163–2167. [CrossRef]
4. Tian, B.; Zhao, Y.; Jiang, Z.; Zhang, L.; Liao, N.; Liu, Y.; Meng, C. Fabrication and structural design of micro pressure sensors for tire pressure measurement systems (TPMS). *Sensors* **2009**, *9*, 1382. [CrossRef]
5. Fragiacomo, G.; Reck, K.; Lorenzen, L.; Thomsen, E.V. Novel designs for application specific MEMS pressure sensors. *Sensors* **2010**, *10*, 9541. [CrossRef]
6. Mitrakos, V.; Hands, P.J.W.; Cummins, G.; Macintyre, L.; Denison, F.C.; Flynn, D.; Desmulliez, M.P.Y. Nanocomposite-based microstructured piezoresistive pressure sensors for low-pressure measurement range. *Micromachines* **2018**, *9*, 43. [CrossRef] [PubMed]
7. Jiang, H.; Yufei, Z.; Runhui, Z.; Lirong, M.; Tao, C.; Wenjie, M.; Caofeng, P. Recent advances of wearable and flexible piezoresistivity pressure sensor devices and its future prospects. *J. Mater.* **2020**, *6*, 86–101. [CrossRef]
8. Yi, Z.; Yang, B.; Zhang, W.; Wu, Y.; Liu, J. Batteryless tire pressure real-time monitoring system driven by an ultralow frequency piezoelectric rotational energy harvester. *IEEE Trans. Ind. Electron.* **2021**, *68*, 3192–3201. [CrossRef]
9. Clementine, M.B.; Yukitoshi, K.; Bob, C.S.; Alex, C.; Anaïs, L.; Zhen, W.; James, C.; Paige, F.; Zhenan, B. A stretchable and biodegradable strain and pressure sensor for orthopaedic application. *Nat. Electron.* **2018**, *1*, 314–321. [CrossRef]
10. Liu, X.; Wei, Y.; Qiu, Y. Advanced flexible skin-like pressure and strain sensors for human health monitoring. *Micromachines* **2021**, *12*, 695. [CrossRef]
11. Tan, Y.; Ivanov, K.; Mei, Z.; Li, H.; Li, H.; Lubich, L.; Wang, C.; Wang, L. A soft wearable and fully-textile piezoresistive sensor for plantar pressure capturing. *Micromachines* **2021**, *12*, 110. [CrossRef]
12. Jian, D.; Zhijian, L.; Heng, J.; Li, S. Monolithic-integrated piezoresistive MEMS accelerometer pressure sensor with glass-silicon-glass sandwich structure. *Microsyst. Technol.* **2017**, *23*, 1563–1574. [CrossRef]
13. Nag, M.; Singh, J.; Kumar, A.; Singh, K. A high sensitive graphene piezoresistive MEMS pressure sensor by integration of rod beams in silicon diaphragm for low pressure measurement application. *Microsyst. Technol.* **2020**, *26*, 2971–2976. [CrossRef]
14. Sheeparamatti, B.G.; Balavalad, K.B. Fabrication and characterization of polysilicon-on-insulator (PolySOI) and a-SOI based micro piezoresistive pressure sensor for harsh environment applications. *Microsyst. Technol.* **2019**, *25*, 4119–4133. [CrossRef]
15. Alcheikh, N.; Coutier, C.; Giroud, S.; Poulain, C.; Rey, P. Characterization and modeling of a piezoresistive three-axial force micro sensor. *Sens. Actuators A Phys.* **2013**, *201*, 188–192. [CrossRef]
16. Yozo, K.; Akio, Y. Optimum design considerations for silicon piezoresistive pressure sensors. *Sens. Actuators A Phys.* **1997**, *62*, 539–542. [CrossRef]
17. Song, P.; Si, C.; Zhang, M.; Zhao, Y.; He, Y.; Liu, W.; Wang, X. A novel piezoresistive MEMS Pressure sensors based on temporary bonding technology. *Sensors* **2020**, *20*, 337. [CrossRef] [PubMed]
18. Jinjin, L.; Qing, W.; Xu, Z.; Yao, L. Flexible strain sensor with good durability and anti-corrosion property based on metal/polymer composite films embedded with silver nanowires. *Arch. Civ. Mech. Eng.* **2020**, *20*, 133. [CrossRef]
19. Won, J.; Choa, S.H.; Yulong, Z. An integrated sensor for pressure, temperature, and relative humidity based on MEMS technology. *J. Mech. Sci. Technol.* **2006**, *20*, 505–512. [CrossRef]
20. Aryafar, M.; Hamedi, M.; Ganjeh, M.M. A novel temperature compensated piezoresistive pressure sensor. *Measurement* **2015**, *63*, 25–29. [CrossRef]
21. Peng, K.H.; Uang, C.M.; Chang, Y.M. The temperature compensation of the silicon piezo-resistive pressure sensor using the half-bridge technique. *Int. Soc. Opt. Eng.* **2003**, *5343*, 292–301. [CrossRef]
22. Yang, C.; Li, C.; Zhang, C. The application of RBF neural network in the compensation for temperature drift of the silicon pressure sensor. In Proceedings of the 2010 International Conference on Computer Design and Applications, Qinhuangdao, China, 25–27 June 2010; IEEE: Qinhuangdao, China, 2010. [CrossRef]
23. Zhou, G.; Zhao, Y.; Guo, F.; Xu, W. A smart high accuracy silicon piezoresistive pressure sensor temperature compensation system. *Sensors* **2014**, *14*, 12174–12190. [CrossRef] [PubMed]
24. Kumar, S.S.; Pant, B.D. Design principles and considerations for the 'ideal' silicon piezoresistive pressure sensor: A focused review. *Microsyst. Technol.* **2014**, *20*, 2303. [CrossRef]
25. Byunghoon, B.; Bruce, R.F.; Kyihwan, P.; Mark, A.S. Design optimization of a piezoresistive pressure sensor considering the output signal-to-noise ratio. *J. Micromech. Microeng.* **2004**, *14*, 1597–1607. [CrossRef]
26. Nallathambi, A.; Shanmuganantham, T.; Sindhanaiselvi, D. Design and analysis of MEMS based Piezoresistive pressure sensor for sensitivity enhancement. *Mater. Today Proc.* **2018**, *5*, 1897–1903. [CrossRef]
27. Song, P.; Ma, Z.; Ma, J.; Yang, L.; Wei, J.; Zhao, Y.; Zhang, M.; Yang, F.; Wang, X. Recent progress of miniature MEMS pressure sensors. *Micromachines* **2020**, *11*, 56. [CrossRef] [PubMed]

28. Sujit, E.S.; Kusuma, N.; Hemalatha, B. Polysilicon piezoresistive MEMS pressure sensor: Study of analytical solutions for diaphragm and design & simulation. In *2017 International Conference on Communication and Signal Processing (ICCSP)*; IEEE: Chennai, India, 2017; pp. 1606–1610. [CrossRef]
29. Haisheng, S.; Hong, Z.; Qiang, Z.; Yuxi, Y.; Xuyuan, C. Silicon–glass-based single piezoresistive pressure sensors for harsh environment applications. *J. Micromech. Microeng.* **2013**, *23*, 075020. [CrossRef]
30. Li, C.; Cordovilla, F.; Jagdheesh, R.; Ocaña, J.L. Design optimization and fabrication of a novel structural SOI piezoresistive pressure sensor with high accuracy. *Sensors* **2018**, *18*, 439. [CrossRef]
31. Asadnia, M.; Kottapalli, A.G.P.; Shen, Z.; Miao, J.M.; Barbastathis, G.; Triantafyllou, M.S. Flexible, zero powered, piezoelectric MEMS pressure sensor arrays for fish-like passive underwater sensing in marine vehicles. In *2013 IEEE 26th International Conference on Micro Electro Mechanical Systems (MEMS)*; IEEE: Taipei, Taiwan, 2013; pp. 126–129. [CrossRef]
32. Wei, C.; Zhou, W.; Wang, Q.; Xia, X.; Li, X. TPMS (tire-pressure monitoring system) sensors: Monolithic integration of surface-micromachined piezoresistive pressure sensor and self-testable accelerometer. *Microelectron. Eng.* **2012**, *91*, 167–173. [CrossRef]

applied sciences

MDPI

Article

A Novel Principal Component Analysis Integrating Long Short-Term Memory Network and Its Application in Productivity Prediction of Cutter Suction Dredgers

Ke Yang, Jun-Lang Yuan, Ting Xiong, Bin Wang and Shi-Dong Fan *

School of Energy and Power Engineering, Wuhan University of Technology, Wuhan 430063, China; k_yang@whut.edu.cn (K.Y.); junlangyuan@whut.edu.cn (J.-L.Y.); cxj_xt@whut.edu.cn (T.X.); wang_bin@whut.edu.cn (B.W.)

* Correspondence: sdfan@whut.edu.cn

Abstract: Dredging is a basic construction for waterway improvement, harbor basin maintenance, land reclamation, environmental protection dredging, and deep-sea mining. The dredging process of cutter suction dredgers is so complex that the operational data show strong characteristics of dynamic, nonlinearity, and time delay, which make it difficult to predict the productivity accurately via basic principles models. In this paper, we propose a novel integrating PCA-LSTM model to improve the productivity prediction of cutter suction dredger. Firstly, multiple variables are reduced in dimension and selected by PCA method based on the working mechanism of cutter suction dredger. Then the productivity is predicted via mud concentration in long short-term memory network with relevant operational time-series data. Finally, the proposed method is successfully applied to an actual case study in China. Also, it performs well in the cross-validation and comparative study for several important characteristics: (i) it involves the operational parameters based on the mechanism analysis; and (ii) it is a deep-learning-based approach that can deal with operation series data with a special memory mechanism. This study provides a heuristic idea for integrating the data-driven method and supervision of human knowledge for application in practical engineering.

Keywords: operational data; productivity prediction; cutter suction dredger; long short-memory network; working mechanism

Citation: Yang, K.; Yuan, J.-L.; Xiong, T.; Wang, B.; Fan, S.-D. A Novel Principal Component Analysis Integrating Long Short-Term Memory Network and Its Application in Productivity Prediction of Cutter Suction Dredgers. *Appl. Sci.* **2021**, *11*, 8159. https://doi.org/10.3390/app11178159

Academic Editor: Dimitris N. Pagonis

Received: 22 July 2021
Accepted: 30 August 2021
Published: 2 September 2021

1. Introduction

Marine-based transportation has always played a critical role in the national economics of China [1], while rivers may suffer from sediment accumulation that obstructs riverways and reduces the carrying activity [2,3]. Cutter suction dredgers are common and useful machines that can remove the mud deposited at the bottom of water and keep transportation routes in good condition [4]. Dredging productivity is one of the most important indexes to evaluate the dredging performance, which is affected by many factors such as soil properties, the power of the pump, the cutter structural parameters, and so on [5]. The process of sand being cut into a mixture of mud and water by a rotating cutter is very complicated. Most of the parameters are dynamically influenced by the uncertain working environment and human operation [6]. Due to the limitations of dredging technology, there are indeed some obstacles for parameters monitoring and real-time prediction, which is a challenge for constructing digital models to describe this process and dredging productivity accurately [7].

Due to the developed sensor technology, more operational data have been gained for analyzing dredging performance. In the literature review, machine learning methods were recently adopted to model the complex and dynamical construction process of CSD (cutter suction dredger) for their excellent learning and mining ability [8]. Generally, the learning-based prediction models can be mainly divided into two main types based on the depth of structure: shallow learning models and deep learning models [9]. The shallow learning methods mainly cover neural-network-based methods such as RBF (radial basis function), ELMs (extreme learning machines), and SVM (support vector machine). As traditional learning models used in productivity prediction, Wang et al. adopted an RBF neural network to deal with the different working conditions and established an accurate nonlinear mathematical model for the instantaneous output prediction with control variables [10]. Guan et al. developed the model of cutter operation parameters using improved ELMs to simulate and predict the productivity distribution in actual construction [11]. Yang et al. predicted the cutter suction dredger production with a double hidden layer BP neural network [12].

The prediction methods on deep learning mainly include: DNN (deep neural network), DBN (deep belief network), CNN (convolutional neural network network), and RNN (recurrent neural network network). DNNs are adopted through multiple auto-encoder (AE) or denoising auto-encoder (DAE) stacking networks, wherein high-dimensional input data are extracted from classless data as the distribution of the original data is represented by deep learning neural networks [13]. Wang et al. developed DNN models for production forecasting wherein the data-driven method was well used to deal with hydraulic fractures and intrinsic complexity [14]. DNN architecture transfers sigmoid function to ReLu and maxout with the purpose of overcoming gradient disappearance, but requires small batch training, which leads to over-fitting and local optimal solution problems. DBN is also a deep network that is stacked with multiple Restricted Boltzman Machines (RBMs) and a classification layer or regression layer. Xu et al. designed a DBN-based model to approximate the function type coefficients of a state-dependent autoregressive model in nonlinear system and realize predictive control [15]. Hu et al. adopted DBN to extract deep hidden features behind the monitoring signals and predict the bearing remaining useful life [16]. Researchers have improved DBN by combining a feed-forward neural network (FNN) to make prediction more accurate [17]. Zhang proposed a multi-target DBN collection method in which the output of multiple DBNs has a certain weight that reveals the final output of the network set; this method performed well on NASA aero-engine data [18]. Furthermore, convolutional neural networks (CNNs) are greatly developed with the excellent characteristics of parameter sharing and spatial pooling, which make it more advantageous in computing speed and accuracy [19]. However, all these ML methods have a limitation in situations that involve a time-series input.

Then recurrent neural networks (RNNs) are proposed with adding a twist where the output from previous time step is fed as input to the current step. The most important feature of RNN is that the hidden state can remember all information calculated from the previous sequence [20]. Thus, it can generate output through prior input (the past memory) and learning in training. It is common to complete the parameter learning of recurrent neural networks by learning over back-propagation, wherein the error is passed forward step by step in the reverse order of time. In [21], a learning-based method is applied to improve the RNN training process while the number of prediction time-steps increases. However, RNNs still suffer from the long-term dependencies, and long short-term memory (LSTM) fills the gap by setting a gate control unit that can choose and keep some useful information in the long-term sequential data. Being different from the traditional RNN, the model is trained by both the stored information of last time step and new input of the current moment, which enhances the prediction accuracy and stability greatly [22].

However, for the practical application in CSD, it is equally significant to analyze the interrelated influencing factors as the productivity prediction. LSTM lacks effective processing for high-dimensional characteristics of large-scale data; it should be used integrating with other methods. Principal component analysis (PCA) is one of the most widely used algorithms for feature reduction, which reconstructs the main k-dimensional feature based on the original n-dimensional feature by deep learning. While PCA is a pure data-driven method that cannot consider the casual relationship and correlation between variables, the procedure of variable analysis based on a working mechanism and human experience is necessary. Yang et al. described a HEPCA model, which made variables supplement based on expert knowledge after the PCA process and generate a more accurate input for the predictive model [23].

Therefore, combining the advantages and characteristics of the different methods described above, this paper presents the long short-term memory integrating principal component analysis model (PCA-LSTM) to predict productivity using the monitoring sensors data. The PCA-LSTM is strutted into four phases. In the first phase, monitoring sensors are analyzed to select related variables according to the working mechanism and knowledge. In the second phase, PCA method is applied to extract the deep feature from the high-dimension dataset and to obtain the correlation of variables. In the third phase, a prediction model is built and trained by the LSTM network. Finally, cross validation and comparative analysis are conducted with a generated model from "Chang Shi 10" in China.

2. Preliminaries

In this section, the related preliminaries regarding PCA and LSTM will be introduced briefly on the basis of the practical application in this study.

2.1. Principal Components Analysis (PCA)

PCA is an important technique that can transform multiple variables into a few main components (comprehensive variables) by means of dimensional reduction, increasing interpretability while minimizing information loss [24]. These main components are usually expressed as linear combinations of the original variables, which can represent most of the information of the whole dataset.

For original data $\boldsymbol{X} = (\boldsymbol{x}_1, \boldsymbol{x}_2, \ldots, \boldsymbol{x}_i, \ldots, \boldsymbol{x}_n)$ and $\boldsymbol{X} \in \mathbb{R}^{k*n}$, we can get the covariance matrix $\boldsymbol{C}_x$:

$$C_x = E[(\boldsymbol{X} - E[\boldsymbol{X}])(\boldsymbol{X} - E[\boldsymbol{X}])^{\mathrm{T}}] \quad (1)$$

After centralizing the data, the mean function $E[\boldsymbol{X}]$ is zero and:

$$C_x = \frac{1}{n}\boldsymbol{X}\boldsymbol{X}^{\mathrm{T}} \quad (2)$$

Assuming that there is a matrix $\boldsymbol{P}$ ($\boldsymbol{P} \in \mathbb{R}^{k'*k}$), through which we can transform the original sample data matrix $\boldsymbol{X}$ into a dimensionality-reduced matrix $\boldsymbol{Y}$ ($\boldsymbol{Y} \in \mathbb{R}^{k'*n}$):

$$\boldsymbol{Y} = \boldsymbol{P}\boldsymbol{X} \quad (3)$$

Then the original data dimension is successfully reduced from k to k', wherein the first k' principal components can explain most of the variance.

For matrix $\boldsymbol{Y}$, its covariance matrix can be expressed through original matrix $\boldsymbol{X}$ as:

$$C_y = \frac{1}{n}\boldsymbol{Y}\boldsymbol{Y}^T = \frac{1}{n}(\boldsymbol{P}\boldsymbol{X})(\boldsymbol{P}\boldsymbol{X})^T = \frac{1}{n}\boldsymbol{P}\boldsymbol{X}\boldsymbol{X}^T\boldsymbol{P}^T = \boldsymbol{P}\frac{1}{n}\boldsymbol{X}\boldsymbol{X}^T\boldsymbol{P}^T = \boldsymbol{P}\boldsymbol{C}_x\boldsymbol{P}^T \quad (4)$$

It is obvious from Equation (4) that C_x is guaranteed to be a non-negative definite matrix and thus is diagonalizable by some unitary matrix. Then the objective optimization is transformed to find an orthonormal transformation matrix $\boldsymbol{P}$. Normally, we can use eigenvalue decomposition or singular value decomposition to solve the matrix $\boldsymbol{P}$, and the first k'-dimensional new features corresponding to k' eigenvalues can represent the whole data best.

2.2. Long Short-Term Memory Network (LSTM)

In this paper, an integrating model of long short-term memory network based on principal components analysis (PCA-LSTM) is explored to analyze the operational time-series data generated from the dredging process. The proposed model is developed on the basis of long short-term memory network (LSTM), which is a special form of recurrent neural networks (RNN) that can address long-distance dependencies and delay in time-series modeling.

The LSTM architecture was firstly proposed by Sepp Hochreiter and Jürgen Schmidhuber in 1997 [25]. There is a special memory cell unit added to the original hidden layer in classic RNN architecture. The cell state is controlled by three gates: Input gate $\boldsymbol{I}_t$, Forget gate $\boldsymbol{F}_t$, and Output gate $\boldsymbol{O}_t$, as shown in Figure 1.

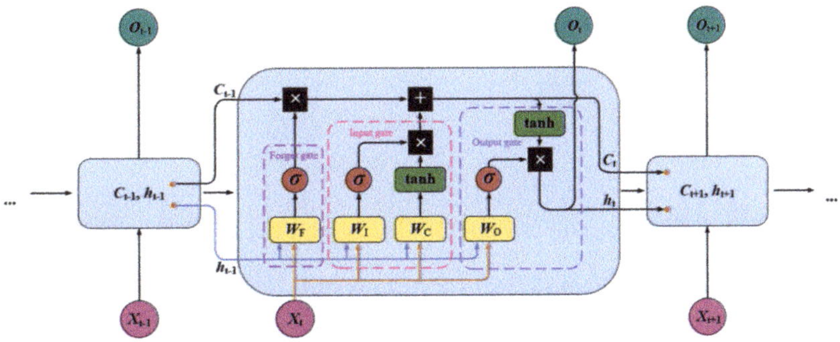

Figure 1. The architecture of classical LSTM.

The forget gate $\boldsymbol{F}_t$ decides which information needs to be kept and which can be forgotten. The information consists of the current input $\boldsymbol{X}_t$ and previous hidden state/short-term memory $\boldsymbol{h}_{t-1}$.

$$\boldsymbol{F}_t = \sigma(\boldsymbol{W}_{Foreget} \cdot [\boldsymbol{h}_{t-1}, \boldsymbol{X}_t]) + \boldsymbol{bias}_{Forget} \tag{5}$$

For every time step, the sigmoid function generates values between 0 and 1 that indicate whether the old information is necessary. 0 denotes forget, and 1 means keep. $\boldsymbol{W}_{Forget}$ is the weight matrix between forget gate and input gate. $\boldsymbol{bias}_{Forget}$ is the connection bias.

The input gate decides what should be stored in the long-term memory in the new information. It works with the current input $\boldsymbol{X}_t$ and previous short memory $\boldsymbol{h}_{t-1}$ through two layers. In the first layer, the short-term memory and current input is passed through a sigmoid function that values from 0 (not important) to 1 (important):

$$\boldsymbol{i}_t = \sigma(\boldsymbol{W}_{Input} \cdot [\boldsymbol{h}_{t-1}, \boldsymbol{X}_t]) + \boldsymbol{bias}_{Input} \tag{6}$$

where $\boldsymbol{W}_{Input}$ is the weight matrix of sigmoid operator between input gate and output gate. $\boldsymbol{bias}_{Input}$ is the bias vector.

The second layer uses the tanh function to regulate the network. The tanh operator creates a vector $\widetilde{\boldsymbol{C}}_t$ with all the possible values between -1 and 1:

$$\widetilde{\boldsymbol{C}}_t = \tanh(\boldsymbol{W}_{Cell} \cdot [\boldsymbol{h}_{t-1}, \boldsymbol{X}_t]) + \boldsymbol{bias}_{Cell} \tag{7}$$

where $\boldsymbol{W}_{cell}$ is the weight matrix of tanh operator between cell state information and network output. $\boldsymbol{bias}_{Cell}$ is the bias vector.

With these two layers input, the cell state updates a new cell state (long-term memory):

$$\boldsymbol{C}_t = \boldsymbol{F}_t \odot \boldsymbol{C}_{t-1} + \boldsymbol{i}_t \odot \widetilde{\boldsymbol{C}}_t \tag{8}$$

where $\odot$ is the Hadamard product.

When it comes to the output gate, the current input $\boldsymbol{X}_t$, the previous short-term memory $\boldsymbol{h}_{t-1}$, and the newly obtained cell state $\boldsymbol{C}_t$ determine the new short-term memory (hidden state) that will be passed on to the cell in the next time step.

$$\boldsymbol{O}_t = \sigma(\boldsymbol{W}_{output} \cdot [\boldsymbol{h}_{t-1}, \boldsymbol{X}_t]) + \boldsymbol{bias}_{output} \tag{9}$$

$$h_t = O_t \odot \tanh(C_t) \tag{10}$$

W_{output} is the weight matrix of the output gate. This hidden state is used for prediction. Both new cell state and hidden state are carried over to the next time step.

3. The Proposed PCA-LSTM Model

As described above, the basic knowledge regarding PCA and LSTM networks was introduced to set up the proposed PCA-LSTM model in this section. Considering mechanism and human experience, the PCA procedure will bring about a more accurate variable analysis for the practical multi-sensors system. The time-series data of effective variables are subsequently learned in LSTM network to output the target prediction.

3.1. PCA Based on Mechanism

The traditional PCA was introduced in Section 2.1. Due to the purely data-driven process, historical data is analyzed in PCA without any prior knowledge, which may bring about that some redundancies may be considered despite the causal relationship. Therefore, human experience will be introduced to interfere with the variable selection procedure ahead of PCA, based on the known mechanism.

The monitoring system always contains a broad range of sensors data related to the target object. Some of the data is on control variables, while some of the data is just on display variables that visualize the parameters.

Assuming the sensor system obtains an initial dataset:

$$X = (x_1, x_2, \ldots, x_{i-m}, x_{i-m+1}, \ldots, x_i, \ldots, x_k) \tag{11}$$

where x_i represents the i-th sensor equipped in the system.

$$x_i = [x_{i1}, x_{i2}, \ldots, x_{ij}, \ldots, x_{in}]^T \tag{12}$$

where x_{ij} represents the j-th data obtained by i-th sensor.

When studying the working mechanism of the target, variables causal relationship will be analyzed and some of the redundant variables will be deleted, as well as some meaningless display parameters. This creates a new sample set:

$$X' = (x_1, x_2, \ldots, x_{i-m}, x_i, \ldots, x_k) \tag{13}$$

PCA based on human experience method obtains a hyperplanar representation of all samples through recent reconstruction, realizing the dimension reduction from k to k' with the least loss.

The samples are centralized firstly as:

$$\sum_i x_i = 0 \tag{14}$$

Then a new coordinate system can be obtained after projection transformation:

$$W = (w_1, w_2, \ldots w_{i-m}, w_i, \ldots, w_k) \tag{15}$$

where w_i is the standard orthogonal basis vector.

$$\|w_i\|_2 = 1 \tag{16}$$

$$w_i{}^T w_j = 0, (i \neq j) \tag{17}$$

If a portion of the coordinate is abandoned, namely the dimension is reduced from k to k' ($k' < k$), the projection of samples x_i in the low-dimensional coordinate system will be:

$$z_{ij} = (z_{i1}, z_{i2}, \ldots, z_{in'}) \tag{18}$$

$$z_{ij} = \boldsymbol{w}_j{}^T \boldsymbol{x}_i \tag{19}$$

where z_{ij} is the j-th coordinate of $\boldsymbol{x}_i$ in low-dimensional space; and $\boldsymbol{x}_i$ can be reconstructed as:

$$\hat{\boldsymbol{x}}_i = \sum_{j=1}^{k'} z_{ij} \boldsymbol{w}_j \tag{20}$$

For the whole training dataset, the distance between original samples $\boldsymbol{x}_i$ and the reconstructed samples $\hat{\boldsymbol{x}}_i$ can then be determined as:

$$\begin{aligned} \sum_{i=1}^{k} \left\| \sum_{j=1}^{n'} z_{ij} \boldsymbol{w}_j - \boldsymbol{x}_i \right\|_2^2 &= \sum_{i=1}^{k} \boldsymbol{z}_i{}^T \boldsymbol{z}_i - 2 \sum_{i=1}^{k} \boldsymbol{z}_i{}^T W^T \boldsymbol{x}_i + const \\ &\propto -tr(\boldsymbol{W}^T (\sum_{i=1}^{k} \boldsymbol{x}_i \boldsymbol{x}_i{}^T) \boldsymbol{W}) \end{aligned} \tag{21}$$

where *const* is the constant item, and $\boldsymbol{W}$ can be obtained by Equation (15).

Because $\sum_{i=1}^{k} \boldsymbol{x}_i \boldsymbol{x}_i{}^T$ is a covariance matrix, the target distance function can be minimized as:

$$\begin{cases} Min_W \; - tr(\boldsymbol{W}^T \boldsymbol{X}' \boldsymbol{X}'^T \boldsymbol{W}) \\ s.t. \; \boldsymbol{W}^T \boldsymbol{W} = \boldsymbol{I} \end{cases} \tag{22}$$

where $\boldsymbol{I}$ is the identity matrix.

With the Langerin multiplier method [26], it can be calculated as:

$$\boldsymbol{X}' \boldsymbol{X}'^T \boldsymbol{w}_i = \lambda_i \boldsymbol{w}_i \tag{23}$$

After eigenvalue decomposition, the eigenvalue λ can be obtained as follows:

$$\lambda = \{\lambda_1, \lambda_2, \ldots, \lambda_{i-m}, \lambda_i, \ldots, \lambda_k) \tag{24}$$

According to the practical demand, reconstruction threshold μ is set to satisfy the condition:

$$\frac{\sum_{i=1}^{k'} \lambda_i}{\sum_{i=1}^{k} \lambda_i} \geq \mu \tag{25}$$

When the maximum threshold μ is satisfied, the eigenvalues can be obtained in turn:

$$\lambda_1 \geq \lambda_2 \geq \ldots \geq \lambda_{k\prime} \tag{26}$$

And the eigenvectors corresponding to the first k' eigenvalues constitute the PCA solution:

$$\boldsymbol{W}^* = (\boldsymbol{w}_1, \boldsymbol{w}_2, \ldots, \boldsymbol{w}_{i-m}, \boldsymbol{w}_i, \ldots, \boldsymbol{w}_{k'}) \tag{27}$$

The variables corresponding to the eigenvectors are:

$$\boldsymbol{X}'' = (\boldsymbol{x}_1, \boldsymbol{x}_2, \ldots, \boldsymbol{x}_{i-m}, \boldsymbol{x}_i, \ldots, \boldsymbol{x}_{k\prime}) \tag{28}$$

Based on the variables obtained by PCA procedure above, the correlation matrix can be calculated as:

$$\boldsymbol{R} = (r_{ij})_{k' * k'} \tag{29}$$

Then the most positively relevant variables to the target will be proceeded in the subsequent prediction model:

$$\boldsymbol{X}^{**} = (\boldsymbol{x}_1, \boldsymbol{x}_2, \ldots, \boldsymbol{x}_p) \tag{30}$$

3.2. The Proposed Methodology

The variables most related to the target were obtained by PCA based on human experience that can be used as inputs in the next LSTM network to obtain prediction results. Namely, with current input being X_t^{**}, the current cell state and hidden state are updated as described in Section 2.2.

$$\begin{aligned} C_t &= F_t \odot C_{t-1} + i_t \odot \widetilde{C}_t \\ &= F_t \odot C_{t-1} + i_t \odot \{\tanh(W_{Cell} \cdot [h_{t-1}, X_t^{**}]) + bias_{cell}\} \\ &= \{\sigma(W_{Forget} \cdot [h_{t-1}, X_t^{**}]) + bias_{Forget}\} \odot C_{t-1} + \{\sigma(W_{Input} \cdot [h_{t-1}, X_t^{**}]) + bias_{Input}\} \odot \{\tanh(W_{Cell} \cdot [h_{t-1}, X_t^{**}]) + bias_{Cell}\} \end{aligned} \tag{31}$$

$$\begin{aligned} h_t &= O_t \odot \tanh(C_t) \\ &= \{\sigma(W_{output} \cdot [h_{t-1}, X_t^{**}]) + bias_{output}\} \odot \tanh(C_t) \end{aligned} \tag{32}$$

Based on the new cell state and hidden state, we define gradient $\delta_h^{(t)}$ and $\delta_c^{(t)}$ to calculate the back propagation error layer by layer:

$$\delta_h^{(t)} = \frac{\partial L_{(t)}}{\partial h_{(t)}} \tag{33}$$

$$\delta_c^{(t)} = \frac{\partial L_{(t)}}{\partial C_{(t)}} \tag{34}$$

where $L(t)$ is the loss function, and at the last sequence index τ, the gradient can be described as follows:

$$\delta_h^{(\tau)} = \frac{\partial L_{(\tau)}}{\partial O_{(\tau)}} \frac{\partial O_{(\tau)}}{\partial h_{(\tau)}} = W_{Output}{}^T(\hat{O}_{(\tau)} - O_{(\tau)}) \tag{35}$$

$$\delta_c^{(\tau)} = \frac{\partial L_{(\tau)}}{\partial h_{(\tau)}} \frac{\partial h_{(\tau)}}{\partial C_{(\tau)}} = \delta_h^{(\tau)} \odot O_{(\tau)} \odot (1 - \tanh^2(C_{(\tau)})) \tag{36}$$

Therefore, for any moment t, $\delta_h^{(t+1)}$ and $\delta_c^{(t+1)}$ can be obtained, deriving from $\delta_h^{(t)}$ and $\delta_c^{(t)}$ as follows:

$$\delta_h^{(t)} = \frac{\partial L_{(t)}}{\partial O_{(t)}} \frac{\partial O_{(t)}}{\partial h_{(t)}} + \frac{\partial L_{(t+1)}}{\partial h_{(t+1)}} \frac{\partial h_{(t+1)}}{\partial h_{(t)}} = W_{Output}{}^T(\hat{O}_{(t)} - O_{(t)}) + W^T \delta_h^{(t+1)} diag(1 - (h_{(t+1)})^2) \tag{37}$$

where W is the coefficient matrix.

Then the reverse gradient error of $\delta_c^{(t)}$ can be obtained through the gradient error of the current layer returned from $h_{(t)}$ and the gradient error of the previous layer $\delta_c^{(t)}$:

$$\delta_c^{(t)} = \frac{\partial L_{(t)}}{\partial C_{(t+1)}} \frac{\partial C_{(t+1)}}{\partial C_{(t)}} + \frac{\partial L_{(t)}}{\partial h_{(t)}} \frac{\partial h_{(t)}}{\partial C_{(t)}} = \delta_c^{(t+1)} \odot F(t+1) + \delta_h^{(t)} \odot O_{(t)} \odot (1 - \tanh^2(C_{(t)})) \tag{38}$$

Then, the gradient of all parameters can be calculated easily using $\delta_h^{(t)}$ and $\delta_c^{(t)}$, and all the parameters can be updated iteratively with the lowest error.

As mentioned above, the proposed method can be run in terms of Figure 2. It mainly consists of two parts: PCA and LSTM. The variables most related to the target are firstly obtained by PCA based on expert knowledge and used as inputs for LSTM network to get the prediction results.

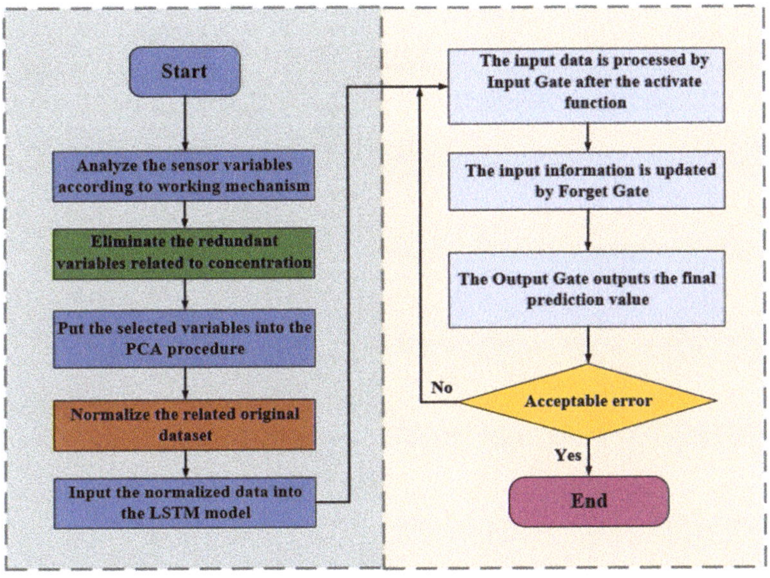

Figure 2. The flowchart of proposed novel PCA-LSTM based on mechanism.

4. Case Study

Cutter Suction Dredger is a special kind of ship that is widely used in dredging engineering. In this section, the proposed method is validated in a real case study of well-equipped 4500 m^3/h cutter suction dredger "Chang Shi 10" that serves in the Yangzi River region.

Mud and sand is cut to mix with water by a rotary cutter during the construction operation of the dredger. Meanwhile, the dredge pump works and creates vacuum pressure at the suction mouth of the cutter. Under the great pumping force, mud is sucked into the dredger pipeline and finally discharged to the dumping area. The primary system according to the dredging procedure was highlighted in Figure 3.

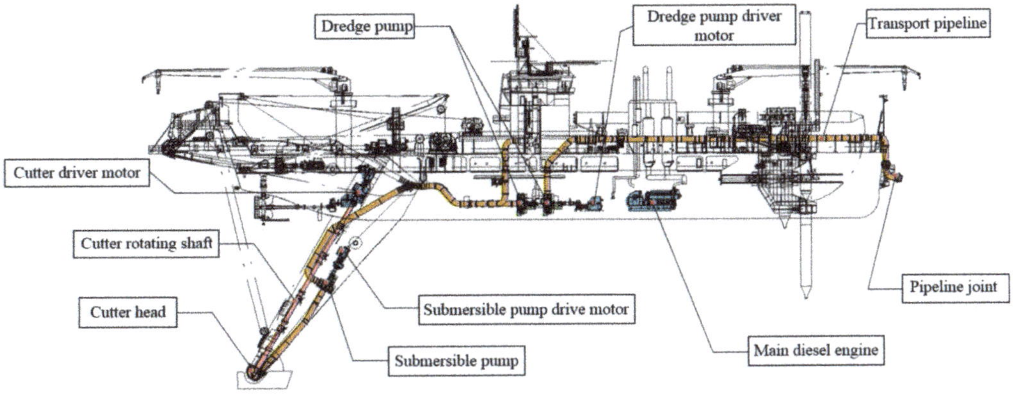

Figure 3. The primary system highlighted in cutter suction dredger (1:150).

4.1. Principal Components Analysis Based on Mechanism and Knowledge

During the construction, mud formation is influenced by many factors such as soil type, mechanical parameters and rotation speed of the cutter, traverse speed of the dredger, dredge pump parameters, and so on. To monitor and control the dredging process, up to 255 specific real-time sensors were arranged to collect the operational data [27]. Figure 4 shows some of the related monitoring parameters and relationship in automatic control system.

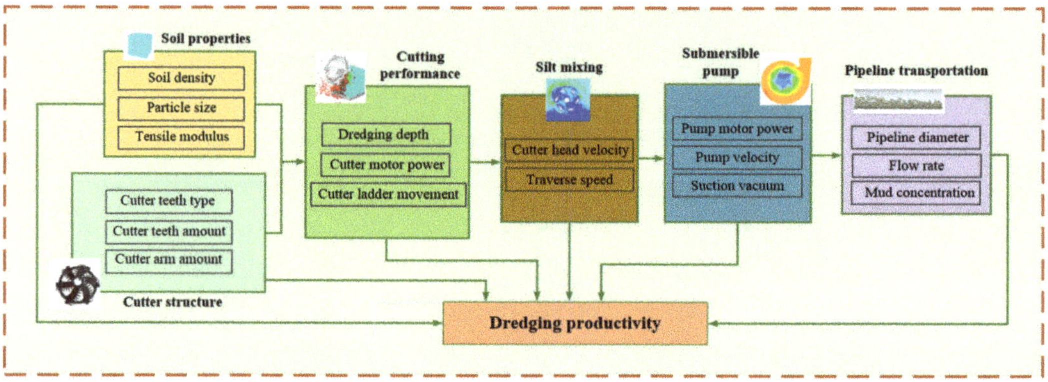

Figure 4. The monitoring parameters related to dredging process and productivity.

As shown in Figure 4, some of the parameters are control variables, while some of them are only display variables that make the data visualization.

Soil property is an important factor affecting the construction process and efficiency of cutter suction dredgers. For different solidity and water-solubility, the mud concentration is limited by cutting performance and silt mixing. The cutter structure, pipeline diameter, and the motor power of pumps are all specific variables (constant), which were demonstrated according to the demand for rated productivity at the beginning of the design. However, the cutter speed, trolley trip, cutter ladder movement, and dredge pump rotation are all control variables that can be adjusted during the operation process in specific construction. When digging hard soil, the dredging depth should be reduced while enhancing the cutter speed to prevent the formation of large diameter mud balls and pipe blocking. When the dredging soil is sediment or silt, the pump velocity should be appropriately increased to reduce the mud concentration and avoid sedimentation or clogging in the pipeline.

According to the actual sensor system of cutter suction dredger "Chang Shi 10", we firstly select 20 variables from the initial operational dataset as shown in Table 1.

Table 1. Initial variables from the operational dataset.

Variable	Description	Unit
S_8	Angle of the cutter ladder	
S_9	Depth of the dredging	m
S_{12}	Rotation speed of the submersible pump	rpm
S_{13}	Rotation speed of the cutter	rpm
S_{20}	Flow	m^3/h
S_{23}	Soil density	kg/m^3

Table 1. *Cont.*

Variable	Description	Unit
S_{79}	Distance of the swing movement	m
S_{80}	Angle of the swing	
S_{100}	Rotation speed of the No.1 dredge pump	rpm
S_{101}	Rotation speed of the No.2 dredge pump	rpm
S_{108}	Power of the cutter	kw
S_{164}	Mud density	kg/m^3
S_{165}	Flow rate	m/s
S_{182}	Trolley trip	m
S_{198}	Discharge pressure of the submersible pump	kPa
S_{199}	Discharge pressure of the No.1 dredge pump	kPa
S_{200}	Discharge pressure of the No.2 dredge pump	kPa
S_{201}	Vacuum	kPa
S_{223}	Water density	kg/m^3
$\mathbf{S_{21}}$	**Mud concentration**	%

Traditionally, the instantaneous productivity of the cutter suction dredger is the product of the flow and mud concentration.

$$P = C_m \cdot Q = C_m \cdot (v \cdot \pi r^2) \tag{39}$$

where $C_m(\%)$ represent the mud concentration; Q is the flow amount per hour; and v is the flow rate.

As shown in Table 1, we choose $\mathbf{S_{21}}$ (mud concentration) as the target variable. In actual dredging construction process, the change of flow rate in the sludge pipeline is one of the important factors affecting the flow. Thus, we delete the redundant variable S_{20} flow in the first step.

Meanwhile, the mud concentration is determined by the density of soil, water, and mud.

$$C_m = \frac{\gamma_m - \gamma_w}{\gamma_s - \gamma_w} \tag{40}$$

where γ_m is the mud density; γ_w is the water density; and γ_s is the soil density.

Then we drop three redundant variables S_{223}, S_{23}, and S_{164} in the second step.

For the study period in this case, the ship works with just No.1 dredge pump. Thus, the variables related to No.2 dredge pump are non-meaningful to the productivity. Namely, S_{101} and S_{200} are dropped according to human analysis and finally we obtain the related variable set as:

$$X' = \{S_8, S_{182}, S_{108}, S_{13}, S_9, S_{201}, S_{12}, S_{198}, S_{100}, S_{199}, S_{165}, S_{79}, S_{80}, S_{21}\} \tag{41}$$

As described in Section 3.1, the selected variables $\mathbf{X'}$ dataset based on human experience will be processed by PCA in this section. In addition, the contribution result is shown in Figure 5.

It is obvious that the top 10 principal components can represent more than 97% of the overall data. For the top two principal components, the dataset can be plot as Figure 6.

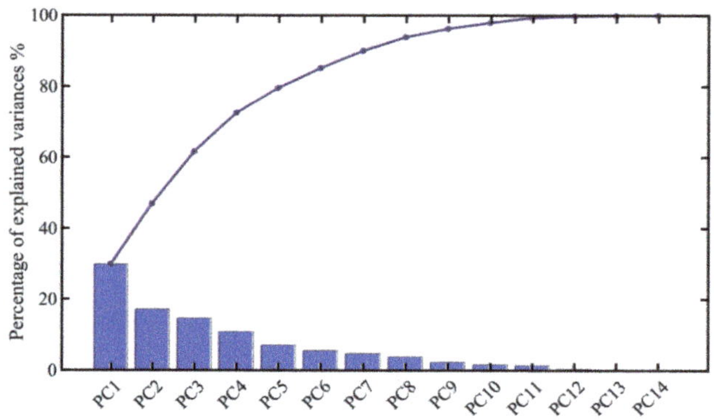

Figure 5. The percentage of explained variance for principal components.

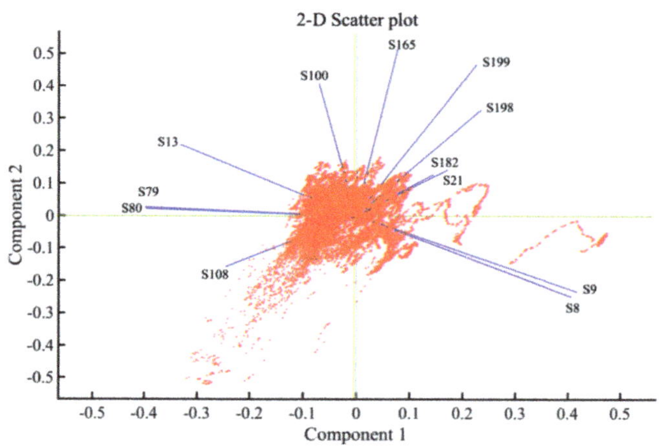

Figure 6. Scatter plot of the dataset after PCA.

From the scatter plot of dataset in 2D figure, the correlation of the variables related to target S_{21} can be set as shown in Equation (28):

$$X'' = \{S_8, S_{182}, S_{108}, S_{13}, S_9, S_{201}, S_{12}, S_{198}, S_{100}, S_{199}, S_{165}, S_{79}, S_{80}\} \quad (42)$$

The most positively relevant variables to target can be further determined by the correlation matrix, as shown in Figure 7.

As the correlation matrix shows, the correlation between S_{21} and S_{199} is 0.48677, which means the discharge pressure of No.1 dredge pump affects the concentration most. It is consistent with the practical production. Pressure will influence the mud and water proportion that is pumped into pipeline. The variable S_{165} shows a correlation of 0.34628, which is also mentioned by other researchers [5,27]. The flow rate may determine the mud sedimentation during pipeline transportation. Furthermore, the vacuum correlation is 0.34152, since the vacuum gauge is installed on the upper part of the cutter, which is sensitive to the change of the mud concentration in the pipeline. Additionally, the angle of cutter ladder, the depth of dredging, and the trolley trip are all factors that affect the mud formation by operation controllers. However, for the discharge of the submersible pump, it is just the indirect factor to display the vacuum condition.

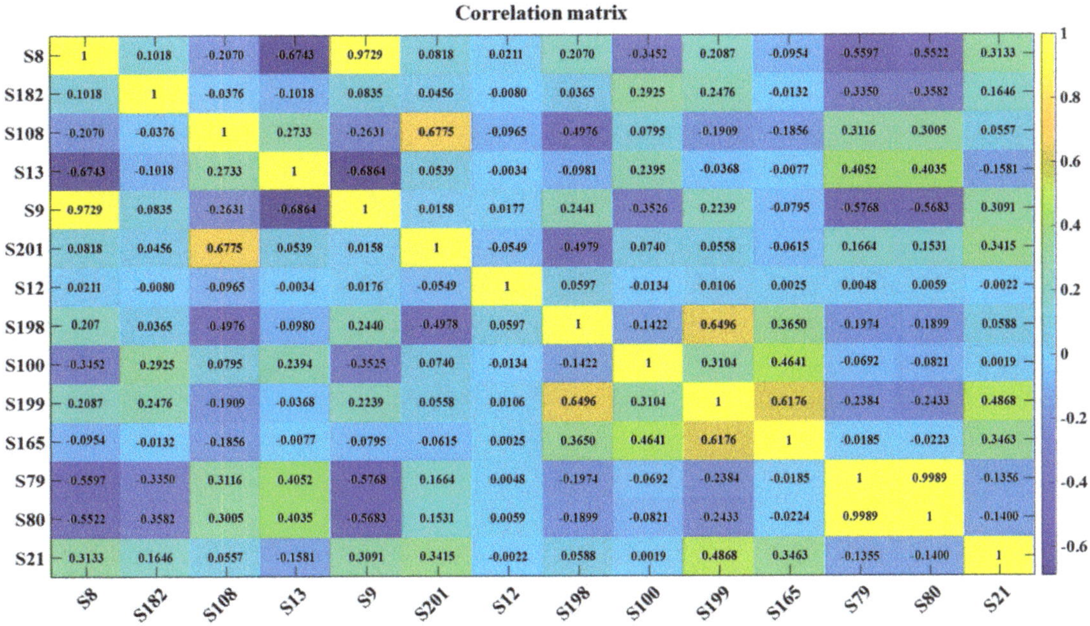

Figure 7. The correlation matrix of the variables.

Meanwhile, it is indicated in the correlation matrix that there are five negative variables to the target. Thus, we obtain the final variables set most positively relevant to the target S_{21} as:

$$X^{**} = \{S_{182}, S_{108}, S_8, S_9, S_{201}, S_{198}, S_{100}, S_{199}, S_{165}\} \quad (43)$$

In general, the mud concentration is mainly inter-influenced by the dredge pump pressure, flow rate, vacuum, cutter ladder angle, dredging depth, and trolley trip.

4.2. Modeling Prediction Analysis

In this section, we choose the first segment series data and follow the steps given in Sections 2.2 and 3.2 to train the proposed model. This segment series data is collected from the monitoring system with a frequency of 100 sample points per minute. We intercept a dataset of 18,000 for 3-hour working time zone and obtain final 16,764 data after the pre-process.

4.2.1. Learning Results Analysis

In terms of the variable's selection process in Section 4.1, we use the most positively related nine parameters as input to predict the target output concentration C_m (%). Considering the data amount effects on the learning ability in data-driven models, we divide the input with a proportion of 6:4 and 7:3 to test the model twice. The learning results are shown in Figures 8 and 9 respectively.

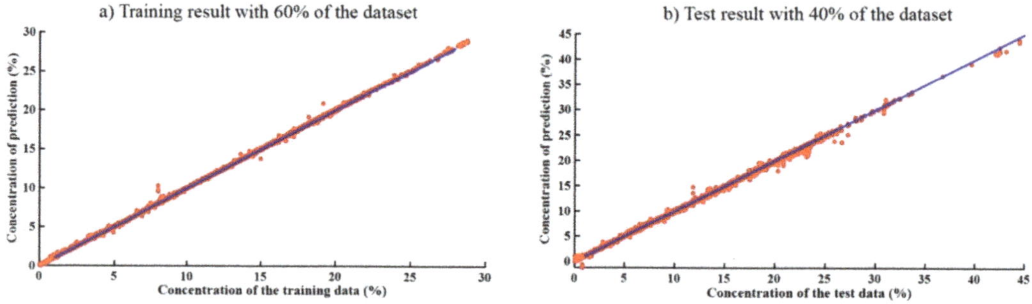

Figure 8. The learning results of 6:4 proportion.

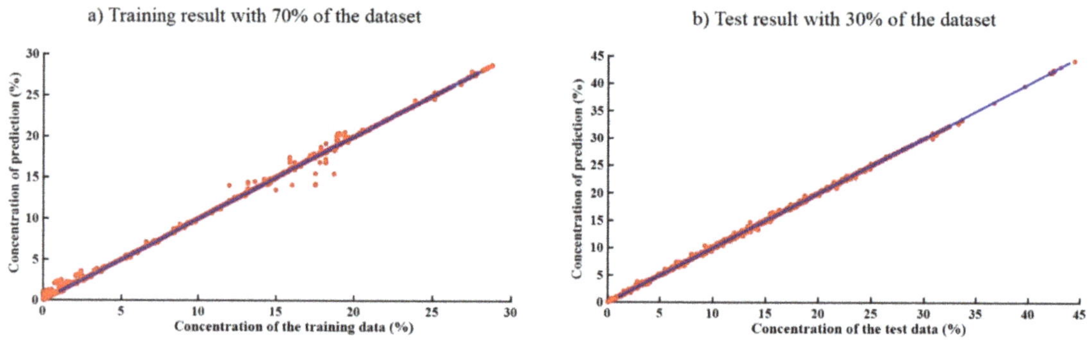

Figure 9. The learning results of 7:3 proportion.

Concentration changes with the working conditions. As shown in the learning results, the normal range of the concentration is from 0 to 45%, which is a comprehensive result of the multiple factors interaction. High concentration is not necessarily good for the production since it may cause sedimentation or clogging in the pipeline. The results in this case are all normal and satisfactory. However, when it comes to the detailed comparison, the learning process with 60% of the dataset performs better than 70% dataset. For the 60% data training, its maximum and minimum error are 0.3091 and 0.0149 respectively. However, the maximum error is 0.526 in the 70% data training process. Also, as shown in Figure 10, for 60% dataset, the loss value decreases and then keeps steady in the training process. The error in testing falls and then keeps steady. However, for the 70% dataset, both training and testing error are less stable and consistent.

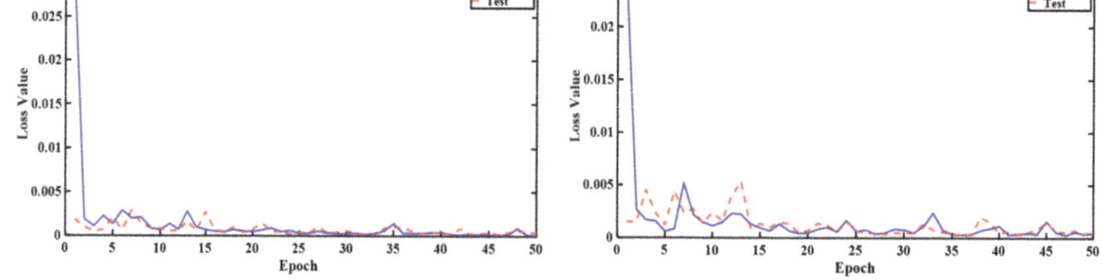

Figure 10. Loss curve of the different datasets.

4.2.2. Cross Validation

Considering the necessary adaptability to dynamical changes, we use another dataset of 36,000 for 6-hour working time zone and obtain a final of 31,304 data samples for further cross validation to illustrate the proposed method's effectiveness and generality. The learning results are shown in Figure 11.

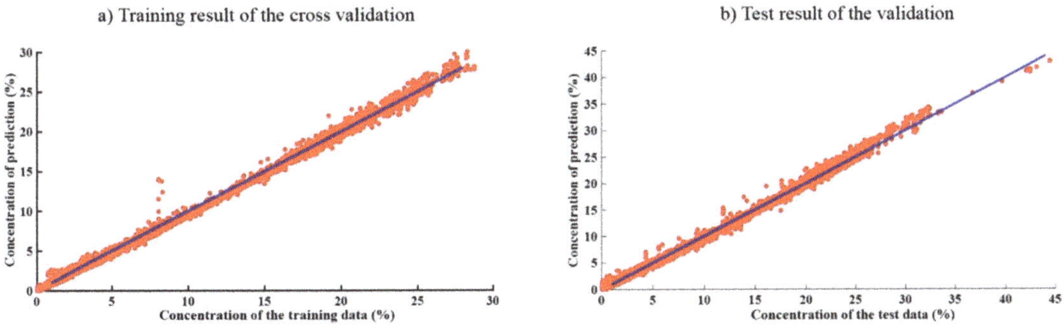

Figure 11. The learning result of cross validation.

It is obvious that the proposed method performs well in both training and testing processes. The average error in cross validation is 1.021%, which decreases since the data amount is becoming larger. In other words, data volume is essential for the deep learning method to function properly. This is just the advantage we explore in this novel model for prediction with operational "big data". Especially, the model can be updated by upcoming new data for more accurate results.

4.3. Comparative Study

This paper presents the novel PCA-LSTM method, which combines advantages of PCA and deep learning algorithm LSTM to manage big time-series data in operation monitoring systems. We compare the proposed method with other prediction methods including traditional PCA-LSTM and LSTM using the same dataset as Section 4.2 for further analysis. The results are shown in Figures 12 and 13.

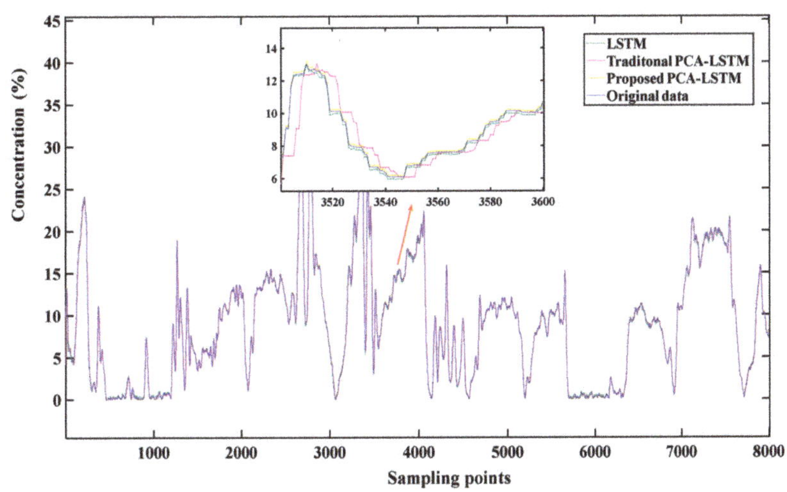

Figure 12. The training result of comparative study.

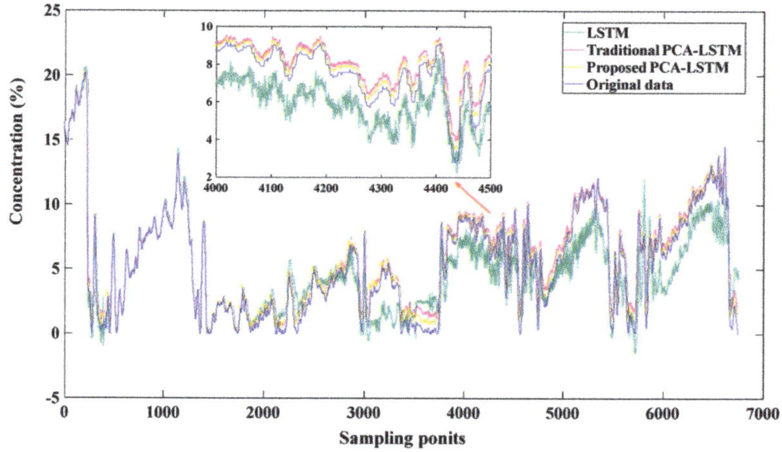

Figure 13. The testing result of comparative study.

It is obvious in Figure 12 that the proposed method works better with a satisfactory error range. LSTM shows the maximum deviation because there is no variable selection before of the prediction process. Although it is a powerful tool to deal with the big series data for its special gate control function, it cannot give consideration to the variable analysis.

In Figure 13, it is easy to find that the novel PCA-LSTM performs better than both traditional PCA-LSTM and LSTM in the test. The yellow line in the figure marks the proposed PCA-LSTM, which has the lowest mean average error of 0.9213%. The green line marks the traditional PCA-LSTM, which shows a mean average error of 1.5301%. However, LSTM shows the worst result with a MAE (mean average error) of 2.0269%. The results differences are caused mainly because of the variable selection for the prediction model. As the input of the data-driven model, variable selection should be more focused with human knowledge and experience.

From a practical point of view, the comparative results are also analyzed by different evaluating indicators such as MAE (mean average error), R^2 (coefficient of determination), and RMSE (root mean square error). As shown in Table 2, all of the models show a good performance with the coefficient of determination, which can explain the LSTM effectiveness. However, for the root mean square error, the proposed method takes on a better stability in the prediction result. The comparative results indicate that the control of input is essential to the machine learning methods.

Table 2. Results analysis of the comparative study.

	MAE	R^2	RMSE
Proposed PCA-LSTM	0.0424	0.9999	0.0925
Traditional PCA-LSTM	0.3063	0.9863	0.4054
LSTM	0.3352	0.9828	0.5010

5. Conclusions

This paper proposes the novel PCA-LSTM method for the productivity prediction of cutter suction dredger, wherein the deep learning process makes good use of the real-time operational monitoring data. The PCA method based on mechanism and knowledge is proposed to analyze the multiple parameters and select relevant variables from the operation process. Then the results are used as input into LSTM model to obtain the target prediction. This approach is also successfully validated by comparison against other methods on a real-world case in China. The productivity of cutter suction dredger is influenced by many correlated factors such as the soil characteristics, cutter parameters, mud pump performance, and pipeline layout. Thus, the mud concentration should be stabilized at a suitable value by comprehensive adjustment to improve its efficiency and productivity.

However, this is still a workable extension of deep learning application in the productivity prediction of cutter suction dredger. In the future, we will further construct the dynamical predictive models according to the changing working condition. When the operational parameters change dynamically under different conditions, the generated data should be classified into status space to study how the operation influences the dredging performance. Additionally, considering the sensors distance in the system, more factors on time-delay should be put up to improve the prediction accuracy.

Author Contributions: Conceptualization, K.Y.; methodology, K.Y.; software, K.Y.; validation, K.Y.; formal analysis, K.Y. and T.X.; data curation, K.Y., J.-L.Y. and B.W.; writing—original draft preparation, K.Y.; writing—review and editing, K.Y. and J.-L.Y.; visualization, K.Y.; supervision, S.-D.F.; funding acquisition, S.-D.F. All authors have read and agreed to the published version of the manuscript.

Funding: This research was funded by National Natural Science Foundation of China under Grant: 51679178 and 52071240.

Institutional Review Board Statement: Not applicable.

Informed Consent Statement: Not applicable.

Data Availability Statement: Not applicable.

Conflicts of Interest: The authors declare no conflict of interest.

References

1. Kawakatsu, H.; Watada, S. Seismic evidence for deep-water transportation in the mantle. *Science* **2007**, *316*, 1468–1471. [CrossRef]
2. Kuehl, S.; DeMaster, D.; Nittrouer, C. Nature of sediment accumulation on the Amazon continental shelf. *Cont. Shelf Res.* **1986**, *6*, 209–225. [CrossRef]
3. Walsh, J.; Nittrouer, C. Contrasting styles of off-shelf sediment accumulation in New Guinea. *Mar. Geol.* **2003**, *196*, 105–125. [CrossRef]
4. Tang, H.; Wang, Q.; Bi, Z. Expert system for operation optimization and control of cutter suction dredger. *Expert Syst. Appl.* **2008**, *34*, 2180–2192. [CrossRef]
5. Wang, B.; Fan, S.; Jiang, P.; Xing, T.; Fang, Z.; Wen, Q. Research on predicting the productivity of cutter suction dredgers based on data mining with model stacked generalization. *Ocean. Eng.* **2020**, *217*, 108001. [CrossRef]
6. Sierhuis, M.; Clancey, W.; Seah, C.; Trimble, J.; Sims, M. Modeling and simulation for mission operations work system design. *J. Manag. Inf. Syst.* **2003**, *19*, 85–128.
7. Blazquez, C.; Adams, T.; Keillor, P. Optimization of mechanical dredging operations for sediment remediation. *J. Waterw. Port Coast. Ocean. Eng.-ASCE* **2001**, *127*, 229–307. [CrossRef]
8. Lai, H.; Chang, K.; Lin, C. A Novel Method for Evaluating Dredging Productivity Using a Data Envelopment Analysis-Based Technique. *Math. Probl. Eng.* **2019**, *2019*, 5130835. [CrossRef]
9. Pei, H.; Hu, C.; Si, X.; Zhang, J.; Pang, Z.; Zhang, P. Review of machine learning based remaining useful life prediction methods for equipment. *J. Mech. Eng.* **2019**, *8*, 1–13. [CrossRef]
10. Wang, L.; Chen, X.; Wang, W. Research and analysis on construction output prediction of cutter suction dredger based on RBF neural network. *China Harb. Eng.* **2019**, *39*, 64–68.
11. Guan, F.; Wang, W. Application of extreme learning machines in productivity prediction of trailing suction hopper dredger. *Sci. Technol. Innov.* **2020**, *8*, 58–61.
12. Yang, J.; Ni, F.; Wei, C. Prediction of cutter suction dredger production based on double hidden layer BP neural network. *Comput. Digit. Eng.* **2016**, *44*, 1234–1237.
13. Ren, L.; Sun, Y.; Cui, J.; Zhang, L. Bearing remaining useful life prediction based on deep autoencoder and deep neural networks. *J. Manuf. Syst.* **2018**, *48*, 71–77. [CrossRef]
14. Wang, S.; Chen, Z.; Chen, S. Applicability of deep neural networks on production forecasting in bakken shale reservoirs. *J. Pet. Sci. Eng.* **2019**, *179*, 112–125. [CrossRef]
15. Xu, W.; Peng, H.; Tian, X.; Peng, X. DBN based SD-ARX model for nonlinear time series prediction and analysis. *Appl. Intell.* **2020**, *50*, 4586–4601. [CrossRef]
16. Hu, C.; Pei, H.; Si, X.; Du, D.; Wang, X. A prognostic model based on DBN and diffusion process for degrading bearing. *IEEE Trans. Ind. Electron.* **2019**, *67*, 8767–8777. [CrossRef]
17. Deutsch, J.; He, D. Using deep learning-based approach to predict remaining useful life of rotating components. *IEEE Trans. Syst. Man Cybern. Syst.* **2017**, *48*, 11–20. [CrossRef]

18. Zhang, C.; Lim, P.; Qin, A.K.; Tan, K.C. Multiobjective deep belief networks ensemble for remaining useful life estimation in prognostics. *IEEE Trans. Neural Netw. Learn. Syst.* **2016**, *28*, 2306–2318. [CrossRef] [PubMed]
19. Wang, Y.; Zhang, Y.; Wu, Z.; Li, H.; Christofides, P. Operational Trend Prediction and Classification for Chemical Processes: A Novel Convolutional Neural Network Method Based on Symbolic Hierarchical Clustering. *Chem. Eng. Sci.* **2020**, *225*, 115796. [CrossRef]
20. Chow, T.; Fang, Y. A recurrent neural-network-based real-time learning control strategy applying to nonlinear systems with unknown dynamics. *IEEE Trans. Ind. Electron.* **1998**, *45*, 151–161. [CrossRef]
21. Malhi, A.; Yan, R.; Gao, R. Prognosis of defect propagation based on recurrent neural networks. *IEEE Trans. Instrum. Meas.* **2011**, *60*, 703–711. [CrossRef]
22. Li, D.; Huang, D.; Yu, G.; Liu, Y. Learning Adaptive Semi-Supervised Multi-Output Soft-Sensors with Co-Training of Heterogeneous Models. *IEEE Access* **2020**, *8*, 46493–46504. [CrossRef]
23. Yang, K.; Liu, Y.; Yao, Y.; Fan, S.; Ali, M. Operational time-series data modeling via LSTM network integrating principal component analysis based on human experience. *J. Manuf. Syst.* **2021**, in press. [CrossRef]
24. D'Agostino, R.B. Principal Components Analysis. In *Handbook of Disease Burdens and Quality of Life Measures*; Preedy, V.R., Watson, R.R., Eds.; Springer: New York, NY, USA, 2020. [CrossRef]
25. Hochreiter, S.; Schmidhuber, J. Long short-term memory. *Neural Comput.* **1997**, *9*, 1735–1780. [CrossRef] [PubMed]
26. Jafari, H.; Alipoor, A. A New Method for Calculating General Lagrange Multiplier in the Variational Iteration Method. *Numer. Methods Partial. Differ. Equ.* **2011**, *27*, 996–1001. [CrossRef]
27. Bai, S.; Li, M.; Kong, R.; Han, S.; Li, H.; Qin, L. Data mining approach to construction productivity prediction for cutter suction dredgers. *Autom. Constr.* **2019**, *105*, 102833. [CrossRef]

applied sciences

MDPI

Article

Measuring Ocean Surface Current in the Kuroshio Region Using Gaofen-3 SAR Data

Yan Li [1,2,3], Jinsong Chong [1,2,3,*], Kai Sun [1,2,3], Yawei Zhao [1,2,3] and Xue Yang [1,2,3]

1 National Key Lab of Microwave Imaging Technology, Beijing 100190, China; liyan1603@mails.ucas.ac.cn (Y.L.); sunkai181@mails.ucas.ac.cn (K.S.); zhaoyawei17@mails.ucas.ac.cn (Y.Z.); yangxue19@mails.ucas.ac.cn (X.Y.)
2 Aerospace Information Research Institute, Chinese Academy of Sciences, Beijing 100190, China
3 School of Electronics, Electrical and Communication Engineering, University of Chinese Academy of Sciences, Beijing 100049, China
* Correspondence: lily@mail.ie.ac.cn; Tel.: +86-10-5888-7125

Abstract: The Kuroshio is the strongest warm current in the western North Pacific, which plays a crucial role in climate and human activities. In terms of this, the accurate acquisition of ocean surface current velocity and direction in the Kuroshio region is of great research value. Gaofen-3 synthetic aperture radar (SAR) provides data support for the study of ocean surface current measurements in the Kuroshio region, but no relevant experimental result has been published yet. In this paper, four available stripmap mode SARs' data acquired by Gaofen-3 in the Kuroshio region are used for measuring the ocean surface current field. In general, the Doppler centroid anomaly (DCA) estimation is a common method to infer ocean surface currents from single-antenna stripmap data, but only the radial velocity component can be retrieved. In order to measure current vectors, a novel method combining the sub-aperture processing and the least squares (LS) technology is suggested and demonstrated by applying to the Gaofen-3 SAR data processing. The experiment's results agree well with model-derived ocean current data, indicating that the Gaofen-3 SAR has the capability to accurately retrieve the ocean surface current field in the Kuroshio region and motivate further research by providing more data.

Keywords: ocean surface current; Gaofen-3 SAR data; the Kuroshio region; sub-aperture processing; the least squares

Citation: Li, Y.; Chong, J.; Sun, K.; Zhao, Y.; Yang, X. Measuring Ocean Surface Current in the Kuroshio Region Using Gaofen-3 SAR Data. *Appl. Sci.* **2021**, *11*, 7656. https://doi.org/10.3390/app11167656

Academic Editor: Dimitrios-Nikolaos Pagonis

Received: 22 July 2021
Accepted: 16 August 2021
Published: 20 August 2021

1. Introduction

Current is ubiquitous in the global ocean, the information about which is valuable in many applications. The circulation current in different regions of the deep ocean will exchange material and energy, which is of great importance for the global climate process. Moreover, monitoring currents in the coastal waters is conducive to navigation, offshore oil and gas field development, fishery resources management, and so forth. According to the data requirements survey report on oceanic variables issued by EuroGOOS, the ocean surface current velocity and direction were the two most requested [1]. Therefore, it is necessary to investigate various methods for measuring the velocity and direction values of the ocean current field. Compared with the in-situ current measurement tools, such as drift buoy and current meter, which are time-consuming and limited in spatial measurement range, the spaceborne synthetic aperture radar (SAR) has become a vital tool for ocean remote sensing mainly due to its day/night and quasi-all-weather capability and its high-spatial resolution [2].

There are two common techniques used for current measurement with SAR data, that is, one is Doppler centroid analysis (DCA) and the other is along-track interferometry (ATI) [3,4]. DCA is suitable for single-antenna SAR data processing and has low requirements for the system. Chapron [5] first proposed the DCA method in 2005 and processed the ENVISAT ASAR data to retrieve the radial velocity of the sea surface in different areas including the current field near the Gulf Stream. In the following years, the DCA method was used many times to measure the ocean surface current field in the Agulhas Current regime [6–8]. These studies show that DCA has good applications in these regions with strong boundary currents. The Kuroshio Current is also one of the strongest western boundary currents, but there have not yet been any relevant SAR experimental results applied to the Kuroshio region. In addition, the DCA method using Doppler centroid shift can only derive the line-of sight (LOS) velocity component of ocean surface current, not including the direction values. The ATI method uses interferometric phase to achieve high-resolution ocean surface current measurements. Moreover, ATI systems can be transformed into ones with squint beams to measure the ocean current vectors [9–11]. However, ATI needs to operate the SAR data of at least two antennas; the related system is complicated, the experimental data are small, and it is far from scale application than DCA using single SAR data.

Like the Gulf Stream in the North Atlantic, the Kuroshio is a powerful western boundary current and forms the western limb of the North Pacific Subtropical Gyre. Owing to the influence of the geostrophic Coriolis force, the remote sensing measurement of Kuroshio current speed can be performed with the spaceborne altimeters [12,13], but the spatial resolution is relatively rough, reaching tens of kilometers, and the data around the land are easily missed. Gaofen-3 is China's first C-band full-polarization synthetic aperture radar imaging satellite, which provides reliable high-resolution SAR image data for ocean remote sensing [14]. So far, Gaofen-3 SAR data have been used to study ocean surface waves [15], ocean winds [16] and internal waves [17]. Gaofen-3 provides data support for the ocean current retrieval in the Kuroshio region, which will promote relevant application research.

This paper uses four available Gaofen-3 SAR datasets to measure the ocean surface current in the Kuroshio region. In order to retrieve current velocity and direction information at the same time, a novel method combining sub-aperture processing technology and least squares technology is proposed and applied. The following second part introduces the principle and process of the data processing method in detail. The third part is to carry out the ocean current measurement experiment of the Gaofen-3 SAR data and verify the experimental results. The fourth part discusses and analyzes the applicability of the method, including the influence of system and environmental factors on the results, and the fifth part presents conclusions.

2. Measurement of the Ocean Surface Current

2.1. Ocean Doppler Information of Single Antenna SAR Data

Due to the relative motion between the SAR platform and the ocean surface, the azimuth Doppler information of the echo signal would change, suggesting that the Doppler frequency can be used to retrieve the ocean surface motion. In fact, considering the complexity of ocean motion and the particularity of SAR for ocean imaging, the estimated Doppler centroid shift in SAR ocean data usually includes the following items [18]:

$$f_{dc} = f_{dc}^{phys} + f_{dc}^{geo} + f_{dc}^{ele} + \triangle f_{dc}. \quad (1)$$

In Equation (1), f_{dc}^{phys} is the geophysical term, f_{dc}^{geo} is the geometric attitude term, f_{dc}^{ele} indicates electronic misdirection of antenna, $\triangle f_{dc}$ represents the residual error in data processing. As one of our concerns, f_{dc}^{phys} is proportional to the ocean surface Doppler velocity in the line-of-sight direction, which contains not only the information of ocean currents, but also the contributions of wind and wave motion. In Bragg theory, wind and wave motion primarily involve Bragg wave phase velocity and large-scale wave orbit velocity. Therefore, f_{dc}^{phys} can be expressed as

$$f_{dc}^{phys} = f_c + f_b + f_o, \quad (2)$$

where f_c, f_b and f_o denote the Doppler centroid frequency variation caused by ocean currents, Bragg wave phase velocity, and large-scale wave orbit velocity, respectively. How to estimate f_{dc}^{phys} from f_{dc} has been studied in detail in reference [8], and the relevant data processing process will not be discussed. The focus of this paper is to separate f_c from f_{dc}^{phys} and further extract the ocean current information.

According to the geometric relationship between ocean current and SAR platform shown in Figure 1, f_c can be expressed as:

$$f_c = \frac{2}{\lambda}(u_x \cos\varphi \sin\theta + u_y \sin\varphi \sin\theta), \tag{3}$$

where u_x and u_y represent the azimuth and range velocity components of the current, respectively. φ is the angle between the beam center and the flight direction in the horizontal plane, which is called the azimuth angle. θ is the incidence angle, and λ is the electromagnetic wavelength. We can also obtain the Doppler frequency caused by platform movement as:

$$f_p = \frac{2v_p}{\lambda} \cos\varphi \sin\theta, \tag{4}$$

where v_p is the radar speed.

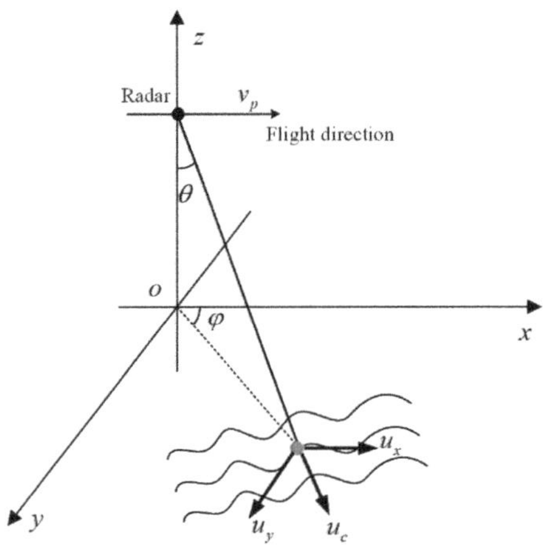

Figure 1. Schematic diagram of the geometric relationship between current and radar.

In Bragg scattering theory, Bragg resonance is the main SAR imaging mechanism for the ocean surface. Furthermore, the phase velocity of the Bragg wave is the main factor affecting the deviation of current measurements, which is expressed as [19,20]:

$$c = \sqrt{g/k_b + Tk_b}, \tag{5}$$

where g is the acceleration of gravity, and T is the ratio of water surface tension to seawater density. $k_b = 2\pi/\lambda_b$ is the wavenumber of Bragg waves, which can be expressed by electromagnetic wavenumber k_e and incidence angle θ_e as

$$k_b = 2k_e \sin\theta_e. \tag{6}$$

Therefore, the Doppler centroid variation caused by the Bragg wave phase velocity is

$$f_b = \frac{2}{\lambda}\sqrt{g/k_b + Tk_b}. \tag{7}$$

In the composite surface model, as one form of Bragg theory, there are not only small-scale Bragg waves but also large-scale modulating waves on the ocean surface. In this situation, the large-scale waves will be displayed in SAR images by modulating the energy distribution of Bragg waves, and the Doppler centroid will also shift with the variation of large-scale wave orbital velocity, that is, there is f_o, which also results in the deviation of current measurements. Generally, the orbital velocity is affected by the weighted average of the local backscatter within the SAR resolution cell and during the SAR integration time [4,5]. In some studies [4,21,22], the empirical model derived from actual data fitting can be used to remove the motion induced by wind and waves, including the effect of orbital speed. This processing method is in good agreement with the actual situation, but is limited by different sensor data and sea surface position, the measurement accuracy will be different. Yu et al. [23] also pointed out that the influence of orbital velocity can be suppressed by space averaging. This operation is simple, but the residual error will be left.

Motion errors will affect SAR imaging and sea surface current field estimation. Reference [24] states that motion errors decrease the image quality in the course of imaging. In principle, the uncertainty caused by motion errors can be corrected with the motion measurement data obtained from the global positioning system (GPS) or other ancillary instrument. In Reference [24], a deep SAR motion compensation scheme is proposed to be combined with the deep SAR imaging algorithm, which could eliminate the influence of motion errors and improve SAR imagery quality in practical applications. Reference [25] also discusses the influence of motion errors on imaging, and points out that classical motion compensation is typically not required for spaceborne systems due to their stable and homogeneous motion. The Gaofen-3 satellite adopts a dual frequency GPS system to realize precise orbit determination, which has high attitude control accuracy and stability [26]. Therefore, the motion errors have a limited impact on the imaging results.

For the application of current measurements, the influence of motion and attitude errors on the predicted Doppler shift is mainly considered, as shown in Reference [8]. Due to errors in satellite position and attitude used to predict the Doppler shift, the ocean current product will deviate. After attitude control, the Gaofen-3 satellite eliminates the change of Doppler center frequency caused by earth rotation, earth ellipticity and satellite orbit oblateness [26]. Therefore, the predicted Doppler shift of SAR data is close to zero. If the azimuth angle error $\triangle\varphi$ is considered in current measurements with actual data, the Doppler frequency caused by platform motion will have errors [27]. Therefore, Equation (8) can be obtained based on Equation (4), and the Doppler frequency caused by current becomes Equation (9):

$$\begin{aligned}\triangle f_a &= \frac{2v_p}{\lambda}[\cos(\varphi+\triangle\varphi)\sin\theta - \cos\varphi\sin\theta] \\ &\approx -\frac{2}{\lambda}v_p\triangle\varphi\sin\varphi\sin\theta\end{aligned} \tag{8}$$

$$\begin{aligned}f_c' &= \frac{2\sin\theta}{\lambda}\left[u_x\cos(\varphi+\Delta\varphi) + u_y\sin(\varphi+\Delta\varphi)\right] \\ &\approx \frac{2\sin\theta}{\lambda}\left[(u_x+\Delta\varphi u_y)\cos\varphi + (u_y-\Delta\varphi u_x)\sin\varphi\right].\end{aligned} \tag{9}$$

2.2. Measurement Method with Single Antenna SAR Data

According to the above analysis, the Doppler centroids in different azimuth angles reflect the current information in different directions. However, single antenna full-aperture SAR data usually only provide the Doppler frequency in one direction. Therefore, the method in this paper firstly carries out sub-aperture processing to acquire sub-apertures image data representing different look directions, and then carries out Doppler centroid estimation, respectively. Furthermore, the equations are established according to the estimated Doppler centroids. Finally, the current vectors are estimated by the least squares method. This method can be divided into three steps: sub-aperture processing, Doppler centroid estimation and vector estimation of current. The flow chart is shown in Figure 2.

In the first step, the input is Gaofen-3 single look complex (SLC) data. After implementing azimuth FFT of the input SLC, the sub-aperture data is extracted by aperture division in the Doppler frequency domain, then IFFT is performed to acquire L sub-aperture SLC in time-domain as the output of the first step. Besides, the azimuth angle φ_k corresponding to each sub-aperture is also the output and used as the input for the third step. The L sub-aperture SLC is inputted in step 2 to estimate the Doppler frequency of each sub-aperture, respectively. In step 2, each input sub-aperture SLC data is divided into $M * N$ blocks, and the Doppler centroid frequency of each block is estimated. The Doppler frequency at the same spatial position block (i, j) with different sub-apertures form a vector $\mathbf{Y}_{i,j}$ as the output. Therefore, azimuth angle φ_k and vector $\mathbf{Y}_{i,j}$ are the inputs of the third part. The coefficient matrix $\mathbf{W}$ is generated by combining the Gaofen-3 system parameters, that is, wavelength, incident angle, flight speed, and the azimuth angle of L sub-apertures. Then, the vector equation of the unknown vector $\mathbf{X}$ is established with vectors $\mathbf{Y}_{i,j}$ and $\mathbf{W}$. The unknown vector $\mathbf{X}$ includes current vector information, so the current vector at each position in space can be calculated by solving the equation with the least squares method. The output of step 3 is the sea surface current field of the entire data. The specific processing process and the calculation of variables are explained respectively in the following subsection.

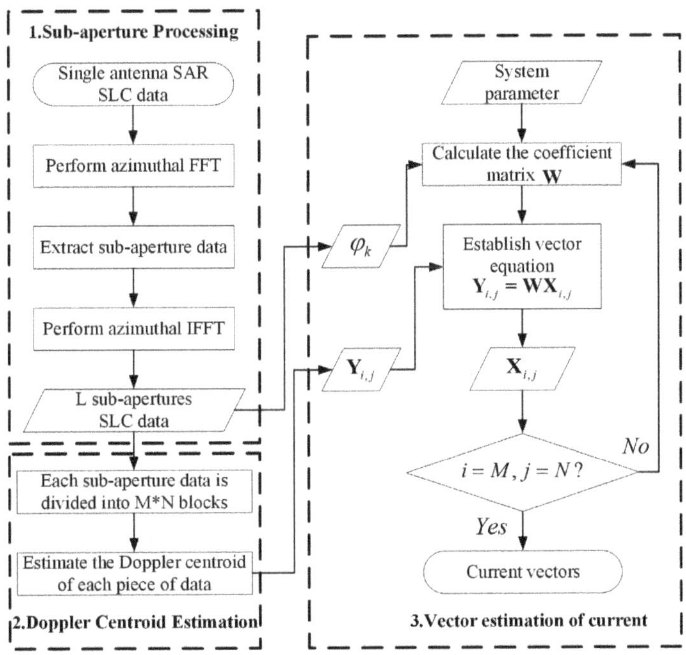

Figure 2. Flow chart of ocean current field measurement using single antenna SAR data.

2.2.1. Sub-Aperture Processing

Generally, the beam of SAR is perpendicular to the direction of flight, and the current velocity component in the range direction is obtained by Doppler centroid estimation with full-aperture data. However, through the above analysis, in order to retrieve the current vectors, it is necessary to derive the Doppler centroids in different directions, so the data under different azimuth angles can be processed through sub-aperture processing.

In the basic theory of SAR, there is a linear relationship between Doppler frequency and azimuth time. By dividing Doppler frequency, the full-resolution image is decomposed into a series of low-resolution images formed by beams in different directions, which is sub-aperture processing technology [28].

In this method, sub-aperture data are divided in the azimuth frequency domain. The divided beam is shown in Figure 3, where the solid line represents the beam center of different sub-aperture S. The sub-aperture bandwidths are same, and the sub-aperture center frequency is f_k, $k = 1, 2, \cdots, L$, where L is the number of sub-apertures. In the ideal side-looking condition, the estimated Doppler centroid should be zero when the full-aperture beam irradiates the stationary target. However, in practical systems, the full-aperture beam is not always perpendicular to the azimuth direction. The full-aperture azimuth angle is calculated from the data recorded by the Inertial Navigation System (INS) or estimated from the data in the stationary area, such as land. According to Equation (4), the azimuth angle φ_k of the sub-apertures beam center line can be expressed as

$$\varphi_k = \arccos\left(\frac{\lambda f_k}{2v_p \sin\theta}\right). \tag{10}$$

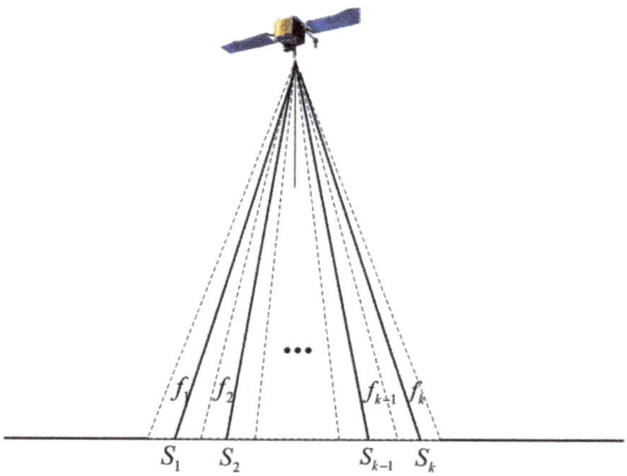

Figure 3. Schematic diagram of sub-aperture processing.

2.2.2. Doppler Centroid Estimation

Accurate Doppler centroid estimation of real data is the basis of subsequent processing through the above analysis. According to Reference [29], the methods used for Doppler centroid estimation include the energy balance method, the power spectrum correlation method, the correlation Doppler centroid estimation method and the optimal estimation method. They use different weighting functions $B(f)$. The results of the optimal estimation method are more accurate, very close to the Cramer–Rao bound, so this paper uses this method to estimate Doppler centroid. The weighting function of optimal estimation method is shown in Equation (11), where $E(f)$ denotes the Doppler power spectrum. According to [29], Figure 4 shows $E(f)$ and $B(f)$, where F_a is pulse repeat frequency (PRF). In Figure 4a, we show the Doppler bandwidth of the sub-aperture image, which corresponds to S in Figure 3. In data processing, we use a band-pass filter to extract the sub-aperture frequency band from the azimuth spectrum. The specific process is referred to in Reference [30]. The frequency point at the center of each sub-aperture bandwidth corresponds to a sub-aperture azimuth angle, and the sub-aperture azimuth angles at both ends are the extreme values, so we can get

$$B(f) = \frac{E'(f)}{E^2(f)}. \tag{11}$$

As shown in Figure 4, the Doppler power spectrum $E(f)$ is an even function, while the weighting function $B(f)$ is an odd function. The convolution of the two functions has only one zero-crossing point, which is then calculated by mathematical methods such as the least squares. The zero-crossing point is the Doppler centroid frequency. We convolute the Doppler spectrum of each sub-aperture with the weighted function in Figure 4b to find the zero-crossing point, so as to estimate the Doppler centroid.

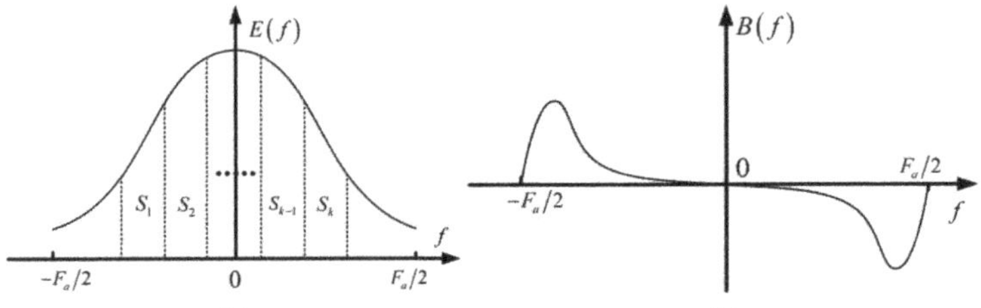

Figure 4. (**a**) Doppler power spectrum with sub-apertures bandwidth. (**b**) Weighting function used for Doppler centroid estimation.

In order to accurately estimate the Doppler centroid in actual data, enough sampling points need to be taken in the azimuth and range direction for calculation. Therefore, it is necessary to block sub-aperture image data for Doppler estimation and eliminate the influence of non-geophysical terms. Finally, an average Doppler frequency for several kilometers' range of ocean surface is acquired, which is expressed as $f_{i,j}$ ($i = 1, 2, \cdots, M$, $j = 1, 2, \cdots, N$, M and N represent the number of blocks in azimuth and range direction of each sub-aperture datapoint , respectively). Generally, there is a time interval between sub-aperture images, and the variation of the resulting ocean surface is usually at the pixel level. However, the Doppler estimation of sub-apertures requires thousands of pixels to obtain the mean motion state of the ocean surface over several kilometers, so it is considered that the changes of Doppler frequency between different sub-apertures are mainly caused by different beam directions, rather than the ocean surface evolution.

It should be noted that the Doppler centroid estimation process in this paper acts on the correctly imaged single look complex data for the purpose of current inversion. In some of the literature, the Doppler estimation process acts on SAR raw data to improve imaging quality, such as in [31,32]. Compared with imaging processing, this paper pays more attention to the application process of imaging data. In addition, the Doppler estimation algorithm in References [31,32] is applied to airborne SAR data with large squint and high contrast. In this paper, the squint angle of the Gaofen-3 satellite is close to zero, and the SAR data scattering is uniform and the contrast is low.

2.2.3. Vector Estimation of Ocean Current

In this section, we assume that the Doppler frequency caused by the Bragg wave phase velocity is the same at the same spatial position in each sub-aperture image. In addition, in order to improve the efficiency of the algorithm, we deal with the influence of orbital velocity through space averaging, and consider that the residual errors in the same Doppler estimation block are the same. Set the Doppler frequency shift caused by the influence of Bragg wave phase velocity and residual orbit velocity to f_B, and combine Equations (8) and (9), then we can get:

$$\begin{aligned} f_{i,j}^k &= \triangle f_a^k + f_c^k(i,j) + f_B(i,j) \\ &= -\frac{2}{\lambda} v_p \triangle\varphi \sin\varphi_k \sin\theta \\ &+ \frac{2}{\lambda}\left(u_x(i,j) + \triangle\varphi u_y(i,j)\right)\cos\varphi_k \sin\theta \\ &+ \frac{2}{\lambda}\left(u_y(i,j) - \triangle\varphi u_x(i,j)\right)\sin\varphi_k \sin\theta + f_B(i,j), \end{aligned} \tag{12}$$

where $f_{i,j}^k$ represents the Doppler results of the data block at the central coordinate (i,j) of the k-th sub-aperture image. Let $\mathbf{Y}_{i,j}$ represent the vector composed of Doppler centroids of all L sub-aperture resolution units at spatial coordinates (i,j), expressed as

$$\mathbf{Y}_{i,j} = [f_{i,j}^1, f_{i,j}^2 \cdots f_{i,j}^k \cdots f_{i,j}^L]^T. \tag{13}$$

Then, the overdetermined linear equations are established and expressed in matrix form as:

$$\mathbf{Y}_{i,j} = \mathbf{W}\mathbf{X}_{i,j}, \tag{14}$$

where the coefficient matrix $\mathbf{W}$ and the unknown vector $\mathbf{X}_{i,j}$ containing the current information are respectively expressed as

$$\mathbf{W} = \frac{2\sin\theta}{\lambda}\begin{bmatrix} -v_p \sin\varphi_1 & \cos\varphi_1 & \sin\varphi_1 & \lambda/2\sin\theta \\ -v_p \sin\varphi_2 & \cos\varphi_2 & \sin\varphi_2 & \lambda/2\sin\theta \\ \vdots & \vdots & \vdots & \vdots \\ -v_p \sin\varphi_L & \cos\varphi_L & \sin\varphi_L & \lambda/2\sin\theta \end{bmatrix} \tag{15}$$

$$\mathbf{X}_{i,j} = [\triangle\varphi, u_x(i,j) + \triangle\varphi u_y(i,j), u_y(i,j) - \triangle\varphi u_x(i,j), f_B(i,j)]^T. \tag{16}$$

Then, after the least squares method is used to solve $\mathbf{X}_{i,j}$, the azimuth velocity component u_x and range velocity component u_y corresponding to the data block at coordinates (i,j) can be expressed as

$$u_x = \frac{X_{i,j}(2) - X_{i,j}(1)X_{i,j}(3)}{1 + X_{i,j}(1)^2} \tag{17}$$

$$u_y = \frac{X_{i,j}(1)X_{i,j}(2) + X_{i,j}(3)}{1 + X_{i,j}(1)^2}. \tag{18}$$

Therefore, the current velocity u_c and direction ϕ_c can be obtained:

$$u_c = \sqrt{u_x^2 + u_y^2} = \sqrt{\frac{X_{i,j}^2(2) + X_{i,j}^2(3)}{1 + X_{i,j}^2(1)}}. \tag{19}$$

$$\phi_c = \arctan\left(\frac{u_x}{u_y}\right) = \arctan\left(\frac{X_{i,j}(2) - X_{i,j}(1)X_{i,j}(3)}{X_{i,j}(1)X_{i,j}(2) + X_{i,j}(3)}\right). \tag{20}$$

The current field corresponding to the whole SAR image can be measured by solving the current vector of each spatial position in turn.

3. Gaofen-3 Data Experiment and Results

3.1. Experiment Description

The experiment site is located in the Kuroshio Current region. The Kuroshio originates from the North Equator, enters the East China Sea via the Philippines, close to the eastern part of Taiwan, and then flows through the Ryukyu Islands, along the southern part of the Japanese archipelago, and ends in the waters near 142° E and 35° N. The Kuroshio has the characteristics of strong flow velocity, large flow, narrow flow range, deep extension, high temperature and high salt, and so forth.

Gaofen-3 is a Chinese spacecraft carrying a C-band SAR, which was launched in August 2016, from Taiyuan (Shanxi Province, China). The Gaofen-3 satellite is capable of high resolution, large imaging width, high radiation accuracy, multiple imaging modes and long-term operation. It can monitor global marine and land information all day and all night, and expand the earth observation range and improve the rapid response capability through left–right attitude maneuver. The satellite has 12 imaging modes, which is the SAR satellite with the most imaging modes in the world. Gaofen-3 has outstanding flexibility, high attitude control accuracy and stability, and has the ability of continuous two-dimensional attitude guidance maneuver. In addition, the satellite adopts an independent health management mechanism to reduce the risk of whole satellite failure. Some Gaofen-3 SAR system parameters are shown in Table 1. For more specific parameters and technical details, please refer to [26].

Table 1. Gaofen-3 system parameters.

Parameters	Values	Units
Radar Frequency	5.4	GHz
Incidence Angle	20–60	deg
Polarization	HH/HV/VH/VV	-
Spatial Resolution	1–500	m
Swath	10–650	km
Platform Speed	7567	m/s
Orbit Altitude	755	km

The four Gaofen-3 SAR datasets are single look complex (SLC) images acquired in ultra-fine strip-map (UFS) mode, of 3 m high resolution within 30 km swath owing to the dual-receive technique. The Gaofen-3 raw data is imaged by a special integrated SAR data processor, and a chirp scaling (CS) algorithm is used for UFS mode data. More details can be found in Reference [33]. The radar wavelength of the Gaofen-3 SAR system is 5.6 cm, the central incident angle is 22.5 degrees in ultra-fine strip-map mode. The GF-3 SAR system is right view, and implemented zero Doppler through attitude control. Further information about the data is shown in Table 2. Figure 5 depicts the geographical locations in the red marks.

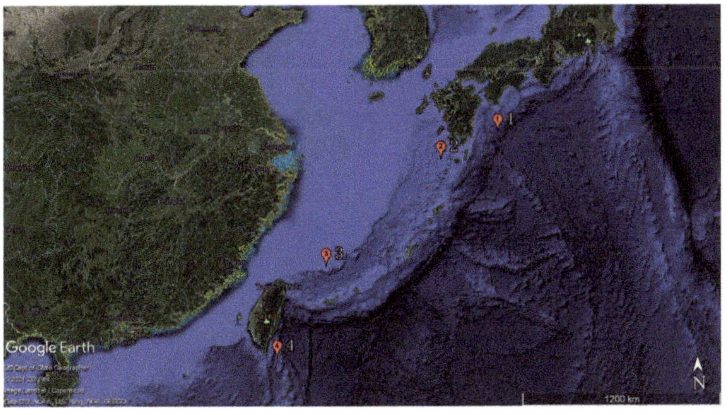

Figure 5. Geographic location of Gaofen-3 SAR data (marked in red).

Table 2. Gaofen-3 experimental data parameters.

Number	Date	Center Coordinates	Orbit	Polarization	Swath
1	8 May 2021	132.79° E, 31.53° N	Descend	HH	30 km
2	23 April 2021	129.70° E, 30.46° N	Descend	HH	30 km
3	15 April 2021	123.86° E, 25.92° N	Ascend	HH	30 km
4	16 November 2020	121.44° E, 21.75° N	Ascend	HH	30 km

Figure 6 shows four SAR intensity images. It can be observed that all the images are relatively uniform in intensity, and are not affected by vortices, internal waves and other sea surface phenomena, which means that the weighted effect of the sea surface normalized radar cross section (NRCS) on Doppler centroid variation is minimal. Note that Johannessen et al. [6] pointed out that the Doppler estimation method works best for the quasi-uniform radar cross section, which is consistent with the image features in this experiment.

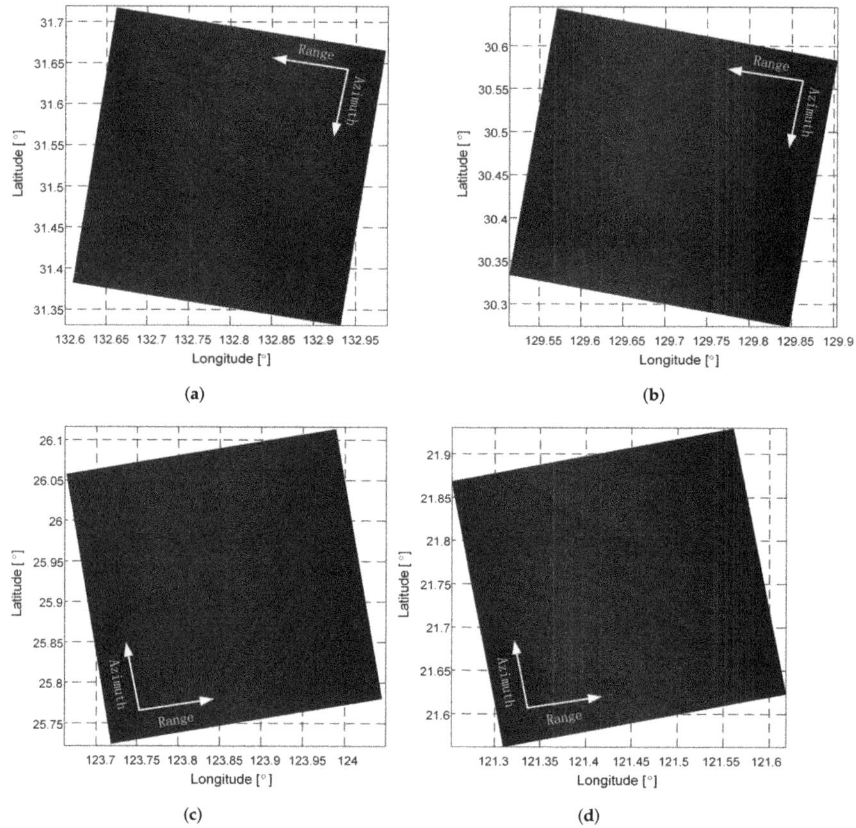

Figure 6. SAR intensity images corresponding to the four positions in Figure 5. (**a**–**d**) correspond to the labels 1, 2, 3, and 4, respectively.

3.2. Retrieved Ocean Surface Current

In the first step of the proposed method, we extracted six sub-apertures for computation. Taking the processing result of label 1 SAR data as an example, after the first step of sub-aperture processing, one of the sub-aperture intensity images is shown in Figure 7. Obviously, the speckle noise of sub-aperture image is more serious and the resolution is reduced. In addition, the maximum azimuth angle of the sub-apertures is 0.3 degrees. In the second part, we divided each sub-aperture dataset into 30 $\times$ 30 blocks, which resulted in a grid spatial resolution of 1 km $\times$ 1 km for each block. Then, after the Doppler centroid estimation, the corresponding Doppler image is shown in Figure 7b.

In the third step, the Doppler image results of each sub-aperture data are calculated according to the flow chart, and finally the two-dimensional current field results are obtained. Corresponding to the four Gaofen-3 SAR datasets in Figure 6, the retrieved ocean surface current field after the proposed method processing is shown in Figure 8, where the color ruler and the arrow point represent the velocity and direction of the ocean current, respectively.

As can be seen from Figure 8, the ocean currents at panel (a), (c) and (d), that is, the positions at labels 1, 3 and 4 in Figure 5, are moving in a direction close to the northeast, which is consistent with the trend of the Kuroshio Current movement. Figure 8a is close to the Japanese islands, with the highest latitude, and the current velocity is the largest. In Figure 8b, the current velocity is the lowest and the direction changes, which is mainly due to the hindrance from the south of Japan island. The current field in Figure 8c is usually affected by the surrounding islands, making the velocity slower and the current direction closer to the east. Figure 8d has the lowest latitude, and the current field is closest to the equator, which is the source of the Kuroshio, so the current velocity is relatively high, and it moves northeast along the Taiwan island.

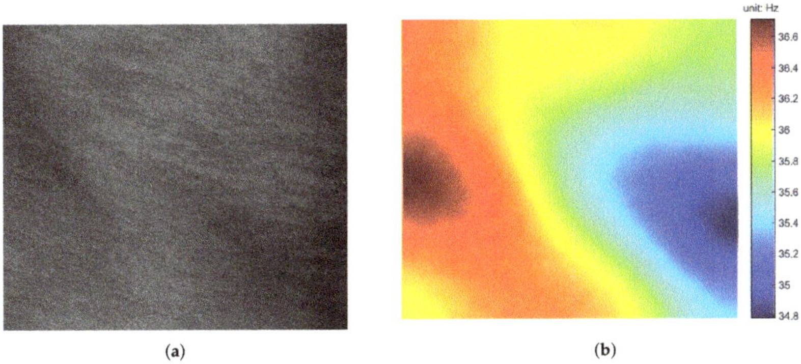

Figure 7. The result of sub-aperture processing: (**a**) one of the sub-apertures intensity image, and the result of Doppler centroid estimation: (**b**) the corresponding Doppler image.

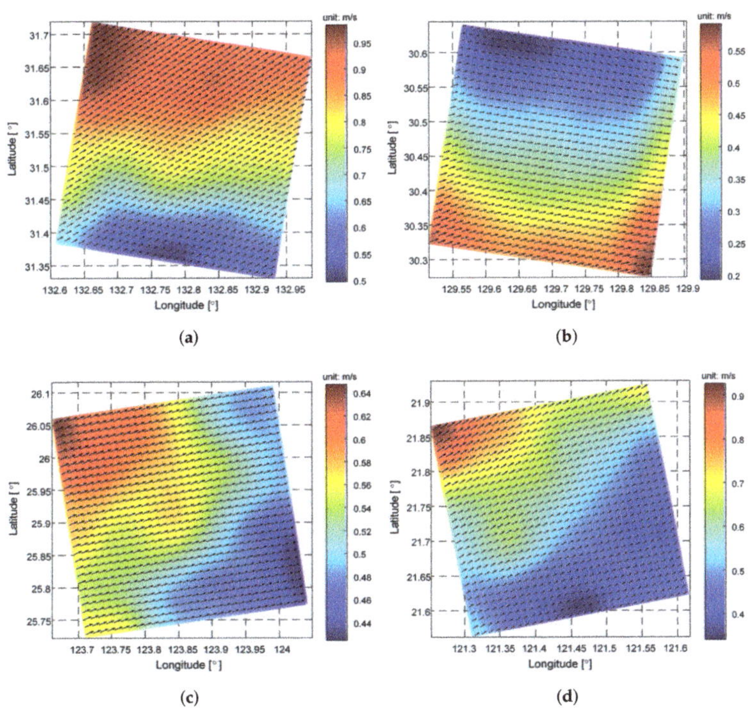

Figure 8. The retrieved ocean surface current field corresponding to the acquisitions in Figure 6. (**a–d**) correspond to the labels 1, 2, 3, and 4, respectively.

3.3. Results Validation

Generally, the in-situ data derived by Argos drifter buoys, current meters, and so forth, is optimally used for the verification of ocean current results. In addition, satellite altimeter data can also be used to verify geostrophic current. Unfortunately, there is no spatio-temporal synchronization in-situ data and altimeter data available for this experiment. In fact, since the ocean current retrieved by the altimeter is usually at a resolution of tens of kilometers, it is not suitable for comparison with the data with a smaller width and higher resolution in this article.

Ocean circulation model data can also be used for ocean current verification. The hybrid coordinate ocean model (HYCOM) has been widely used in a variety of ocean current research projects [34–36]. Therefore, the HYCOM model data are used to verify the results in this article. In this section, HYCOM's global reanalysis dataset with a spatial resolution of 1/12° and a temporal resolution of 1 day is acquired for comparison. Figure 9 shows the HYCOM ocean current model data consistent with the SAR acquisition time and location. The black box represents the boundary of the experimental area. The length and color of the arrow represent the current velocity, and the direction of the arrow indicates the current direction. It is obvious that the ocean current state in the black box compares qualitatively well with the experimental results in Figure 8.

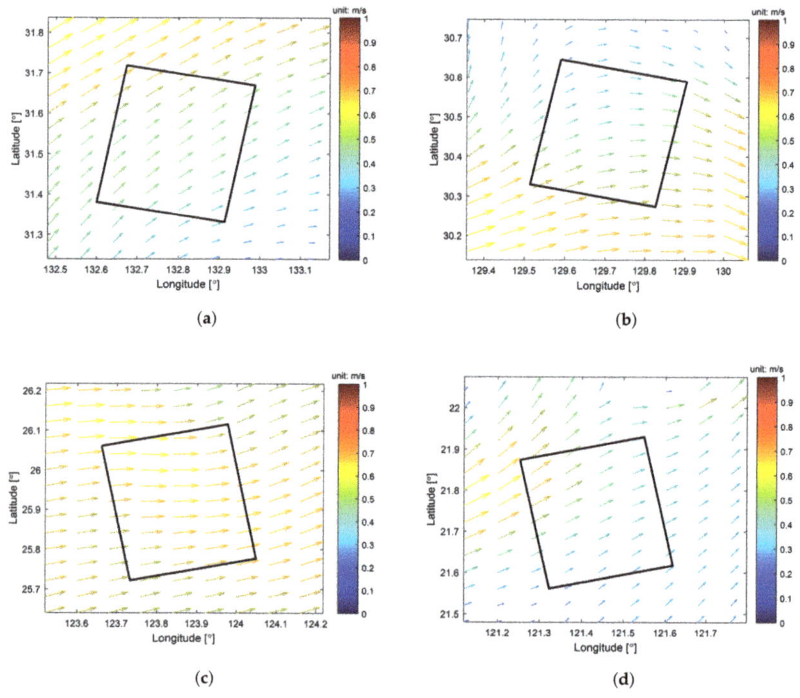

Figure 9. The HYCOM model data corresponding to the ocean current results in Figure 8. (**a**–**d**) correspond to the labels 1, 2, 3, and 4, respectively.

In order to quantitatively analyze the relationship between experimental results and the model data, the results in Figure 8 were down-sampled and matched with the data in Figure 9, and the current velocity and direction scatter plots are shown in Figures 10 and 11, respectively. Figure 10 illustrates the statistical comparison of current velocity, where the minimum correlation coefficient is close to 0.9 and the maximum correlation coefficient is 0.98. Both root mean square error (RMSE) and bias are less than 0.1 m/s, which meets the requirements for accuracy of current velocity measurement in marine environment applications.

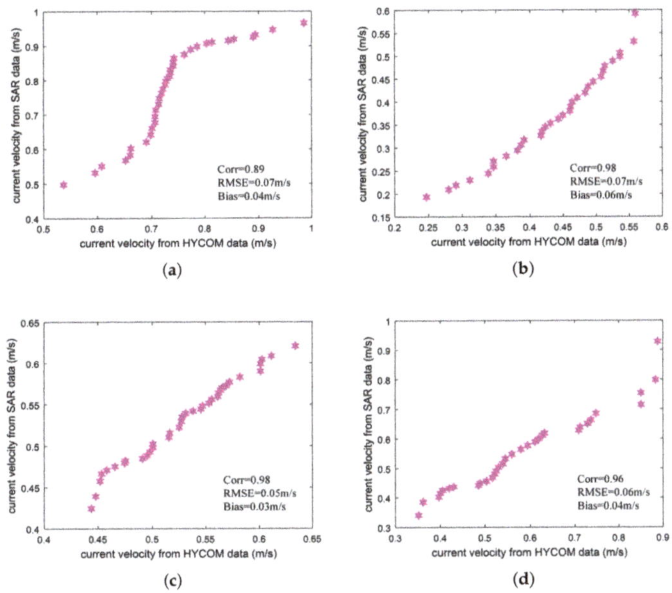

Figure 10. The scatter plot of the current velocity after comparing the experimental results in Figure 8 with the model data in Figure 9. (**a**–**d**) correspond to the labels 1, 2, 3, and 4, respectively.

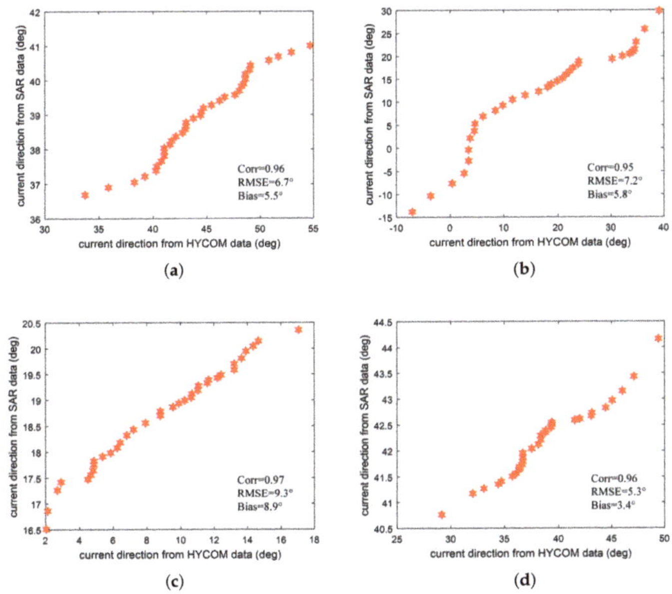

Figure 11. The scatter plot of the current direction after comparing the experimental results in Figure 8 with the model data in Figure 9. (**a**–**d**) correspond to the labels 1, 2, 3, and 4, respectively.

Figure 11 shows the statistical comparison for the current direction results, where a direction of zero means the ocean current is going eastward. The current direction statistics of the four plots show that the correlation coefficients between the experimental results and the model data are all greater than 0.95. Both RMSE and bias are less than 10°. Although the HYCOM data represent the average state of the current field in the day, the Kuroshio current does not change much in one day. The results in Figures 10 and 11 suggest that, after using the proposed method to process Gaofen-3 data, the ocean surface current fields in the Kuroshio region are measured accurately.

The current characteristics in this paper and HYCOM need to be further compared. The proposed method has the advantages of flexible data acquisition time and high spatial sampling rate. The Gaofen-3 system can image the sea area of interest and obtain the current information at specific times, with high spatial resolution of only a couple of kilometers. However, the disadvantage is that the acquisition of current field in the case of complex sea conditions and uneven sea surface scattering is not applicable, so it needs to be further studied in combination with other information. The advantage of the current field obtained from HYCOM data is that the model integrates a variety of sensor data and is applicable to a wide range of sea environment, but the disadvantage is that it only represents the average motion state of one day, and the spatial resolution is only tens of kilometers, and there is a lack of data in the nearshore sea area. Therefore, in terms of the factors that lead to the difference between the two models in retrieving the current field, the input data type and accuracy, the calculation and processing process, and the temporal and spatial characteristics of the current field products will make them different.

4. Discussion

Considering that the marine environment is complex and full of many unknowns and uncertainties, many works about ocean surface current field measurement are mainly based on data-driven studies, especially in the application of using Doppler centroids to estimate ocean currents. As mentioned in the introduction, ENVISAT ASAR data has been used to study the sea surface current field in different regions including Gulf Stream and Agulhas Stream [5,6,8]. The SAR data acquired by different systems, such as ERS [37], TerraSAR-X [38] and Sentinel-1 [39], are also used in sea surface current measurement research, respectively, demonstrating the ability of currents retrieval. The work in this paper focuses on using the Gaofen-3 SAR data to carry out application research on the ocean surface current measurement in the Kuroshio region. In terms of this, the promising experimental results have proved the current measurement capability of the Gaofen-3 system. However, it should be noted that the results will also be affected by the system, environment, and processing methods. Since there is little knowledge about the actual sea surface state, it is hard to accurately and quantitatively analyze the dynamic process of the sea surface. Nevertheless, the main influencing factors and processes of the system and environment on the experimental results will be discussed below in this section, and the applicable scope of the proposed method will be given.

The incidence angle will affect the SAR ocean surface imaging mechanism. At a medium incidence angle range from 20° to 60°, which is consistent with the Gaofen-3 system design, the sea surface is dominated by Bragg scattering, and Bragg waves are the main scatterers on the sea surface, which is in line with the principles of the proposed method. At steep incidence angles, the specular scattering will increase imaging nonlinearity, which is not conductive to current measurement. Moreover, if the incidence angle is too large, the sea surface will be affected by wave breaking. In the experiment, the azimuth beam width determines the number of sub-apertures, which in turn also affects the current measurement results. When the azimuth beam width of the system is large, the number of sub-apertures will be more, and the number of azimuth angles will increase, which will lead to more equations used for least squares estimation to suppress the influence of random noise. However, in the case of the designed azimuth beam width, an increase in the number of sub-apertures will result in a decrease in sub-aperture time illuminating the ocean surface, which results in a decrease in the signal-to-noise ratio and current accuracy. Therefore, the number of sub-apertures is a compromise selection based on the data processing process. In this paper, six sub-apertures are used in the Gaofen-3 data experiment.

Generally, in order to improve the accuracy of Doppler estimation, the size of the resolution cell for ocean current products exceeds 1 km. When the system resolution increases, the amount of data that can be used to estimate the Doppler center increases, which will also increase the accuracy of Doppler estimation. The resolution of the Gaofen-3 UFS mode data used in this paper is about 2 m, and the spatial resolution of the current field is 1 km.

Usually, the wind field will cause the variation of sea surface wave motion, resulting in the deviation of current measurement results. Therefore, in traditional current measurement methods, such as DCA and ATI, it is necessary to first estimate the deviation value using the known wind field information, and then subtract it from the current measurement results. This process makes the current measurement process more complex, and the accuracy of wind field information and the difference of deviation estimation methods will affect the current measurement accuracy. The method used in this paper does not need prior wind field information to estimate wave motion, but reasonably introduces variables representing Bragg wave phase velocity and residual orbital velocity into the equations, and finally directly solves the current information. The experimental results show that this procedure is feasible. Note that the actual sea surface environment is complex and changeable, and it is difficult to consider all the characteristics of the marine environment in one method. In this article, the experiment is located in the Kuroshio region, which has the characteristics of high current velocity and uniform ocean surface backscattering, and is less affected by strong sea surface motion phenomena such as hurricanes and vortices. These are the basic environmental features for the success of the experiment.

In summary, the method proposed in this paper is suitable for SAR systems with a medium incidence angle and high resolution, and the appropriate number of sub-apertures is selected according to the azimuth beam width and processing process. The sea surface current field with high velocity and uniform backscattering, such as Kuroshio, Agulhas current and Gulf Stream, and so forth, is an ideal measurement object. The marine environment is required to be stable without being affected by complex ocean motion phenomena.

5. Conclusions

The ocean surface current field affects global climate change and human activities, and its speed and direction information is one of the most valuable parameters in ocean research. At present, using the Doppler centroid anomaly to process a single SAR image is a commonly used method of remote sensing current measurement, and the SAR data of different systems have retrieved relatively satisfactory ocean surface radial velocities in the Gulf Stream, Agulhas Stream and other sea areas. The Kuroshio Current is also one of the largest western boundary currents in the world, but there are no more relevant experimental reports on the application of SAR data current measurement. Gaofen-3 SAR data have played an important role in retrieving ocean waves, ocean winds and other ocean parameters, but its ability to measure ocean currents needs to be verified. Therefore, this paper uses four available Gaofen-3 SAR datasets to carry out the ocean surface current measurement experiment in the Kuroshio region.

In order to derive the ocean surface current vectors, a Doppler-based current measurement method combining sub-aperture processing and the least squares technology is suggested. Instead of prior wind information, the method considers the influence of wave motion in the equations, and then estimates the current field directly. The ocean current results derived from Gaofen-3 SAR data are compared with those from ocean model data. It is found that the current velocity and direction both show good agreement. The velocity accuracy is within 0.1 m/s and the direction accuracy is within 10°, which can meet the application requirements and demonstrate the current measurement capability of Gaofen-3.

Finally, the influencing factors of the experimental process and the applicable conditions of the method are discussed. System parameters, such as incidence angle and resolution, sea conditions, and changes in the processing process, will all have an impact on the current measurement results. The proposed method is suitable for SAR data with medium incidence angle, high resolution and uniform sea surface scattering. The data processing procedures under more complex sea conditions need further research and analysis in future work.

Author Contributions: Conceptualization, Y.L. and J.C.; methodology, Y.L.; software, Y.L. and K.S.; validation, Y.L. and J.C.; formal analysis, Y.L., Y.Z. and X.Y.; investigation, Y.L.; writing—original draft preparation, Y.L.; writing—review and editing, J.C. and Y.L.; supervision, J.C. All authors have read and agreed to the published version of the manuscript.

Funding: This research received no external funding.

Institutional Review Board Statement: Not applicable.

Informed Consent Statement: Not applicable.

Data Availability Statement: Not applicable.

Conflicts of Interest: The authors declare no conflict of interest.

References

1. Fischer, J.; Flemming, N.C. *Operational Oceanography: Data Requirements Survey*; Southampton Oceanography Centre: Southampton, UK, 1999.
2. Elyouncha, A.; Eriksson, L.E.B.; Romeiser, R.; Ulander, L.M.H. Measurements of Sea Surface Currents in the Baltic Sea Region Using Spaceborne Along-Track InSAR. *IEEE Trans. Geosci. Remote Sens.* **2019**, *57*, 8584–8599. [CrossRef]
3. Romeiser, R. The future of SAR-based oceanography: High-resolution current measurements by along-track interferometry. *Oceanography* **2013**, *26*, 92–99. [CrossRef]
4. Elyouncha, A.; Eriksson, L.E.B.; Romeiser, R.; Ulander, L.M.H. Empirical Relationship between the Doppler Centroid Derived from X-Band Spaceborne InSAR Data and Wind Vectors. *IEEE Trans. Geosci. Remote Sens.* **2021**, in press. [CrossRef]
5. Chapron, B.; Collard, F.; Ardhuin, F. Direct measurements of ocean surface velocity from space: Interpretation and validation. *J. Geophys. Res. Oceans* **2005**, *110*, C07008. [CrossRef]
6. Johannessen, J.A.; Chapron, B.; Collard, F.; Kudryavtsev, V.; Mouche, A.; Akimov, D.; Dagestad, K.F. Direct ocean surface velocity measurements from space: Improved quantitative interpretation of Envisat ASAR observations. *Geophys. Res. Lett.* **2008**, *35*, L22608. [CrossRef]
7. Rouault, M.J.; Mouche, A.; Collard, F.; Johannessen, J.; Chapron, B. Mapping the Agulhas Current from space: An assessment of ASAR surface current velocities. *J. Geophys. Res. Oceans* **2010**, *115*, C10026. [CrossRef]
8. Hansen, M.W.; Collard, F.; Dagestad, K.F.; Johannessen, J.A.; Fabry, P.; Chapron, B. Retrieval of sea surface range velocities from Envisat ASAR Doppler centroid measurements. *IEEE Trans. Geosci. Remote Sens.* **2011**, *49*, 3582–3592. [CrossRef]
9. Frasier, S.J.; Carswell, J.R.; Capdevila, J. A pod-based dual-beam interferometric radar for ocean surface current vector mapping. In Proceedings of the IEEE International Geoscience and Remote Sensing Symposium (IGARSS), Sydney, NSW, Australia, 9–13 July 2004; pp. 561–563.
10. Toporkov, J.V.; Perkovic, D.; Farquharson, G.; Sletten, M.A.; Frasier, S.J. Sea surface velocity vector retrieval using dual-beam interferometry: First demonstration. *IEEE Trans. Geosci. Remote Sens.* **2005**, *43*, 2494–2502. [CrossRef]
11. Martin, A.C.H.; Gommenginger, C. Towards wide-swath high-resolution mapping of total ocean surface current vectors from space: Airborne proof-of-concept and validation. *Remote Sens. Environ.* **2017**, *197*, 58–71. [CrossRef]
12. Hwang, C.; Kao, R. TOPEX/POSEIDON-derived space–time variations of the Kuroshio Current: Applications of a gravimetric geoid and wavelet analysis. *Geophys. J. Int.* **2002**, *151*, 835–847. [CrossRef]
13. Chao, M.; Wu, D.; Lin, X. Variability of surface velocity in the Kuroshio Current and adjacent waters derived from Argos drifter buoys and satellite altimeter data. *Chin. J. Oceanol. Limnol.* **2009**, *27*, 208–217.
14. Yang, J.; Yuan, X.; Han, B.; Zhao, L.; Sun, J.; Shang, M.; Wang, X.; Ding, C. Phase Imbalance Analysis of GF-3 Along-Track InSAR Data and Ocean Current Measurements. *Remote Sens.* **2021**, *13*, 269. [CrossRef]
15. Yang, J.S.; Ren, L.; Wang, J. The First Quantitative Remote Sensing of Ocean Surface Waves by Chinese GF-3 SAR Satellite. *Oceanol. Limnol. Sin.* **2017**, *48*, 207–209.
16. Ren, L.; Yang, J.; Mouche, A.A.; Wang, H.; Zheng, G.; Wang, J.; Zhang, H.; Lou, X.; Chen, P. Assessments of Ocean Wind Retrieval Schemes Used for Chinese Gaofen-3 Synthetic Aperture Radar Co-Polarized Data. *IEEE Trans. Geosci. Remote Sens.* **2019**, *57*, 7075–7085. [CrossRef]
17. Yang, J.; Wang, J.; Lin, R. The first quantitative remote sensing of ocean internal waves by Chinese GF-3 SAR satellite. *Acta Oceanol. Sin.* **2017**, *36*, 118. [CrossRef]
18. Moiseev, A.; Johnsen, H.; Hansen, M.W.; Johannessen, J.A. Evaluation of Radial Ocean Surface Currents Derived From Sentinel-1 IW Doppler Shift Using Coastal Radar and Lagrangian Surface Drifter Observations. *J. Geophys. Res. Oceans* **2020**, *125*, 1–18. [CrossRef]
19. Plant, W.J. Bragg scattering of electromagnetic waves from the air/sea interface. In *Surface Waves and Fluxes*; Springer: Amsterdam, The Netheriands, 1990; pp. 41–108.

20. Plant, W.J.; Keller, W.C.; Hayes, K. Measurement of river surface currents with coherent microwave systems. *IEEE Trans. Geosci. Remote Sens.* **2005**, *43*, 1242–1257. [CrossRef]
21. Yurovsky, Y.Y.; Kudryavtsev, V.N.; Grodsky, S.A.; Chapron, B. Sea Surface Ka-Band Doppler Measurements: Analysis and Model Development. *Remote Sens.* **2019**, *11*, 839. [CrossRef]
22. Mouche, A.A.; Collard, F.; Chapron, B.; Dagestad, K.-F.; Guitton, G.; Johannessen, J.A.; Kerbaol, V.; Hansen, M.W. On the Use of Doppler Shift for Sea Surface Wind Retrieval From SAR. *IEEE Trans. Geosci. Remote Sens.* **2012**, *50*, 2901–2909. [CrossRef]
23. Yu, X.Z.; Chong, J.S.; Hong, W. An Iterative Method for Ocean Surface Current Retrieval by Along-track Interferometric SAR. *J. Electron. Inf. Technol.* **2012**, *34*, 2660–2665. [CrossRef]
24. Pu, W. Deep SAR Imaging and Motion Compensation. *IEEE Trans. Image Process.* **2021**, *30*, 2232–2247. [CrossRef]
25. Romeiser, R.; Schwäbisch M.; Schulz-Stellenfleth J. *Study on Concepts for Radar Interferometry from Satellites for Ocean (and Land) Applications (KoRIOLiS)*; University of Hamburg: Hamburg, Germany, 2002.
26. Zhang, Q. Design and Key Technologies of the GF-3 Satellite. *Acta Geod. Cartogr. Sin.* **2017**, *46*, 269–277.
27. Pan, X.; Liao, G.; Yang, Z.; Dang, H. Sea surface current estimation using airborne circular scanning SAR with a medium grazing angle. *Remote Sens.* **2018**, *10*, 178. [CrossRef]
28. Ferro-Famil, L.; Reigber, A.; Pottier, E.; Boerner, W.M. Scene characterization using sub-aperture polarimetric SAR data. *IEEE Trans. Geosci. Remote Sens.* **2003**, *41*, 2264–2276. [CrossRef]
29. Bamler, R. Doppler frequency estimation and the Cramer-Rao bound. *IEEE Trans. Geosci. Remote Sens.* **1991**, *29*, 385–390. [CrossRef]
30. Cumming, I.; Wong, F. *Digital Signal Processing of Synthetic Aperture Radar Data: Algorithms and Implementation*; Artech House Incorporated: London, UK, 2005.
31. Li, W.; Huang, Y.; Yang, J.; Wu, J.; Kong, L. An Improved Radon-Transform-Based Scheme of Doppler Centroid Estimation for Bistatic Forward-Looking SAR. *IEEE Geosci. Remote Sens. Lett.* **2011**, *8*, 379–383. [CrossRef]
32. Kong, Y.; Cho, B.; Kim, Y. Ambiguity-free Doppler centroid estimation technique for airborne SAR using the Radon transform. *IEEE Trans. Geosci. Remote Sens.* **2005**, *43*, 715–721. [CrossRef]
33. Han, B.; Ding, C.; Zhong, L.; Liu, J.; Qiu, X.; Hu, Y.; Lei, B. The GF-3 SAR Data Processor. *Sensors* **2018**, *18*, 835. [CrossRef] [PubMed]
34. Backeberg, B.C.; Bertino, L.; Johannessen, J.A. Evaluating two numerical advection schemes in HYCOM for eddy-resolving modelling of the Agulhas Current. *Ocean Sci.* **2009**, *5*, 173–190. [CrossRef]
35. Seo, S.; Park, Y.-G.; Park, J.-H.; Lee, H.J.; Hirose, N.J.O.; Research, P. The Tsushima Warm Current from a high resolution ocean prediction model, HYCOM. *Ocean Polar Res.* **2013**, *35*, 135–146. [CrossRef]
36. Wang, M.; Liu, Z.; Zhu, X.; Yan, X.; Zhang, Z.; Zhao, R. Origin and formation of the Ryukyu Current revealed by HYCOM reanalysis. *Acta Oceanol. Sin.* **2019**, *38*, 1–10. [CrossRef]
37. Kim, J.-E.; Kim, D.-J.; Moon, W.M. Enhancement of Doppler centroid for ocean surface current retrieval from ERS-1/2 raw SAR. In Proceedings of the IEEE International Geoscience and Remote Sensing Symposium (IGARSS), Anchorage, AK, USA, 20–24 September 2004; pp. 3118–3120.
38. Rossi, C.; Runge, H.; Breit, H.; Fritz, T. Surface current retrieval from TerraSAR-X data using Doppler measurements. In Proceedings of the IEEE International Geoscience and Remote Sensing Symposium (IGARSS), Honolulu, HI, USA, 25–30 July 2010; pp. 3055–3058.
39. Johnsen, H.; Nilsen, V.; Engen, G.; Mouche, A.A.; Collard, F. Ocean doppler anomaly and ocean surface current from Sentinel 1 tops mode. In Proceedings of the IEEE International Geoscience and Remote Sensing Symposium (IGARSS), Beijing, China, 10–15 July 2016; pp. 3993–3996.

applied sciences

MDPI

Article

Engine Malfunctioning Conditions Identification through Instantaneous Crankshaft Torque Measurement Analysis

Konstantinos-Marios Tsitsilonis [1,2,*] and Gerasimos Theotokatos [2]

1 Datum Electronics Ltd., 37 Castle St, East Cowes PO32 6EZ, UK
2 Department of Naval Architecture, Ocean and Marine Engineering, University of Strathclyde, 100 Montrose St, Glasgow G4 0LZ, UK; gerasimos.theotokatos@strath.ac.uk
* Correspondence: konstantinos.tsitsilonis@datum-electronics.co.uk

Abstract: In this study a coupled thermodynamics and crankshaft dynamics model of a large two-stroke diesel engine was utilised, to map the relationship of the engine Instantaneous Crankshaft Torque (ICT) with the following frequently occurring malfunctioning conditions: (**a**) change in Start of Injection (SOI), (**b**) change in Rate of Heat Release (RHR), (**c**) change in scavenge air pressure, and (**d**) blowby. This was performed using frequency analysis on the engine ICT, which was obtained through a series of parametric runs of the coupled engine model, under the various malfunctioning and healthy operating conditions. This process demonstrated that engine ICT can be successfully utilised to identify the distinct effects of malfunctions (**c**) or (**d**), as they occur individually in any cylinder. Furthermore by using the same process, malfunctions (**a**) and (**b**) can be identified as they occur individually for any cylinder, however there is no distinct effect on the engine ICT among these malfunctions, since their effect on the in-cylinder pressure is similar. As a result, this study demonstrates the usefulness of the engine ICT as a non-intrusive diagnostic measurement, as well as the benefits of malfunctioning conditions mapping, which allows for quick and less resource intensive identification of engine malfunctions.

Keywords: instantaneous crankshaft torque; frequency analysis; two-stroke diesel engine; engine harmonic frequencies; malfunctioning conditions mapping; thermodynamics model; crankshaft dynamics model; degradation; diagnostics

Citation: Tsitsilonis, K.-M.; Theotokatos, G. Engine Malfunctioning Conditions Identification through Instantaneous Crankshaft Torque Measurement Analysis. *Appl. Sci.* **2021**, *11*, 3522. https://doi.org/10.3390/app11083522

Academic Editor: Dimitrios-Nikolaos Pagonis

Received: 30 March 2021
Accepted: 13 April 2021
Published: 14 April 2021

1. Introduction

Large two-stroke diesel engines are highly efficient and robust internal combustion engines, and as a result they are utilised as main engines in the majority of the world ocean-going fleet, operating under a wide range of conditions [1]. However, main engine breakdown and engine components degradation still present major challenges, leading to significant costs due to vessel down time and engine inefficiencies [2,3], as well as compromising crew safety [4]. Consequently, early diagnosis of engine abnormalities, degradation and faults is essential in maintaining the vessel's safety, as well as avoiding higher operational expenditure associated with fuel, maintenance, costly downtime and increased repair costs.

In order to diagnose the engine's condition, measurement and monitoring of vital performance parameters are required. Amongst the most important parameters indicating the engine health status is the in-cylinder pressure [5,6], which can effectively be used to analyse combustion quality [7], diagnose cylinder blowby and compression faults [8], and diagnose the fuel injector's condition [9]. However, obtaining the in-cylinder pressure measurement can be intrusive [10] and requires the deployment of large networks of expensive sensors to perform continuous measurement [11], thus increasing costs, creating data synchronisation issues and introducing multiple failure points.

To overcome the above shortcomings, other peripheral engine measurements are used to accurately determine the cylinders condition. In specific, a combination engine working media pressures and temperatures are used, however these are not as revealing of the engine cylinders' health status as the in-cylinder pressure [6]. Thus, the Instantaneous Crankshaft Speed (ICS) measurement has been utilised in automotive applications to obtain combustion phasing information [12], identify imbalanced cylinders [13], and cylinders with a decreased mean effective pressure [14]. Regarding larger engines, the ICS signal has been utilised to detect misfiring for medium [15] and slow-speed engines [16], in addition to cylinder imbalances [17]. Thus, the ICS measurement can be utilised due to the low cost of speed sensors, however the raw ICS signal from the angle encoder requires significant processing and high sample rates to provide sufficient detail, which is often not achieved in practice [18]. Hence, by using ICS measurements faults in the engine cylinders are identified, but their specific causes are often not revealed.

The Instantaneous Crankshaft Torque (ICT) can be used to acquire detailed information of the cylinders' condition, which can be obtained at sufficiently high sample rates and accuracy [19]. In specific, for automotive applications the peak pressure position [20], and properties such as combustion phasing and the burned mass fraction were estimated for each cylinder [21], by utilising measurements from an ICT sensor. Furthermore, progress was made in reconstructing the in-cylinder pressure diagram, however singularities were encountered during the torque signal inversion [11]. Regarding larger engines the ICT measurement was utilised for a two-stroke diesel engine of a land-based power plant, to determine the indicated power for each cylinder using an Artificial Neural Network (ANN) [18]. The ICT measurement is thus able to provide greater insight into the cylinders condition using just a single sensor, however its utilisation in the marine industry remains a novel concept as more traditional means of monitoring and diagnostics employed [22,23].

In the shipping industry, the most common form of diagnostics relies on human judgement combined with information on acceptable degradation tolerances of engine parts and performance parameters, which is provided by the manufacturer [24]. However, this is a manual approach which is error-prone due to the involvement of the human factor. As a result, statistical techniques such as regression [25], Support Vector Machines (SVMs) [26], Bayesian networks [27], and ANNs [28] have been implemented which utilise data from a multitude of engine parameters to perform diagnostics on the main engine holistically or its individual components. These techniques rely on the collection of training and validation data sets from large sensor networks in both healthy and faulty conditions, which are hard to obtain. This is due to the lack of such sensor networks and high quality data for the majority of the world fleet [29], as well as data sharing issues [30]. As a result, first-principles modelling is employed to reduce the required number of measured parameters and identify the cylinders condition. Specifically, thermodynamics modelling has been used in conjunction with some measured parameters in order to identify engine faults such as excessive peak combustion pressure [31], fuel injectors condition [32,33] and cylinder compression as well as blowby [34]. These techniques also demonstrated their usefulness by using engine models to derive baseline conditions, which facilitate comparisons with the current engine operation, making it easier to reveal faults and degradation effects.

From the above literature review, the following research gaps were identified: Despite the usefulness of the ICT measurement, there is no study for large two-stroke diesel engines that utilises said measurement to identify specific types of engine faults. Subsequently, the aim of this study is to provide a framework for assessing the performance large two-stroke diesel engines, by mapping the relationship of specific malfunctioning conditions with the engine ICT, through the comparison with engine healthy (baseline) conditions. This is accomplished by utilising a novel combination of less data-intensive first-principles models, which implies coupling a thermodynamics and crankshaft dynamics model to predict the engine's ICT at various malfunctioning conditions. The following four most frequently encountered malfunctioning conditions are considered:

1. Change in the Start of Injection (SOI); occurs due to fuel injector wear and engine detuning [32].
2. Change in the Rate of Heat Release (RHR); occurs due to poor fuel quality, and fuel injector or fuel pump wear [33,35].
3. Change in the scavenge air pressure; occurs due to turbocharger compressor or scavenge air cooler fouling [34].
4. Cylinder blowby; occurs due to wear in the cylinder liner or piston rings [34,36].

An initial study on a modelling approach for predicting a marine engine's shaft dynamics was presented by the authors [37]. The present study is an extension of that, and includes improved thermodynamics model formulations, and additional analysis on the modulus and phase angle of the most affected frequency components of the engine ICT signal.

2. Methods and Tools

2.1. Reference System & Methodology

The reference system considered in this study is a 10-cylinder two-stroke diesel engine, driving a generator operating at a constant speed of 125 rev/min. This engine was utilised for the development of a coupled thermodynamics and crankshaft dynamics model, which provides the in-cylinders pressure and flywheel ICT as output. The technical characteristics and measured data used to develop and calibrate the coupled model were obtained from [18,38]. In specific, the measured data include the in-cylinder pressure for a single cylinder and the flywheel ICT for four loads as shown in Figure 1, whilst a summary of the engine technical details is listed in Table 1.

Table 1. Reference system basic technical specifications [18].

Maximum Continuous Rating	15.50 MW
Number of cylinders	10
Nominal rotational speed	125 rev/min
Piston bore	0.67 m
Stroke	1.70 m
Firing order	1-9-4-6-3-10-2-7-5-8

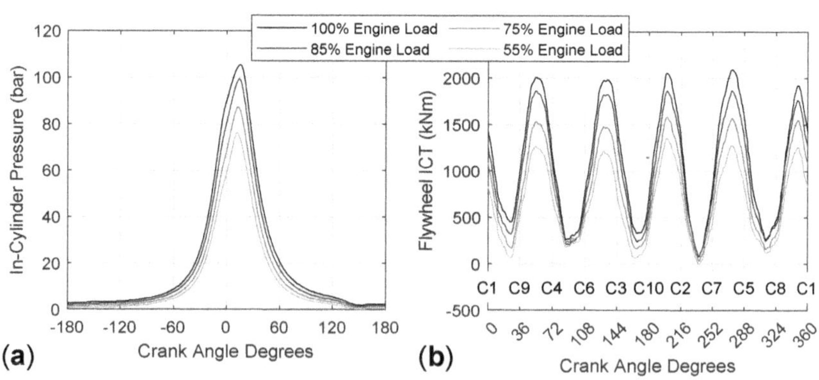

Figure 1. Measured (**a**) in-cylinder pressure and (**b**) flywheel ICT, at four engine loads of 100%, 85%, 75% and 55%. Firing cylinders are marked as C1, C2, ... C10.

Initially, the engine thermodynamic model and the engine crankshaft dynamics model were developed and coupled with each other. These models and their coupling are described in Sections 2.2 and 2.3. Subsequently, the thermodynamic model parameters calibration was carried out by employing a nonlinear data-fitting Levenberg-Marquardt algorithm from MATLAB [39]. The objective employed was to minimise the least-squares difference between the measured and simulated in-cylinder pressure from the coupled model, by assuming that all cylinders have identical in-cylinder pressure. The state of the calibrated model was considered to represent the engine healthy conditions.

Following the calibration process, four different engine malfunctioning conditions were examined at 100% load as listed in Section 1, and four corresponding parameters within the coupled model were used to simulate the engine malfunctioning operation. To test individually all malfunctioning conditions, one parameter was varied 6 times in equal steps from its minimum to its maximum values (defined in Section 2.4), while the rest of the parameters were assigned a constant value corresponding to the healthy engine conditions. The above procedure was performed for each cylinder for three model parameters that correspond to each of the following malfunctioning conditions; change in SOI, change of RHR, and blowby. The remaining parameter corresponding to the change in scavenge air pressure malfunction, was varied for the entire engine and not for each individual cylinder, as a drop in the scavenge air pressure affects all cylinders simultaneously. Therefore, a total of six parametric runs were performed for the three malfunctioning conditions of each cylinder, providing a total of 180 ($3 \times 6 \times 10$) parametric runs. In addition to these parametric runs, six parametric runs were performed for the scavenge air pressure malfunction, which affects all cylinders uniformly, raising the total to 186 runs.

For each parametric run, a frequency analysis by using a Fast Fourier Transform (FFT) was performed to the output of the engine ICT, and the top three most affected signal frequencies and phase angles where identified. These were utilised to map the relationship between the engine ICT and the specific malfunctioning conditions considered in this study.

2.2. Thermodynamic Model Description

An in-house thermodynamics model was developed to predict the engine in-cylinder pressure under the various operating conditions examined. The engine scavenging process was modelled using a two-zone approach, which is considered necessary in order to capture the complexities during the fresh air and exhaust gas mixing [40]. The remaining part of the engine cycle from Exhaust Valve Close (EVC) to Scavenge Port Open (SPO) was modelled using a one-zone approach which is able to provide sufficient accuracy for the in-cylinder pressure diagram prediction [41].

The formulation of each approach is developed using the energy the working medium as homogeneous. In specific, it involves a system of non-linear differential equations that is solved for the pressure, temperature and burnt fuel fraction state variables for each zone. For the one-zone approach the following system of ordinary differential equations (ODE) is obtained:

$$\begin{bmatrix} \dot{p} \\ \dot{T} \\ \dot{\xi} \end{bmatrix} = \begin{bmatrix} m\dfrac{\partial u}{\partial p} & m\dfrac{\partial u}{\partial T} & 0 \\ V - mT\dfrac{\partial R}{\partial p} & -m\left(R + T\dfrac{\partial R}{\partial T}\right) & 0 \\ 0 & 0 & 1 \end{bmatrix}^{-1} \begin{bmatrix} \dot{Q}_{\Sigma} - p\dot{V} + \dot{H}_{\Sigma} - u\dot{m} - \dfrac{\partial u}{\partial \phi} m\dot{\phi} \\ T\left(\dfrac{\partial R}{\partial \phi} m\dot{\phi} + R\dot{m}\right) - p\dot{V} \\ F_s\dot{\phi}(1-\xi)^2 \end{bmatrix} \quad (1)$$

where p, T and ξ are the pressure, temperature and burnt fuel fraction respectively; u, R, F_s and ϕ denote the internal energy, gas constant, stoichiometric fuel-air ratio, and equivalence ratio respectively; and finally, V, m and $\dot{Q}_{\Sigma}$ are the volume, mass, sum of heat release and heat transfer rates respectively, whereas $\dot{H}_{\Sigma}$ denotes the sum of enthalpy flow rates entering and exiting the cylinder.

The internal energy u and gas constant R, along with their corresponding partial derivatives with respect to p, T and ϕ are calculated through combustion gas properties [42]. The stoichiometric fuel-air ratio F_s is set to 0.0694 for marine diesel fuel [43], whilst the equivalence ratio ϕ is calculated considering the burnt fuel fraction ξ in accordance to the following equation [5]:

$$\phi = \frac{\xi}{F_s(1-\xi)} \tag{2}$$

Furthermore, the mass conservation applied for the one-zone approach provides the following equation:

$$\dot{m} = \dot{m}_i - \dot{m}_o + \dot{m}_{fb} \tag{3}$$

where the subscripts i and o denote inlet and outlet mass flowrates respectively, which are calculated by utilising the valve isentropic flow equations [5]. The burnt mass rate is denoted by $\dot{m}_{fb}$ and is calculated using the single-Wiebe combustion model according to the following equation [7]:

$$\dot{m}_{fb} = m_f\, \dot{\theta}\, \frac{a\,(w+1)}{\Delta\theta} \left(\frac{\theta-\theta_{SOC}}{\Delta\theta}\right)^{w} \exp\left\{-a\left(\frac{\theta-\theta_{SOC}}{\Delta\theta}\right)^{w+1}\right\} \tag{4}$$

where m_f is the total fuel injected in the cylinder, θ denotes the crank angle, $\Delta\theta$ is the duration of combustion, w is the Wiebe shape parameter, whereas a is set to 6.9078 to maintain a combustion efficiency of 99.9% [44]. In addition, θ_{SOC} denotes the start of combustion crank angle, which is calculated in accordance to the following equation:

$$\theta_{SOC} = \theta_{SOI} + \Delta\theta_{IGD} \tag{5}$$

where θ_{SOI} is the start of injection crank angle which is provided as input, and $\Delta\theta_{IGD}$ is the ignition delay calculated in accordance to the Sitkey equation [7].

The sum of heat transfer and heat release rates is calculated by considering the burnt mass rate from Equation (4) above, and the sum of enthalpy flowrates is calculated by considering the valve inlet and outlet mass flowrates, in accordance to the Equations (6) and (7) respectively:

$$\dot{Q}_{\Sigma} = \dot{m}_{fb} h_{form} - C_w \sum_j A_j (T - T_j) \tag{6}$$

$$\dot{H}_{\Sigma} = \dot{m}_i h_i - \dot{m}_o h_o \tag{7}$$

where h_{form} is the fuel enthalpy of formation, C_w is the Woschni heat transfer coefficient [7], and A_j, T_j are the heat transfer areas and surface temperatures respectively for the jth cylinder part. The subscript j denotes the piston, cylinder head, liner and exhaust valve.

The scavenging period (from the SPO to EVC) is modelled by employing the two-zone approach [41], which considers that the cylinder is divided into two zones, namely, the charge air zone, and the exhaust gas zone (the latter contains a mixture of the exhaust gas and charge air). By applying the energy and mass conservation equations as well as the ideal gas law and considering the homogeneous mixture, the following system of ODE is obtained:

$$\begin{bmatrix} \dot{p} \\ \dot{V}_1 \\ \dot{T}_1 \\ \dot{V}_2 \\ \dot{T}_2 \\ \dot{\xi}_2 \end{bmatrix} = \begin{bmatrix} 0 & 1 & 0 & 1 & 0 & 0 \\ V_1 & p & -m_1 R_1 & 0 & 0 & 0 \\ 0 & p & m_1 \dfrac{\partial u_1}{\partial T_1} & 0 & 0 & 0 \\ V_2 & 0 & 0 & p & -m_2 R_2 & 0 \\ 0 & 0 & 0 & p & m_2 \dfrac{\partial u_2}{\partial T_2} & 0 \\ 0 & 0 & 0 & 0 & 0 & 1 \end{bmatrix}^{-1} \begin{bmatrix} \dot{V} \\ \dot{m}_1 R_1 T_1 \\ \dot{Q}_1 + \dot{m}_i h_i - \dot{m}_{1,2} h_1 - \dot{m}_1 u_1 \\ T_2\left(\dot{m}_2 R_2 + \dfrac{\partial R_2}{\partial \phi_2} m_2 \dot{\phi}_2\right) \\ \dot{Q}_2 + \dot{m}_{1,2} h_1 - \dot{m}_o h_o - \dot{m}_2 u_2 - \dfrac{\partial u_2}{\partial \phi_2} m_2 \dot{\phi}_2 \\ F_s \dot{\phi}_2 (1-\xi_2)^2 \end{bmatrix} \tag{8}$$

where the subscripts 1 and 2 denote the charge air and mixture zones respectively, and the subscript 1,2 denotes the charge air flow from zone 1 to zone 2. Furthermore, $\dot{Q}_1$ and $\dot{Q}_2$

denote the heat transfer rates for each zone respectively, which are calculated using the Woschni heat transfer coefficient C_w in accordance to the following equations:

$$\dot{Q}_1 = -C_w \left[A_p(T - T_p) + A_l \frac{V_1}{V}(T - T_l) \right] \tag{9}$$

$$\dot{Q}_2 = -C_w \left[\sum_q A_q(T - T_q) + A_l \frac{V_2}{V}(T - T_l) \right] \tag{10}$$

where the subscripts p and l denote the piston and cylinder liner respectively, and the subscript q denotes the valve and cylinder head.

The mass flow rate from zone 1 to zone 2 is calculated according to the following equation:

$$\dot{m}_{1,2} = c_s \dot{m}_i \tag{11}$$

where c_s is the scavenging mixing factor calibrated at every load point [41].

The application of the mass and volume conservation provide the following equations:

$$\dot{m}_1 = (1 - c_s)\dot{m}_i \tag{12}$$

$$\dot{m}_2 = c_s \dot{m}_i - \dot{m}_o \tag{13}$$

$$\dot{V} = \dot{V}_1 + \dot{V}_1 \tag{14}$$

The switch between the one-zone and two-zone approaches takes place after the scavenge ports open, when the charge air zone volume V_1 is sufficiently large, such that singularities during the matrix inversion in Equation (8) are avoided [45]. In specific, when the scavenge ports open, the one-zone approach equations from Equation (1) continue to be integrated, along with the following additional differential equation:

$$\dot{V}_1 = \dot{m}_1 R_1 T_1 p^{-1} \tag{15}$$

Subsequently, the zone volume fraction is utilised as the criterion which performs the switch from the one-zone to the two-zone approach, according to whether the following inequality is satisfied:

$$\frac{V_1}{V} \geq \mathrm{c} \tag{16}$$

where c is set to 10^{-3}.

2.3. Crankshaft Dynamics Model

The thermodynamics model was coupled to the crankshaft dynamics model, which is configurable to predict the ICT for either stationary generators or propulsion applications. The engine crankshaft dynamics was modelled using a lumped parameter model, the detailed development of which is described in [37]. The lumped parameter model represents the entire engine crankshaft in a series of 16 discrete inertia disks or Degrees of Freedom (DOFs), as shown for the reference system in Figure 2. The DOFs 4–8 and 10–14 correspond to each of the 10 cylinders, whilst DOFs 1–3, 9, 15, and 16 correspond to the engine vibration dampers, chaindrive, flywheel and generator armature respectively. Each DOF is connected to its neighbouring DOF(s) using a spring and a damper element, that simulate the shaft's rigidity and structural damping. Furthermore, each DOF is connected with a damper element to a non-rotating frame of reference, simulating the friction in the shafting system's bearings.

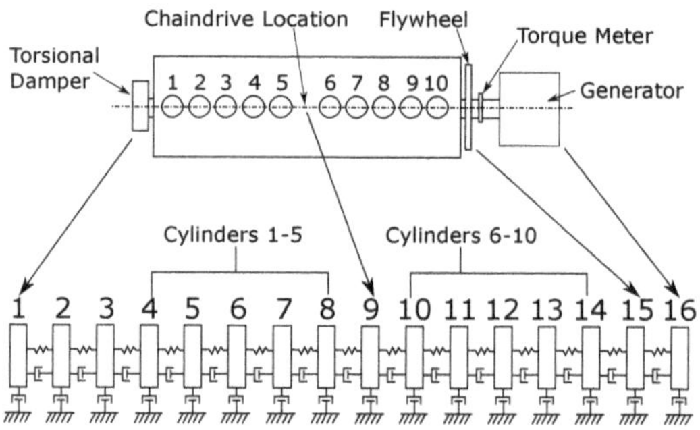

Figure 2. Crankshaft dynamics lumped mass model schematic [46].

The following crankshaft dynamics model governing equation was derived by applying the angular momentum conservation on the engine shafting system:

$$\boldsymbol{J}\ddot{\boldsymbol{\Theta}} + \boldsymbol{C}\dot{\boldsymbol{\Theta}} + \boldsymbol{K}\boldsymbol{\Theta} = \boldsymbol{T}_J + \boldsymbol{T}_C + \boldsymbol{T}_F + \boldsymbol{T}_L \tag{17}$$

where $\boldsymbol{\Theta}$, $\dot{\boldsymbol{\Theta}}$ and $\ddot{\boldsymbol{\Theta}}$ are 16×1 vectors of the angular displacement and its respective first and second derivatives; $\boldsymbol{J}$ is a 16×16 diagonal inertia matrix; $\boldsymbol{C}$ and $\boldsymbol{K}$ are 16×16 tridiagonal damping and stiffness matrices respectively; and $\boldsymbol{T}_J$, $\boldsymbol{T}_C$, $\boldsymbol{T}_F$ and $\boldsymbol{T}_L$ are 16×1 vectors of the variable inertia, cylinder combustion, engine friction, and engine load torques respectively.

The inertia variation throughout the engine rotation is a result of the rotating and reciprocating motion of the engine pistons and the connecting rods [20]. As a result, the variable inertia vector elements corresponding to the engine cylinders (DOFs 4–8 and 10–14) are calculated using Equation (18) below, whilst all other elements corresponding to DOFs 1–3, 9, and 15–16 are set to zero [46]. Thus, the variable inertia torque is calculated according to the following equation:

$$\boldsymbol{T}_{J,n} = M_{rec}\left[r^2 - \left(\frac{\mathrm{d}s_n}{\mathrm{d}\theta_n}\right)^2\right]\ddot{\theta}_n + M_{rec}\,\frac{\mathrm{d}^2 s_n}{\mathrm{d}\theta_n^2}\,\frac{\mathrm{d}s_n}{\mathrm{d}\theta_n}\dot{\theta}_n^2 \tag{18}$$

where M_{rec} and r are the reciprocating piston mass and crank radius respectively, s is the instantaneous piston displacement with respect to the Top Dead Center (TDC), and the subscript n refers to the DOFs corresponding to each cylinder.

The cylinder combustion torque vector is arranged in a similar way as the variable inertia torque vector, where its elements corresponding to the engine cylinders are calculated using Equation (19), and all other elements are set to zero. The combustion torque for individual cylinders is therefore calculated according to the following equation:

$$T_{C,n} = \frac{\mathrm{d}s_n}{\mathrm{d}\theta_n}\,p_n A_p \tag{19}$$

where by using the same notation as in the thermodynamics model, p_n is the in-cylinder pressure for the nth DOF corresponding to each cylinder, and A_p corresponds to the piston area.

The engine friction torque is calculated as a function of the engine rotational speed, utilising the friction coefficients in accordance to the following equation [18]:

$$T_F = \boldsymbol{F}\dot{\boldsymbol{\Theta}} \tag{20}$$

where $\boldsymbol{F}$ is a 16×16 diagonal matrix containing the friction coefficients.

The engine load torque is calculated as a constant for the case where the engine is configured for a stationary generator [18], or as a function of the engine speed for the case where the engine is configured for propulsion applications [47]. In specific, the load torque is implemented at the last element of the engine load vector, such that $T_L = [0,\ 0, \cdots, \tau_L]^T$, and for both of the above configurations the following equations are used respectively:

$$\tau_{L,gen} = k_g \frac{P_{MCR}}{\dot{\theta}_{MCR}} \tag{21}$$

$$\tau_{L,prop} = k_d \dot{\theta}^N \quad \text{where } k_d = \frac{P_{MCR}}{\dot{\theta}^{N+1}_{MCR}} \tag{22}$$

where P_{MCR} and $\dot{\theta}_{MCR}$ is the engine power and rotational speed at MCR respectively, k_g is the generator load constant, which is obtained from on-site bus voltage measurements [18], and $N \approx 2$ is the propeller law exponent [47].

The torque at the flywheel is calculated considering the shaft stiffness at the flywheel location, and angle of twist between the flywheel and the generator armature according to the following equation:

$$\tau_{fly} = k_{15}(\theta_{16} - \theta_{15}) \tag{23}$$

where k_{15} is the shaft stiffness at the 15th DOF, and θ_{16}, θ_{15} are the angular displacements of DOFs 15 and 16, corresponding to the flywheel and generator armature, respectively.

2.4. Malfunctioning Conditions Modelling

In total, 4 malfunctioning conditions are investigated in this study, as listed in Section 1. The first one is the change in SOI, which can be caused due to the fuel injector wear and tear or engine detuning. In specific, engine detuning can occur after a large number of operating hours, as the Variable Injection Timing (VIT) of individual cylinders may require retuning [33]. In addition, detuning occurs due to the wear and tear of the injector pump or the camshaft lobe, the latter occurs to a large percentage of the world fleet, which still operates using camshaft-driven engines. The SOI angle variation for each cylinder are taken into account by using the θ_{SOI} term in Equation (5), which is expected to vary by $\pm 2.5°$ with respect to the healthy engine condition SOI [48].

The second malfunctioning condition considered is the change in RHR. In specific, this can occur due to a number of reasons including poor fuel quality, and worn or clogged fuel injectors [33,34]. To change the RHR, this study considers the variation of the Weibe shape parameter w emplyed in Equation (4). This impacts the fuel burn rate, and as a results the combustion rate. Since the model is very sensitive to changes of the Weibe shape parameter, its range was considered ± 0.5 with respect to the healthy engine's parameter value.

The third malfunctioning condition is a change in the scavenge air pressure. A decrease in the scavenge air pressure can occur due to scavenge air cooler fouling, turbocharger compressor fouling, air filter fouling, turbine wear or a combination of those. On the contrary, an increase in the scavenge air pressure can occur due to fouling in the turbine nozzle, or due to higher exhaust gas temperatures resulting from the incomplete combustion, late combustion of poor quality fuel [29]. This malfunction is implemented to the thermodynamic model as a change in the scavenge air pressure, which is an input parameter affecting the air flow rate though the scavenging ports that are utilised to calculate the cylinder inlet mass m_i. Ths study considers changes in the scavenge air receiver pressure by ± 0.2 bar with respect value for the healthy engine condition [32] .

The fourth malfunctioning condition is the engine blowby, which occurs due to wear in the cylinder liner or piston rings [34,36]. This was modelled by simulating an additional isentropic orifice, namely the blowby orifice, which connects the scavenge air receiver with the cylinder. The effective area profile of the blowby orifice was set as constant throughout the entire engine revolution, and its value was adjusted manually such that a range of typical blowby mass flowrates can be obtained [36,49].

The modelled malfunctioning conditions and their respective parameter changes are summarised in Table 2.

Table 2. Malfunctioning condition parameters variation.

Malfunctioning Condition	Changed Parameter	Parameter Range	Parameter Step
Change in SOI	Start of Combustion (θ_{SOI})	$\pm 2.5°$	1.25°
Change in RHR	Weibe shape parameter (w)	± 0.5	0.25
Change in scavenge air pressure	Scavenge air pressure	± 0.3 bar	0.15 bar
Blowby	Blowby effective area	$0 \sim 28.5$ mm^2	7.12 mm^2

3. Results & Discussion

3.1. Healthy Engine Conditions

To simulate the engine performance for the baseline conditions, the coupled thermodynamic and crankshaft dynamics model was calibrated at 100% engine load using the measured in-cylinder pressure as described in Section 2.1. In specific, the calibration parameters employed were the start of injection (Equation (5)), duration of combustion and Weibe shape parameter (Equation (4)), as well as the scavenging mixing factor (Equations (11) and (12)). Subsequently, by comparing to the measured data, the coupled model was validated at 100%, 85%, 75%, and 55% engine loads as shown in Figure 3.

The calibration process resulted to the simulated in-cylinder pressure having a Normalised Root Mean Square Error (NRMSE) of 9.2% at 100% engine load, as shown in Figure 3a. Through the validation of the in-cylinder pressure for the remaining three engine loads in Figurs 3c,e,g, it is observed that the NRMSE is increasing to 10.8%, 11.8% and 13.9% respectively. This error increase occurs as these loads are further away from the calibration load of 100%. In addition, the simulated peak combustion pressure is close to the measurements, with a maximum difference of 2.3 bar which occurs at 85% engine load. Also, the crank angle corresponding to the simulated peak combustion pressure is always within 2 crank angle degrees from the measurements, with the exception at 85% engine load where the simulated peak pressure is at 3.5 crank angle degrees less than the measurements. Finally, it is observed that the compression part of the cycle is captured accurately by the simulated pressure. However the expansion from the peak pressure to approximately 60°CA, appears to occur quicker in the case of the simulated in-cylinder pressure as compared to the measurements, for all engine loads.

The NRMSE error in the simulated flywheel ICT in Figure 3b,d,f,h is observed to follow the same pattern as the error for the in-cylinder pressure, since it steadily increases at the engine loads further way from the calibration load. Furthermore, the error in the simulated flywheel ICT is observed to be greater than that of the in-cylinder pressure for all loads, which is due to the assumptions made in the crankshaft dynamics model. In specific, since the in-cylinder pressure from only one cylinder was available, all the cylinders in the crankshaft dynamics model were set to have identical in-cylinder pressures, which would also represent best the healthy engine conditions. As a result, the flywheel ICT was simulated using identical in-cylinder pressures, whilst in the physical engine the in-cylinder pressures were most likely to be different [38], which would result to a larger error in the flywheel ICT. Further analysis on the errors induced in the simulated flywheel ICT can be found in [46].

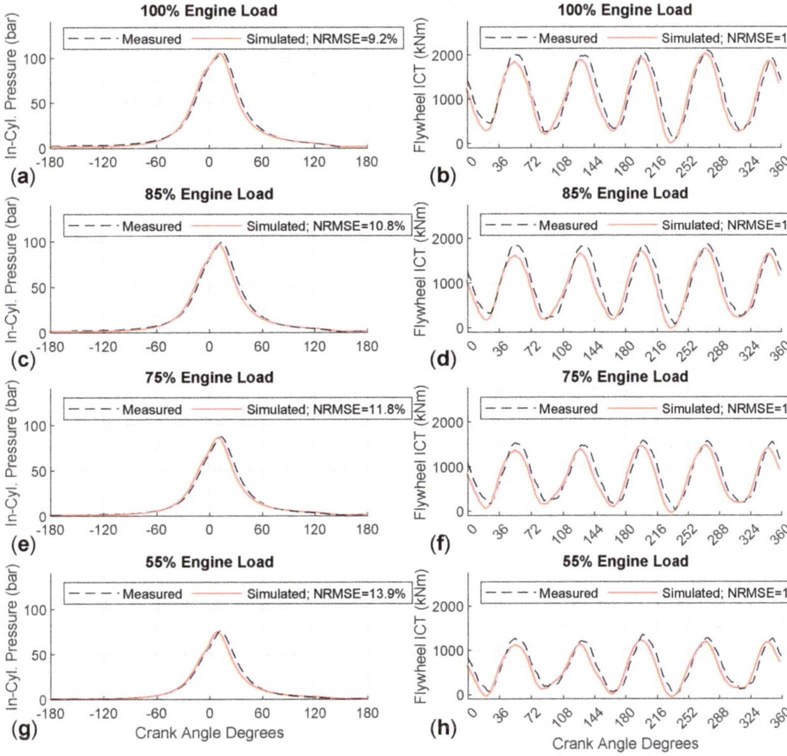

Figure 3. In-cylinder pressure at (**a**) 100%, (**c**) 85%, (**e**) 75% and (**g**) 55% engine load. Flywheel ICT at (**b**) 100%, (**d**) 85%, (**f**) 75% and (**h**) 55% engine load.

3.2. *Malfunctioning Engine Conditions*

The malfunctioning conditions were simulated using the coupled thermodynamics and crankshaft dynamics model, in accordance to the procedure mentioned in Section 2.1. The change of the in-cylinder pressure for a malfunctioning cylinder is shown in Figure 4, whereas the change in the flywheel ICT for the representative case of cylinder 2 malfunctioning is shown in Figure 5. The flywheel ICT change was normalised with respect to the average torque of the healthy engine conditions.

Regarding the change in SOI, it is observed from Figure 4a that as the start of injection is retarded relative to the healthy conditions, the peak in-cylinder pressure decreases. This occurs since combustion takes place later in the cylinder expansion process compared to the healthy conditions, which can have adverse effects on engine efficiency. Conversely, as the injection timing is advanced the pear in-cylinder pressure increases as combustion takes place closer to the TDC. Furthermore, as shown in Figure 5a, the flywheel ICT is observed to achieve its maximum difference compared to the healthy condition at approximately 35°CA after the TDC of cylinder 2, which is followed by an oscillatory behaviour that attenuates throughout the engine revolution. This delay in the torque response after the malfunctioning cylinder is attributed to the crankshaft's flexibility, hence the malfunction of cylinder 2 on the flywheel torsional stresses do not appear instantly. Moreover, depending on whether the peak in-cylinder pressure is greater or less than the respective one of the engine healthy conditions, the amplitude of the flywheel ICT difference varies accordingly.

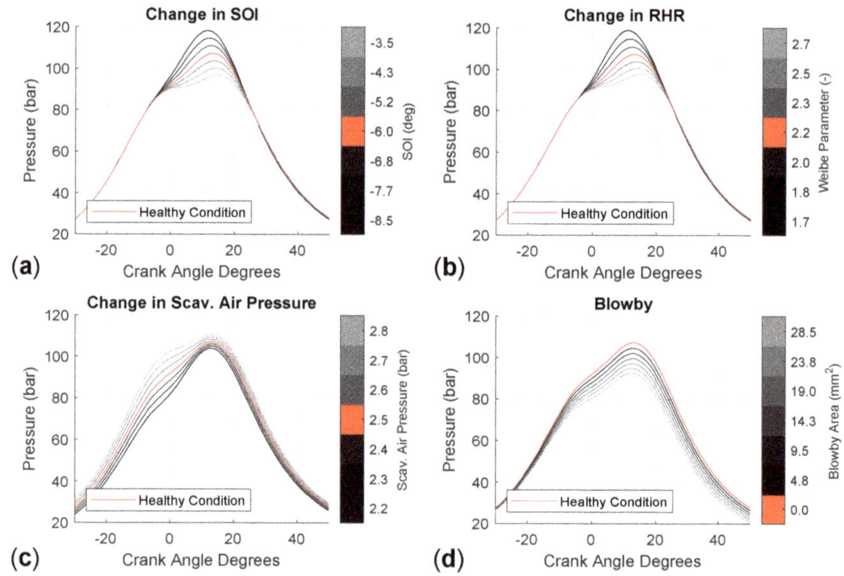

Figure 4. In-cylinder pressure for the healthy and malfunctioning conditions of (**a**) change in SOI, (**b**) change in RHR, (**c**) change in scavenge air pressure, and (**d**) blowby.

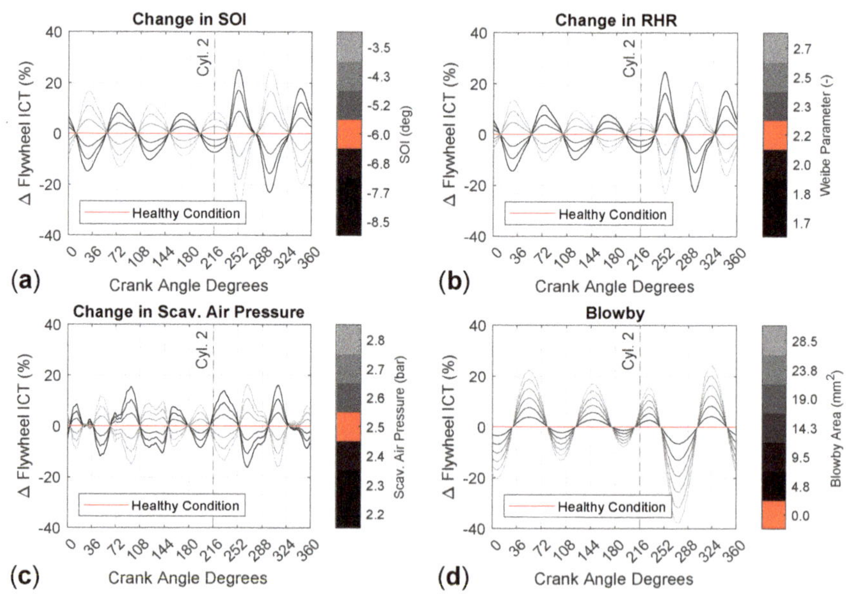

Figure 5. Normalised difference in flywheel ICT of the healthy engine conditions and the cylinder 2 malfunctioning conditions of (**a**) change in SOI, (**b**) change in RHR, (**c**) change in scavenge air pressure, and (**d**) blowby.

Regarding the change in RHR, as shown in Figures 4b and 5b, it appears to have a very similar effect in both the in-cylinder pressure and the flywheel ICT as compared to the change in SOI malfunctioning condition. In specific, as the Weibe shape parameter increases the RHR becomes faster causing the in-cylinder pressure to peak higher as compared to the healthy condition.

The change in the scavenge air pressure effects the entire in-cylinder pressure curve, but has the greater impact in the compression part of the cycle as observed from Figure 4c. In specific, a decrease of 0.3 bar in the scavenge air pressure from the healthy conditions, can result in a decrease of up to 10 bar in the in-cylinder pressure at TDC, and decrease of 5 bar in the peak combustion pressure. Moreover, since all cylinders are affected with this malfunctioning condition, a decrease of 0.3 bar with respect to the healthy conditions, results to a 0.5% decrease in the flywheel ICT. As a result, small changes in the scavenge air pressure can result to significant changes in the maximum compression and peak combustion pressure of all cylinders, which subsequently affects significantly the engine torque. Hence, this underpins the importance of maintenance in the scavenge air system. Furthermore, after cylinder 2 fires there is no significant change in the flywheel ICT oscillations as shown in Figure 5c, since all cylinders are affected.

Finally, when the engine cylinder experiences blowby, this decreases the combustion and expansion part of the in-cylinder pressure diagram, as shown in Figure 4d. As compared to the SOI and RHR malfunctioning conditions, the change in the peak combustion pressure although significant, may not be as large, however the expansion of the in-cylinder pressure diagram is affected, thus reducing the individual piston torque significantly. As a result the normalised difference in the flywheel ICT experiences the largest oscillations compared to all other malfunctioning conditions reaching a minimum of 39% and a maximum of 23%, as shown in Figure 5d. Therefore, this malfunctioning condition can potentially introduce large torsional strain on the engine shaft. Similarly to the SOI and RHR malfunctioning conditions, a delay of 54°CA between the TDC of the malfunctioning cylinder 2 and the maximum flywheel ICT oscillation is observed from Figure 5d, which occurs as a result of the shafting system's flexibility. However, in contrast to the SOI and RHR malfunctioning conditions, the phase in the difference of the flywheel ICT does not change, as during blowby the in-cylinder pressure diagram remains always lower compared with the respective one of the healthy engine conditions.

3.3. Frequency Analysis and Malfunctioning Conditions Mapping

In accordance to the methodology in Section 2.1, by performing a frequency analysis on the flywheel ICT, the engine harmonic frequencies with the most affected modulus and phase angle are determined as a result of the malfunctioning conditions. In specific, according to Figure 6, it is deduced that the change in SOI and RHR for the malfunctioning cylinder no. 2, affects mostly the modulus of the 28th and the 34th harmonic frequencies, and the phase angle of the 14th harmonic frequency. The rest harmonic frequencies affected are observed to be identical for both malfunctioning conditions, which as mentioned in Section 3.2 confirms that both the change in SOI and RHR have a similar effect on the flywheel ICT. Furthermore, the change in scavenge air pressure and blowby modulus is on average 14.3 Nm less for their respective harmonic frequencies, compared to the change in SOI and RHR malfunctioning conditions. However, the harmonic frequencies with the most affected modulus for the change in scavenge air pressure and blowby, are lower by an average of 12 harmonics as compared to the change in SOI and RHR. Therefore, it can be distinguished from noise, and thus it can readily measure their modulus with adequate accuracy.

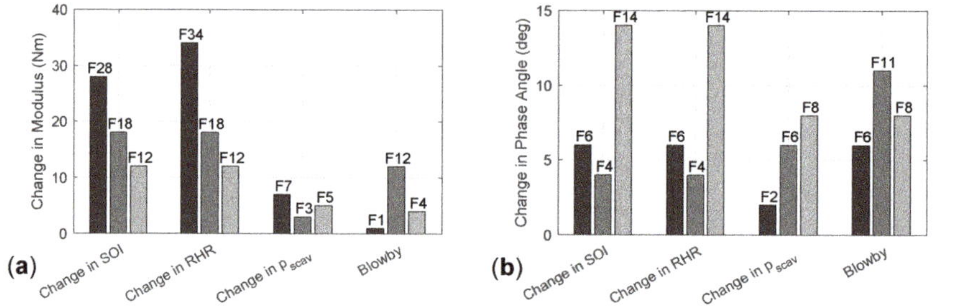

Figure 6. Change in the (**a**) modulus and (**b**) phase angle of the top three most affected engine harmonic frequencies from a malfunction of cylinder 2, at 100% load. Most affected frequencies are denoted with the letter F at the top of each bar.

The results shown in Figure 6 can be condensed to perform the malfunctioning conditions mapping, which includes the harmonic frequencies of the top three most affected frequency moduli and phase angles for every malfunctioning cylinder, as shown in Figure 7a,b respectively. It is visible in Figure 7, that specific malfunctioning conditions have a distinct effect on the flywheel ICT harmonic frequencies for all cylinders, which makes them easily identifiable by employing this mapping process. In specific, as shown in Figure 7a, the modulus of distinct harmonic frequencies is affected for the case of blowby. Hence, the modulus of the 1st harmonic frequency for malfunctioning cylinders 2–10, and the modulus of the 4th harmonic frequency for malfunctioning cylinders 1, 2, 4, and 10 are affected. Similarly, for the change in scavenge air pressure the modulus of the 3rd and 5th harmonic frequencies is affected distinctly, for all engine cylinders malfunctioning. As a result, the above suffices to diagnose blowby or a change in scavenge air pressure if they occur individually, to all 10 cylinders respectively.

However, the change in SOI and RHR have a very similar impact on the flywheel ICT, hence the modulus of the nearly identical harmonic frequencies is affected, as shown in Figure 7a. This makes it difficult to distinguish weather a change in SOI or RHR is responsible when a specific cylinder is malfunctioning. In specific both malfunctioning conditions affect the modulus of the 4th harmonic frequency for malfunctioning cylinders 3, 5, 6, and 8, as well as the 8th and 9th harmonic frequencies for malfunctioning cylinders 4 and 8 respectively. For the remaining cylinders, the affected moduli are exceeding the 10th harmonic frequency which can make them difficult to distinguish from noise in the flywheel ICT signal. However, as shown in Figure 7b, the most affected phase angle of the 4th harmonic frequency corresponds to both malfunctioning conditions, for cylinders 1, 2, 7, 9 and 10. As a result, since phase angle of a lower harmonic frequency is affected as compared to the effected modulus harmonic frequencies, the effect of both malfunctioning conditions is easier to identify for all 10 cylinders respectively.

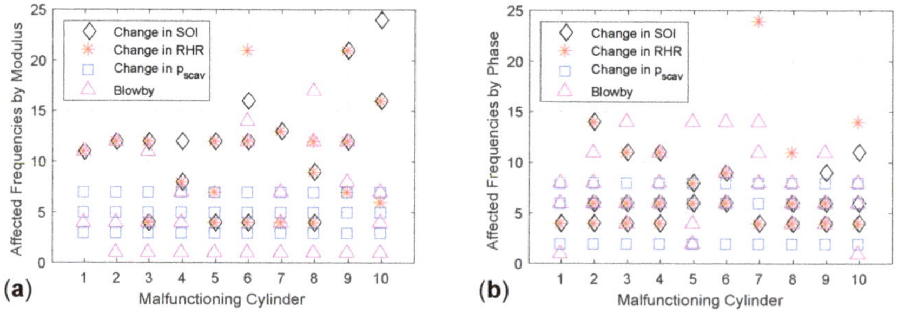

Figure 7. Mapping of individually occurring malfunctioning conditions for each cylinder showing the harmonic frequencies with the most affected (**a**) modulus and (**b**) phase angle, at 100% load.

4. Conclusions

In this study, the mapping of four frequently occurring malfunctioning conditions was performed for a large two-stroke diesel engine, by simulating and identifying their effect on the engine's flywheel ICT. The malfunctioning conditions examined were (a) change in SOI, (b) change in RHR, (c) change in scavenge air pressure and (d) blowby, which were simulated using a validated thermodynamics and crankshaft dynamics coupled engine model.

In specific, changes in SOI and RHR have a nearly identical impact on the engine's in-cylinder pressure diagram, which primarily affects the maximum combustion pressure and its location. Conversely, changes in the scavenge air pressure primarily affect the compression part of the cycle, with smaller effects observed in the peak combustion pressure and its location. Most importantly changes in the scavenge air pressure affect all cylinders uniformly rather than individually, hence the impact of this malfunctioning condition on the engine performance is greater. This also differs from the effect that blowby has on the in-cylinder pressure diagram, where the combustion and expansion part of the cycle are primarily effected.

As a result, the changes in SOI and RHR appear to have similar effects in the flywheel ICT, where as compared to the healthy engine conditions, a sudden change in the torque fluctuations is observed at approximately 35°CA after the malfunctioning cylinder reaches TDC. The delay in the change of the ICT fluctuations measured at the flywheel is a result of the crankshaft flexibility, demonstrating the importance in considering this phenomenon when considering engine dynamics to identify malfunctions in specific cylinders. The change in scavenge air pressure does not cause similar fluctuations in the flywheel ICT as all cylinders are affected uniformly, thus the crankshaft torsional stresses do not change drastically. However, when blowby occurs the flywheel ICT fluctuations are observed to change significantly at 54°CA after the faulty cylinder reaches TDC, and they attenuate for the rest of the engine's rotation.

Subsequently, using the flywheel ICT of the malfunctioning engine conditions, a frequency analysis was performed to identify and map the harmonic frequencies with the most affected modulus and phase angle, with their corresponding malfunctioning cylinder. From this mapping process it was determined that the change in scavenge air pressure, and blowby for any malfunctioning cylinder, have a distinct effect on the flywheel ICT harmonic frequencies, and can thus be identified easily using the malfunctioning conditions map. However, the change in SOI and RHR display mostly similar effects on the flywheel ICT. Thus according to the malfunctioning conditions map it is possible to detect weather a cylinder is experiencing a change in SOI or RHR, but cannot distinguish between the two.

The benefit of the malfunctions mapping approach developed in this study is that the engine model is employed only once to develop the malfunctions map, and simulate the flywheel ICT under healthy engine conditions. Subsequently, a computationally cheap and simple algorithm can be deployed in-situ which performs a frequency analysis on the flywheel ICT signal, and determines the harmonic frequencies of the most effected modulus and phase angle. As a result, in a future study the results of this process can then be readily compared with the malfunctions map to diagnose the engine's condition. Furthermore, malfunctioning conditions occurring in a combination of different cylinders can be examined and mapped in a more extensive database, that includes a larger range of malfunctioning and encompasses better the engine's realistic operating conditions.

Author Contributions: Conceptualization, K.-M.T. and G.T.; methodology, K.-M.T. and G.T.; software, K.-M.T.; validation, K.-M.T.; formal analysis, K.-M.T.; investigation, K.-M.T.; resources, G.T.; data curation, K.-M.T. and G.T.; writing—original draft preparation, K.-M.T.; writing—review and editing, G.T.; visualization, K.-M.T.; supervision, G.T.; project administration, G.T.; funding acquisition, G.T. All authors have read and agreed to the published version of the manuscript.

Funding: The authors gratefully acknowledge the financial support of Innovate UK, through the Knowledge Transfer Partnership project (project number 11577), for the work reported in this article.

Acknowledgments: The authors from MSRC greatly acknowledge the funding from DNV AS and RCCL for the MSRC establishment and operation. The opinions expressed herein are those of the authors and should not be construed to reflect the views of Innovate UK, DNV AS and RCCL.

Conflicts of Interest: The funders had no role in the design of the study; in the collection, analyses, or interpretation of data; in the writing of the manuscript, or in the decision to publish the results.

Abbreviations

The following abbreviations are used in this manuscript:

ANN	Artificial Neural Network
DOF	Degree of Freedom
EVC	Exhaust Valve Close
FFT	Fast Fourier Transform
ICS	Instantaneous Crankshaft Speed
ICT	Instantaneous Crankshaft Torque
MCR	Maximum Continuous Rating
NRMSE	Normalised Root Mean Squared Error
RHR	Rate of Heat Release
SOI	Start of Injection
SPO	Scavenge Port Open
SVM	Support Vector Machine
TDC	Top Dead Center
VIT	Variable Injection Timing

References

1. Kyrtatos, N.P. From Hercules A-B-C to Hercules-2: Cutting Edge R&D in Ship Engines. *Transp. Res. Procedia* **2016**, *14*, 1581–1590. [CrossRef]
2. The Sweedish Club. *Main Engine Damage*; The Sweedish Club: Göteborg, Sweden, 2018.
3. Legović, D.; Dejhalla, R. An Overview of Measures for Ship's Energy Efficiency Improvement. In *Zbornik Radova 22. Simpozija Teorija i Praksa Brodogradnje, in Memoriam Prof. Leopold Sorta (Sorta 2016)*; Sveučilište u Zagrebu: Zagreb, Croatia, 2016.
4. Matulić, N.; Radica, G.; Nižetić, S. Engine model for onboard marine engine failure simulation. *J. Therm. Anal. Calorim.* **2020**, 141, 119–130.
5. Heywood, J.B. *Internal Combustion Engine Fundamentals*; McGraw-Hill: New York, NY, USA, 1988.
6. Martyr, A.J.; Plint, M.A. *Engine Testing: The Design, Building, Modification and Use of Powertrain Test Facilities*; Elsevier: Amsterdam, The Netherlands, 2012.
7. Merker, G.P.; Schwarz, C.; Stiesch, G.; Otto, F. *Simulating Combustion: Simulation of Combustion and Pollutant Formation for Engine-Development*; Springer Science & Business Media: Berlin/Heidelberg, Germany, 2005.
8. Polanowski, S.; Pawletko, R. Acquisition of diagnostic information from the indicator diagrams of marine engines using the electronic indicators. *J. KONES* **2011**, *18*, 359–366.
9. Lin, T.R.; Tan, A.C.; Mathew, J. Condition monitoring and diagnosis of injector faults in a diesel engine using in-cylinder pressure and acoustic emission techniques. *Dyanmics Sustain. Eng.* **2011**, *1*, 454–463.
10. Woodyard, D. *Pounder's Marine Diesel Engines and Gas Turbines*; Butterworth-Heinemann: Oxford, UK, 2009.
11. Aulin, H.; Tunestal, P.; Johansson, T.; Johansson, B. Extracting cylinder individual combustion data from a high precision torque sensor. In Proceedings of the ASME 2010 Internal Combustion Engine Division Fall Technical Conference, San Antonio, TX, USA, 12–15 September 2010; pp. 619–625.
12. Gao, J.; Shen, T. Cylinder pressure sensor-based real-time combustion phase control approach for SI engines. *IEEJ Trans. Electr. Electron. Eng.* **2017**, *12*, 244–250. [CrossRef]
13. Geveci, M.; Osburn, A.W.; Franchek, M.A. An investigation of crankshaft oscillations for cylinder health diagnostics. *Mech. Syst. Signal Process.* **2005**, *19*, 1107–1134. [CrossRef]
14. Taraza, D.; Henein, N.; Bryzik, W. The frequency analysis of the crankshaft's speed variation: A reliable tool for diesel engine diagnosis. *J. Eng. Gas Turbines Power* **2001**, *123*, 428–432. [CrossRef]
15. Charles, P.; Sinha, J.K.; Gu, F.; Lidstone, L.; Ball, A. Detecting the crankshaft torsional vibration of diesel engines for combustion related diagnosis. *J. Sound Vib.* **2009**, *321*, 1171–1185. [CrossRef]
16. Dereszewski, M.; Charchalis, A.; Polanowski, S. Analysis of diagnostic utility of instantaneous angular speed fluctuation of diesel engine crankshaft. *J. KONES* **2011**, *18*, 123–128.
17. Desbazeille, M.; Randall, R.; Guillet, F.; El Badaoui, M.; Hoisnard, C. Model-based diagnosis of large diesel engines based on angular speed variations of the crankshaft. *Mech. Syst. Signal Process.* **2010**, *24*, 1529–1541. [CrossRef]

18. Guerrero, D.P.; Jiménez-Espadafor, F.J. Torsional system dynamics of low speed diesel engines based on instantaneous torque: Application to engine diagnosis. *Mech. Syst. Signal Process.* **2019**, *116*, 858–878. [CrossRef]
19. Espadafor, F.J.J.; Villanueva, J.A.B.; Guerrero, D.P.; García, M.T.; Trujillo, E.C.; Vacas, F.F. Measurement and analysis of instantaneous torque and angular velocity variations of a low speed two stroke diesel engine. *Mech. Syst. Signal Process.* **2014**, *49*, 135–153. [CrossRef]
20. Schagerberg, S.; McKelvey, T. *Instantaneous Crankshaft Torque Measurements-Modeling and Validation*; Technical Report; SAE: Warrendale, PA, USA, 2003.
21. Thor, M.; Egardt, B.; McKelvey, T.; Andersson, I. Using combustion net torque for estimation of combustion properties from measurements of crankshaft torque. *Control Eng. Pract.* **2014**, *26*, 233–244. [CrossRef]
22. Acciaro, M.; Hoffmann, P.N.; Eide, M.S. The energy efficiency gap in maritime transport. *J. Shipp. Ocean Eng.* **2013**, *3*, 1.
23. Kitada, M.; Ölçer, A. Managing people and technology: The challenges in CSR and energy efficient shipping. *Res. Transp. Bus. Manag.* **2015**, *17*, 36–40. [CrossRef]
24. MAN B&W. Instruction Book 'Operation' for 46-108MC/MC-C Engines. Hyundai-MAN B&W. Available online: https://seatracker.ru/viewtopic.php?t=1516 (accessed on 13 April 2021).
25. Capezza, C.; Coleman, S.; Lepore, A.; Palumbo, B.; Vitiello, L. Ship fuel consumption monitoring and fault detection via partial least squares and control charts of navigation data. *Transp. Res. Part D Transp. Environ.* **2019**, *67*, 375–387. [CrossRef]
26. Lazakis, I.; Gkerekos, C.; Theotokatos, G. Investigating an SVM-driven, one-class approach to estimating ship systems condition. *Ships Offshore Struct.* **2019**, *14*, 432–441. [CrossRef]
27. Dikis, K.; Lazakis, I.; Turan, O. Probabilistic risk assessment of condition monitoring of marine diesel engines. In Proceedings of the ICMT 2014, York, UK, 21–22 July 2014.
28. Cheliotis, M.; Lazakis, I.; Theotokatos, G. Machine learning and data-driven fault detection for ship systems operations. *Ocean Eng.* **2020**, *216*, 107968. [CrossRef]
29. Tsitsilonis, K.M.; Theotokatos, G. A novel systematic methodology for ship propulsion engines energy management. *J. Clean. Prod.* **2018**, *204*, 212–236. [CrossRef]
30. Lind, M.; Michaelides, M.; Ward, R.; Herodotou, H.; Watson, R. *Boosting Data-Sharing to Improve Short Sea Shipping Performance: Evidence from Limassol Port Calls Analysis*; Technical Report; Cyprus University of Technology: Limassol, Cyprus, 2019.
31. Kouremenos, D.; Hountalas, D. Diagnosis and condition monitoring of medium-speed marine diesel engines. *Tribotest* **1997**, *4*, 63–91. [CrossRef]
32. Hountalas, D.; Kouremenos, D.; Sideris, M. A diagnostic method for heavy-duty diesel engines used in stationary applications. *J. Eng. Gas Turbines Power* **2004**, *126*, 886–898. [CrossRef]
33. Lamaris, V.; Hountalas, D. A general purpose diagnostic technique for marine diesel engines—Application on the main propulsion and auxiliary diesel units of a marine vessel. *Energy Convers. Manag.* **2010**, *51*, 740–753. [CrossRef]
34. Watzenig, D.; Sommer, M.; Steiner, G. Engine state monitoring and fault diagnosis of large marine diesel engines. *E I Elektrotechnik Inf.* **2009**, *126*, 173–179. [CrossRef]
35. Benvenuto, G.; Campora, U. Performance prediction of a faulty marine diesel engine under different governor settings. In Proceedings of the International Conference on Marine Research And Transportation, Naples, Italy, 28–30 June 2007; pp. 35–44.
36. Kouremenos, D.; Rakopoulos, C.; Hountalas, D.; Kouremenos, A. The maximum compression pressure position relative to top dead centre as an indication of engine cylinder condition and blowby. *Energy Convers. Manag.* **1994**, *35*, 857–870. [CrossRef]
37. Tsitsilonis, K.; Theotokatos, G.; Habens, M. A modelling approach for predicting marine engines shaft dynamics. In Proceedings of the International Naval Engineering Conference & Exhibition 2020, London, UK, 5–9 October 2020.
38. Guerrero, D.P. Aportaciones al Mantenimiento Predictivo de Plantas de Potencia: Aplicación a Motores Diesel Lentos de 2 Tiempos. Ph.D. Thesis, University of Seville, Sevilla, Spain, 2017.
39. MATLAB. *9.10.0.1602886 (R2021a)*; The MathWorks Inc.: Natick, MA, USA, 2021.
40. Foteinos, M.I.; Papazoglou, A.; Kyrtatos, N.P.; Stamatelos, A.; Zogou, O.; Stamatellou, A.M. A three-zone scavenging model for large two-stroke uniflow marine engines using results from CFD scavenging simulations. *Energies* **2019**, *12*, 1719. [CrossRef]
41. Guan, C.; Theotokatos, G.; Chen, H. Analysis of two stroke marine diesel engine operation including turbocharger cut-out by using a zero-dimensional model. *Energies* **2015**, *8*, 5738–5764. [CrossRef]
42. Gardiner, W.C.; Burcat, A. *Combustion Chemistry*; Springer: Berlin/Heidelberg, Germany, 1984.
43. CIMAC HFO Working Group. Recommendations Regarding Fuel Quality for Diesel Engines. 2003. Available online: https://www.cimac.com/cms/upload/Publication_Press/Recommendations/Recommendation_21_rev1.pdf (accessed on 13 April 2021).
44. Xiang, L.; Theotokatos, G.; Cui, H.; Xu, K.; Ben, H.; Ding, Y. Parametric knocking performance investigation of spark ignition natural gas engines and dual fuel engines. *J. Mar. Sci. Eng.* **2020**, *8*, 459. [CrossRef]
45. Öberg, P. A Dae Formulation for Multi-Zone Thermodynamic Models and Its Application to Cvcp Engines. Ph.D. Thesis, Linköping University, Linköping, Sweden, 2009.
46. Tsitsilonis, K.M.; Theotokatos, G.; Xiros, N.; Habens, M. Systematic investigation of a large two-stroke engine crankshaft dynamics model. *Energies* **2020**, *13*, 2486. [CrossRef]

47. MAN Diesel & Turbo. Basic Principles of Ship Propulsion. 2011. Available online: https://spain.mandieselturbo.com/docs/librariesprovider10/sistemas-propulsivos-marinos/basic-principles-of-ship-propulsion.pdf?sfvrsn=2 (accessed on 6 February 2021).
48. Hountalas, D.T. Prediction of marine diesel engine performance under fault conditions. *Appl. Therm. Eng.* **2000**, *20*, 1753–1783. [CrossRef]
49. Wolff, A. Influence of piston ring pack configuration on blowby and friction losses in a marine two-stroke engine. *Combust. Engines* **2017**, *56*, doi:10.19206/CE-2017-328. [CrossRef]

applied sciences

MDPI

Article

A Novel Methodology for Hydrocarbon Depth Prediction in Seabed Logging: Gaussian Process-Based Inverse Modeling of Electromagnetic Data

Hanita Daud [1], Muhammad Naeim Mohd Aris [1,*], Khairul Arifin Mohd Noh [2,*] and Sarat Chandra Dass [3]

1 Department of Fundamental and Applied Sciences, Universiti Teknologi PETRONAS, Seri Iskandar 32610, Perak, Malaysia; hanita_daud@utp.edu.my
2 Department of Geosciences, Universiti Teknologi PETRONAS, Seri Iskandar 32610, Perak, Malaysia
3 School of Mathematical and Computer Sciences, Heriot-Watt University Malaysia, Putrajaya 62200, Malaysia; s.dass@hw.ac.uk
* Correspondence: muhammad_16000991@utp.edu.my (M.N.M.A.); khairula.nmoh@utp.edu.my (K.A.M.N.)

Abstract: Seabed logging (SBL) is an application of electromagnetic (EM) waves for detecting potential marine hydrocarbon-saturated reservoirs reliant on a source–receiver system. One of the concerns in modeling and inversion of the EM data is associated with the need for realistic representation of complex geo-electrical models. Concurrently, the corresponding algorithms of forward modeling should be robustly efficient with low computational effort for repeated use of the inversion. This work proposes a new inversion methodology which consists of two frameworks, namely Gaussian process (GP), which allows a greater flexibility in modeling a variety of EM responses, and gradient descent (GD) for finding the best minimizer (i.e., hydrocarbon depth). Computer simulation technology (CST), which uses finite element (FE), was exploited to generate prior EM responses for the GP to evaluate EM profiles at "untried" depths. Then, GD was used to minimize the mean squared error (MSE) where GP acts as its forward model. Acquiring EM responses using mesh-based algorithms is a time-consuming task. Thus, this work compared the time taken by the CST and GP in evaluating the EM profiles. For the accuracy and performance, the GP model was compared with EM responses modeled by the FE, and percentage error between the estimate and "untried" computer input was calculated. The results indicate that GP-based inverse modeling can efficiently predict the hydrocarbon depth in the SBL.

Keywords: seabed logging; electromagnetic data; hydrocarbon depth; inverse modeling; Gaussian process; gradient descent; computer simulation technology

Citation: Daud, H.; Mohd Aris, M.N.; Mohd Noh, K.A.; Dass, S.C. A Novel Methodology for Hydrocarbon Depth Prediction in Seabed Logging: Gaussian Process-Based Inverse Modeling of Electromagnetic Data. *Appl. Sci.* **2021**, *11*, 1492. https://doi.org/10.3390/app11041492

Received: 21 December 2020
Accepted: 14 January 2021
Published: 7 February 2021

1. Introduction

Seabed logging (SBL) is an application of the marine controlled-source electromagnetic (CSEM) surveying technique for discovering and characterizing high-resistive bodies (e.g., hydrocarbon-saturated reservoirs) remotely in the deep marine environment based on contrasts of electrical resistivity of the subsurface. Its capability of identifying the fluid content inside the potential reservoirs is undoubted. This emerging application uses a powerful electric dipole transmitter to continuously emit low-frequency electromagnetic (EM) waves, normally between 0.1 and 10 Hz [1], in all directions through the seawater column and subsurface beneath the seabed. The SBL exploits the fact that hydrocarbon-saturated reservoirs have higher electrical resistivity (approximately 30–500 Ωm) than its surroundings, such as seawater (approximately 0.5–2.0 Ωm) and sedimentary rocks (approximately 1.0–2.0 Ωm) [2]. The conductivities (i.e., inversely proportional to resistivities) of hydrocarbon-filled reservoirs can reach about 0.01–0.1 S/m or lower in certain depths underneath the seabed. The main component of an SBL is the utilization of a towed EM source known as a horizontal electric dipole (HED) and stationary EM receivers for recording the returned EM signals [1]. Studies related to the commercial SBL application can be found in [1–9]. Figure 1 is adopted from [10] to depict the illustration of the SBL application in the offshore environment.

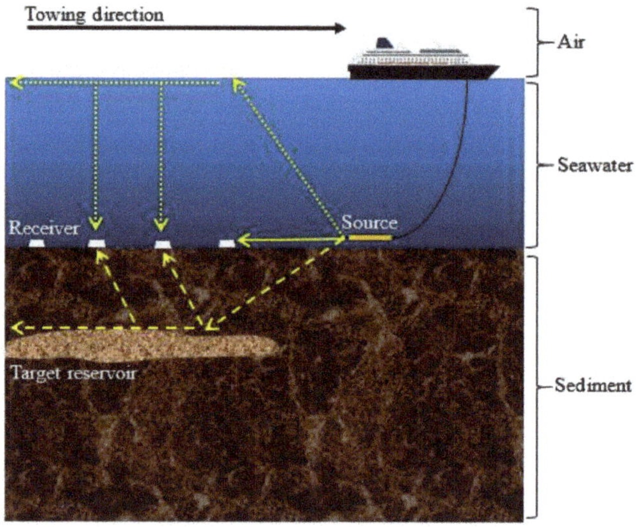

Figure 1. Illustration of the seabed logging (SBL) application in the deep marine environment. The source is towed by a ship and the electromagnetic (EM) receivers are placed on the seafloor beforehand. The target reservoir is in the sedimentary rock. The low-frequency EM waves travel in all directions (straight arrow: direct wave; dotted arrow: air wave; dashed arrow: guided wave).

Numerical modeling is a crucial task in geophysics problems since it provides prior knowledge and information of the potential reservoirs. Many powerful computational modeling techniques have been exploited for processing the EM responses, such as finite element (FE), finite difference (FD), finite volume (FV), integral equations (IE), etc. These numerical techniques are capable of modeling the subsurface realistically in the marine environment by defining grids or meshes. According to [11], the FE technique is highly preferred due to its ability to support unstructured meshes. These meshes are able to provide a quality geological representation by placing small grid elements in the areas where better resolution is highly needed, such as in large conductivity contrast areas [11]. However, as the consequence, the mesh-based algorithms require a very lengthy computational time to solve the linear system; in addition, these forward modeling techniques involve very complicated deterministic mathematical equations. According to [12,13], the most time-consuming task in the algorithms is solving the integral and multiple linear equations, not to mention for the inversion scheme that requires multiple EM forward solutions [13]. Here, the corresponding forward modeling should be robust and capable of operating the EM simulation rapidly, especially for repeated use in the large number of iterations involved in the inversion scheme as well as for multiple source–receiver positions in the SBL application [14].

To address these problems, this work aimed to develop a novel Gaussian process-based inversion methodology to provide meaningful information where the hydrocarbon potentially resides underneath the seabed. For the optimization procedure, this inverse modeling utilized the gradient descent (GD) method to find the best estimate (i.e., hydrocarbon depth) based on the forward Gaussian process (GP) regression model. As mentioned before, GP acts as the alternative forward modeling technique as it is capable of being a surrogate model for the complicated mathematical model that requires high computational time to solve [15]. Researchers in [16] also stated that GP is an elegant and simple approach, where it performs well in modeling nonlinear problems. Mean squared error (MSE) was used as the objective function (i.e., loss function) in the optimization. The main focus of this work was to find the depth of hydrocarbon of observational EM responses, whose hydrocarbon layer was assumed to be present but with an unknown depth. This work is novel in two ways. First, it presents the integration of a stochastic process (the GP regression model acts as the EM forward model) and an indirect search algorithm (the iterative GD method as the acquirer of minimizer) in one inversion methodological flow. Second, applying this proposed inverse modeling for predicting the input parameter of the potential high-resistive bodies (e.g., hydrocarbon depth) is a new procedure or approach in SBL data processing. This attempt could be very beneficial to de-risk hydrocarbon exploration in the offshore environment. Details of the GP and GD approaches are discussed in the next section.

2. Mathematical Background

Gaussian process (GP) and gradient descent (GD) algorithms are commonly exploited in machine learning (ML) problems. In order to seek the most favorable fairness between the performance of the modeling and the need of low computational time, considering the strengths and integrating these two mathematical approaches could be very advantageous to any applications that struggle with the same concern. Here, Figure 2 shows the common ML process to predict outputs given a set of inputs, its corresponding outputs, and desired/query inputs. It represents the general idea on which this work is focusing. The proposed inversion methodology is believed capable of providing meaningful and useful information to the SBL application with an acceptable accuracy and low computational efforts. In this section, the background and examples from the literature of GP and GD approaches are thoroughly discussed in Sections 2.1 and 2.2, respectively.

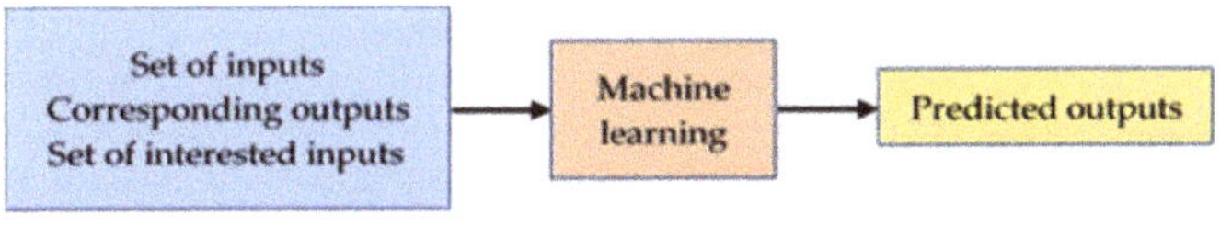

Figure 2. Machine learning (ML) process. ML utilizes sets of inputs, the corresponding outputs, and interested inputs to yield the predictive outputs.

2.1. Gaussian Process

Mathematically, the Gaussian process (GP) can be defined as an infinite collection of random variables where any of the finite subsets can be treated as Gaussian (i.e., normal) distribution [17,18]. A GP model is fully expressed or specified by a mean function $m(x)$ in a vector and a covariance $k(x, x')$ in the form of a matrix. GP is a probabilistic and nonparametric method for fitting any functional forms based on observations of the domain. This method differs from most of the other black-box identification approaches. GP does not try to approximate the model by trying to fit the parameters of the chosen basis function, but it searches the relationship among the given knowledge or information. This stochastic approach can provide predictive mean and variance. The mean represents the most likely outputs, whereas the variance plays a role as the uncertainty's quantifier. The magnitude of variance is incorporated into the GP prediction and inference. It does reflect the amount and quality of the data distribution, and it is an essential numerical feature when it comes to differentiating the GP regression from any other computational intelligence methods.

GP regression is a well-known technique and has been adopted in many scientific and engineering fields, such as machine learning, computer experiments, geostatistics, electronics, etc. Its convenient properties in modeling make GP successful in demonstrating outstanding performances in numerous applications. Researchers in [19] employed GP regression for predicting the concentration of ozone in an area of Bourgas, Bulgaria. The prediction was based on the measurements of the concentration of ozone, phenol, sulphur dioxide, and benzene in the air (hourly). Researchers in [20] studied the capability of GP in estimating load-bearing pile capacity. This research work compared the performance of GP prediction with other approaches, and GP worked very well in predicting the pile capacity. In a wireless networking application, researchers in [21] inferred that, besides resulting in low-cost outcomes, GP was also capable of obtaining low complex proximity reports for a received-signal-strength (RSS) threshold. Next, GP was also used by researchers in [22] to explain the uncertainty in estimating the state of health (SOH) of lithium-ion batteries. Additionally, according to researchers in [23], GP can successfully decrease the complexity of the computer vision with a better performance of reconstructions.

In [24], GP was hybridized with particle swarm optimization (PSO) to do advanced prediction of roadway broken road zone based on data collected in China. The study inferred that the hybrid model showed the best performance with higher variance accounted for (VAF), higher coefficient of determination (R^2), and lower root mean squared error (RMSE) compared to the other methods such as multiple linear regression (MLR) and artificial neural network (ANN). For geophysics problems, researchers in [25] mentioned that GP was able to yield reliable information of permeability and porosity in well analysis. GP was also used by researchers in [26] to predict the presence of a thin hydrocarbon layer by utilizing its predictive variance for quantifying the uncertainties of EM responses. Furthermore, GP was exploited for seismic signal detection as well. According to [27], the rate of detection when using GP was higher compared to other methods and the rate of false alarm was lower especially for the weak seismic signals. Researchers in [28] proved that GP was successfully capable of being exploited in processing EM responses by calibrating the stochastic and the computer experiment models for magnitude versus offset (MVO) analysis.

2.2. Gradient Descent

Gradient descent (GD) is a popular iterative process used to update the model parameters of a function by minimizing the objective function (i.e., loss function) based on a first-derivative (i.e., slope) basis. The derivative of a function is fully employed to determine whether to increase or decrease the model parameters (i.e., direction to move). Moreover, finding the slope can also tell how big of a step should be taken to reach the optimal point. Conceptually, this approach optimizes the function by iteratively moving in the negative direction of the gradient, also known as steepest descent. The smoothness of the moving process to reach the minimum depends on the size of steps, which in the GD method is called the learning rate, i.e., α. A higher learning rate can extend over a larger function area, but there is a risk of overshooting the optimal point. Meanwhile, a smaller learning rate is more precise since the slope is frequently re-calculated. However, if it is not wisely parameterized, the optimization needs more time to converge and obtain the minima. Thus, choosing the suitable step size implies a high efficiency of the optimization algorithm.

This well-known method has been employed and exercised in many scientific applications, mostly in optimization problems such as in machine learning [29]. Researchers in [30] exploited GD to establish an operation model of nodes for the classification of heart conditions based on a deep neural network. Furthermore, researchers in [31] adopted the GD method to produce phase-only computer-generated holograms (CGHs). This study inferred that the GD required low time to optimize the phase-only CGH distribution and provided higher-precision images compared to the Gerchberg–Saxton (GS) algorithm. Next, researchers in [32] used the GD method to optimize the vector centers of the consequent layer functions and receptive field matrices in a neuro-fuzzy model based on the standard criterion of mean square error. Furthermore, GD was also used by [33] for minimizing the equation of error used in the study by updating the W_1 and W_2 matrices in order to solve the predicting student performance (PSP) problem. Researchers in [34] also utilized stochastic GD as one of the machine learning methods to be compared for the inversion of a stochastic skin optical model. The researchers also used multi-layer perceptron (MLP), linear support vector regressor (SVR), convolutional neural network (CNN), and lasso regressor in the comparative study.

3. Methodology

This work aimed to predict the depth of hydrocarbon in the seabed logging (SBL) application by using a novel methodology, namely Gaussian process (GP)-based inverse modeling. Here, four subsections are discussed: 3.1. Preprocessing computer outputs, 3.2. Evaluating electromagnetic (EM) profile using the GP, 3.3. Predicting depth of hydrocarbon, and 3.4. Estimate and GP model validation.

3.1. Preprocessing Computer Outputs

Computer simulation technology (CST) software, which uses finite element (FE), is exploited in this work to model and evaluate prior information of EM responses (e.g., magnitudes of electric (E)-field) in the SBL for the proposed GP-based inversion methodology. CST is part of SIMULIA, a Dassault Systèmes brand, sourced from Darmstadt, Germany. According to [35], CST Studio Suite® exclusively uses the FE method as the EM simulation solver for low-frequency application. Nine SBL models with the same input properties, except the hydrocarbon depth, were simulated using the CST software. Details of the SBL data acquisition are presented in Appendix A. For demonstration, one SBL model replicated by the CST is depicted in Figure A1. The SBL models, known as target models, were replicated with three background layers such as air, seawater, and sediment. The nine different "tried" depths were between 200 m and 1000 m from the seafloor with an increment of 100 m each. All the computer input properties and parameters used in the SBL modeling are indicated in Table 1.

Table 1. Input properties and parameters of the seabed logging (SBL) models.

Input Property or Parameter	Numerical Values in Unit
Electrical conductivity (Air)	1.0×10^{-11} S/m
Electrical conductivity (Seawater)	1.63 S/m
Electrical conductivity (Sediment)	1.00 S/m
Electrical conductivity (Hydrocarbon)	2.0×10^{-3} S/m
Thickness of layer (Air)	300 m
Thickness of layer (Seawater)	1000 m
Total thickness of layer (Sediment)	3500 m
Thickness of layer (Hydrocarbon)	200 m
Transmission frequency	0.125 Hz
Current strength	1250 A
Transmitter length	270 m
Depth of elevation (Transmitter)	30 m
Length of SBL models	20,000 m
Width of SBL models	20,000 m
Height of SBL models	5000 m

As discussed by researchers in [28], interpretation of the SBL data relies on an assumption that the normalized magnitude versus offset (MVO) between the target responses (i.e., with presence of hydrocarbon) and reference responses (i.e., without presence of hydrocarbon) can pre-identify the presence of high-resistive bodies. The idea is that the magnitudes of target response are divided by the magnitudes of reference response along the offsets (i.e., source–receiver separation distances). Ratio of the normalization should be greater than one, which is associated with the existence of the potential reservoirs. Thus, all the generated CST outputs were preprocessed before being used for the forward GP modeling. By considering the ratio measurements and the symmetrical property of the developed SBL models, we chose nine different datasets with 73 datapoints per set, which are from an approximate distance of 11,840.80 m to an approximate distance of 19,004.97 m, for the data processing presented in the next subsection.

3.2. Evaluating Electromagnetic Profile Using the Gaussian Process

This work uses Gaussian process machine learning (GPML) MATLAB codes for the GP regression, previously written by Rasmussen and Nickisch, which can be found in [36]. Note that the MATLAB version used by this work is R2019b. As mentioned before, the Gaussian process (GP) acts as the alternative approach to the numerical forward modeling in the SBL application for evaluating EM profiles at various "untried" hydrocarbon depths. Suppose all nine CST datasets, $\mathcal{D}_1, \mathcal{D}_2, \ldots, \mathcal{D}_9 = \{(x_i, y_i)\}_{i=1}^{N}$, where $N = 73$, are treated as the training datasets for the GP modeling. It is also possible to simply write the d-dimensional input variable vector and its corresponding output, which is magnitude of E-field vector, as $x_i \in X$ and $y_i \in Y$, respectively. Generally, in this work, X is the N-by-d matrix and Y is the N-by-1 column vector, where $x \in \mathbb{R}^{d=2}$ and $y \in \mathbb{R}^{d=1}$. In other words, $x = (s, h)$, where s and h represent the offset (i.e., source–receiver separation distance) and the depth of hydrocarbon, respectively.

In order to evaluate or estimate the EM profiles at various "untried" depth inputs conforming to the testing input, $x_* = (s_*, h_*)$, the posterior distribution of the testing output y_* is defined in Equation (1). This work's interest was to predict the depth parameter of hydrocarbon along the offsets from 11,840.80 m to 19,004.97 m. Thus, here, only the testing input variable of hydrocarbon depth, h_*, is varied accordingly, while the s_* remains unchanged. Note that the interested hydrocarbon depths in the GP regression are from 200 m to 1000 m with an increment of 20 m each (including the "tried" depths).

$$(y_*|X, Y, x) \sim G(m_*, \kappa_*), \tag{1}$$

where the predictive mean, m_*, and the predictive variance, κ_*, are expressed in Equations (2) and (3), respectively.

$$m_* = K(X, x_*)^T \left[K(X, X) + \sigma_N^2 I_N\right]^{-1} Y, \tag{2}$$

and

$$\kappa_* = k(x_*, x_*) - K(X, x_*)^T \left[K(X, X) + \sigma_N^2 I_N\right]^{-1} K(X, x_*). \tag{3}$$

Based on Equations (2) and (3), $K(X, X)$ is the covariance matrix with a dimension of N-by-N amongst the X, $K(X, x_*)$ is the covariance matrix with a dimension of N-by-1 between the X and x_*, $k(x_*, x_*)$ is the covariance of the x_*, I_N is an N-dimensional unit matrix, and $\sigma_N^2 I_N$ is the noise covariance matrix. All these covariance matrices are composed of squared exponential (SE) terms, where each element of the matrices is computed using Equation (4) as follows:

$$k(x, x') = \sigma_f^2 \left[exp\left(-\frac{|x - x'|^2}{2\ell^2}\right)\right], \tag{4}$$

where σ_f represents the signal variance and ℓ represents the isotropic length scale.These two parameters are known as the hyperparameters of the SE covariance function, $\theta = \left\{\sigma_f, \ell\right\}$. SE is a very famous covariance function used in GP regression. According to [37], GP with SE is infinitely differentiable. This feature is very important for the derivative-based approach.

Here, it must be noted that all the computations of Equations (1)–(4) require the estimate of θ. These hyperparameters are numerically estimated by minimizing the negative log marginal likelihood (NLML) using a gradient-based optimizer. The NLML equation is defined in Equation (5) as follows:

$$NLML = -log[p(Y|X, \theta)] = \frac{N \log(2\pi)}{2} + \frac{\log\left|K(X, X) + \sigma_N^2 I_N\right|}{2} + \frac{Y^T \left[K(X, X) + \sigma_N^2 I_N\right]^{-1} Y}{2}. \tag{5}$$

In Equation (5), a logarithmic identifier is used to simplify the integral expression involved in the marginal likelihood method [25]. Once the hyperparameters are successfully estimated, the estimation of the GP regression is then defined using Equations (2) and (3). The m_* is a mean value of the estimation that gives the best estimate of y_* (i.e., magnitude of E-field at x_*), and the κ_* is the predictive variance of the testing input in terms of two standard deviations (95% confidence interval), which reflects the reliability of the predictive mean. These two equations are the basis in the forward GP modeling.

3.3. Predicting Depth of Hydrocarbon

The fitted GP model is denoted as $Y_{fit}(s, h)$, where h here includes all "tried" and "untried" depths. This means that, for every h (i.e., 200 m–1000 m with a gap of 20 m), the GP gives estimates of the EM profile, $Y_{fit}(s, h)$, at distances from 11,840.80 m to 19,004.97 m. Assume that the observational dataset $Y_{obs}(s, \blacksquare)$ is observed at $N = 73$ distance points and the hydrocarbon resides underneath the seabed but with "unknown" depth. To find out the depth of hydrocarbon of the observational dataset Y_{obs}, the least square criterion is formed between the Y_{fit} and Y_{obs}. Mean squared error (MSE) is used as the loss function in the optimization algorithm. In general, MSE measures the average amount of squared differences between Y_{fit} and Y_{obs}. The loss function $L(h_j)$ is expressed in Equation (6) as follows:

$$\{L(h_j)\}_{j=1}^{M} = \frac{1}{N}\sum_{i=1}^{N}\left[Y_{fit}(s_i, h_j) - Y_{obs}(s_i)\right]^2, \tag{6}$$

where $L(h_j)|j = 1, 2, \ldots, M$ and $M = 41$.

Equation (6) is a function of h. The loss function is minimized over the h with the minimizer serving as an estimate of the hydrocarbon depth. By using Equation (6), the $L(h_j)$ between the Y_{fit} and Y_{obs} are calculated. Here, the least $L(h_j)$ is then identified. As for now, the minimizer of the least $L(h_j)$ serves as the best estimate of hydrocarbon depth for Y_{obs}. Thus, in this work, the respective minimizer is treated as the initial guess of depth. Then, the iteration index j is set as zero and the initial guess is denoted as h_0.

Next, for the optimization using the GD method, the first-partial derivative of the $L(h_j)$ with respect to h_j is required. The partial derivative of the loss function is defined in the equation below. The derivation of Equation (7) is provided in Appendix B (in Equation (A1)):

$$L'(h_j) = \frac{2}{N}\left(\sum_{i=1}^{N}\left[Y_{fit}(s_i, h_j) - Y_{obs}(s_i)\right]\right)\cdot Y'_{fit}(s_i, h_j), \tag{7}$$

where Y'_{fit} is the derivative of the GP using the SE covariance function with respect to h_j. It must be noted here that the forward GP model is defined by estimations composed by the predictive mean as shown in Equation (2). To find out the Y'_{fit}, the predictive mean m_* is differentiated with respect to h_j as follows:

$$m'_* = K'(X, x_*)^T\left[K(X, X) + \sigma_N^2 I_N\right]^{-1} Y, \tag{8}$$

where $K'(X, x_*)$ is the derivative of covariance function between X and x_* with respect to h_j. As mentioned earlier, each element of the covariance matrix is computed using the SE covariance function (Equation (4)). Thus, the derivative of SE between X and x_* with respect to h_j is defined in order to find the $K'(X, x_*)$, as expressed in Equation (9). Its derivation is provided in Appendix B (in Equation (A2)):

$$k'(x, x_*) = \sigma_f^2\cdot\left(\frac{|x - x_*|}{\ell^2}\right)\cdot exp\left(-\frac{|x - x_*|^2}{2\ell^2}\right). \tag{9}$$

Next, the following step is finding the correction or update of h_j. It is denoted as h_{j+1}. By referring to the GD algorithm, the equation of h_{j+1} is expressed as follows:

$$h_{j+1} = h_j - \alpha \cdot L'(h_j), \tag{10}$$

where α is the learning rate, which governs the step size of the optimization. Normally, as a start, a geometric sequence with a common ratio of 10 is used to indicate which minimizer can give a significant optimization.

After getting the h_{j+1}, the EM profile at h_{j+1}, denoted as $Y_{fit}(s_i, h_{j+1})$, is then estimated using the GP regression by defining the h_{j+1} as its testing input and the GP modeling steps are repeated. Next, the new loss function for h_{j+1} between Y_{fit} and Y_{obs} is calculated using Equation (11) as follows:

$$L(h_{j+1}) = \frac{1}{N}\sum_{i=1}^{N}\left[Y_{fit}(s_i, h_{j+1}) - Y_{obs}(s_i)\right]^2. \tag{11}$$

The tolerance threshold ε, which acts as the stopping criterion for the iterations, is defined for verifying the $L(h_{j+1})$ before deciding the optimal hydrocarbon depth for the Y_{obs}. In the context of this work, it was decided that the changes between each subsequent minimizer or depth had to be less than 0.05. As presented in Section 4, the specific limit leads to acceptable results. Equation (12) shows the equation of relative percentage change (RPC) used in this work:

$$RPC_{j+1} = \frac{|h_{j+1} - h_j|}{|h_j|} \times 100\%. \tag{12}$$

The next conditions used in this work are the following: the maximum number of iterations is 100 and the loss function (i.e., MSE) must converge. In other words, $L(h_{j+1})$ must be always lower than $L(h_j)$. Each iteration must satisfy these three threshold conditions. The optimal depth of hydrocarbon for Y_{obs} is believed to be successfully predicted if the process does not meet the conditions. Otherwise, it is returned to the step of evaluating the $L'(h_j)$ onwards by defining a new iteration index, $j = j + 1$. All the flows are repeated until the ε is neglected, then the iterations stop. The methodological flow presented in this section is shown in Figure 3.

3.4. Estimate and GP Model Validation

Error measurement is very important to determine the reliability of estimations. In this work, for estimate validation, the percentage error between the computer input (i.e., hydrocarbon depth used in the computer simulation) and the depth predicted by the GP-based inverse modeling was calculated using Equation (13). Meanwhile, for GP model validation, three error validators were used in this work, namely mean absolute deviation (MAD), mean absolute percentage error (MAPE), and root mean squared error (RMSE). The errors of EM profiles provided by the GP model were calculated by assuming the EM responses modeled by the FE (through the CST software) as the true EM values. All equations of the MAD, MAPE, and RMSE are defined in Equations (14)–(16), respectively, as follows:

$$Percentage\ Error = \left|\frac{h_{true} - h_{estimate}}{h_{true}}\right| \times 100\%, \tag{13}$$

$$MAD = \frac{1}{N}\sum|y_{FE} - y_{GP}|, \tag{14}$$

$$MAPE = \frac{1}{N}\sum\frac{|y_{FE} - y_{GP}|}{|y_{FE}|} \times 100\%, \tag{15}$$

$$RMSE = \sqrt{\frac{1}{N}\sum(y_{FE} - y_{GP})^2}, \tag{16}$$

where h_{true} is the CST depth input, $h_{estimate}$ is the depth estimated by the GP-based inverse modeling, y_{FE} is the EM response modeled by FE method, y_{GP} is the EM response evaluated by the forward GP modeling, and N is the total number of observations.

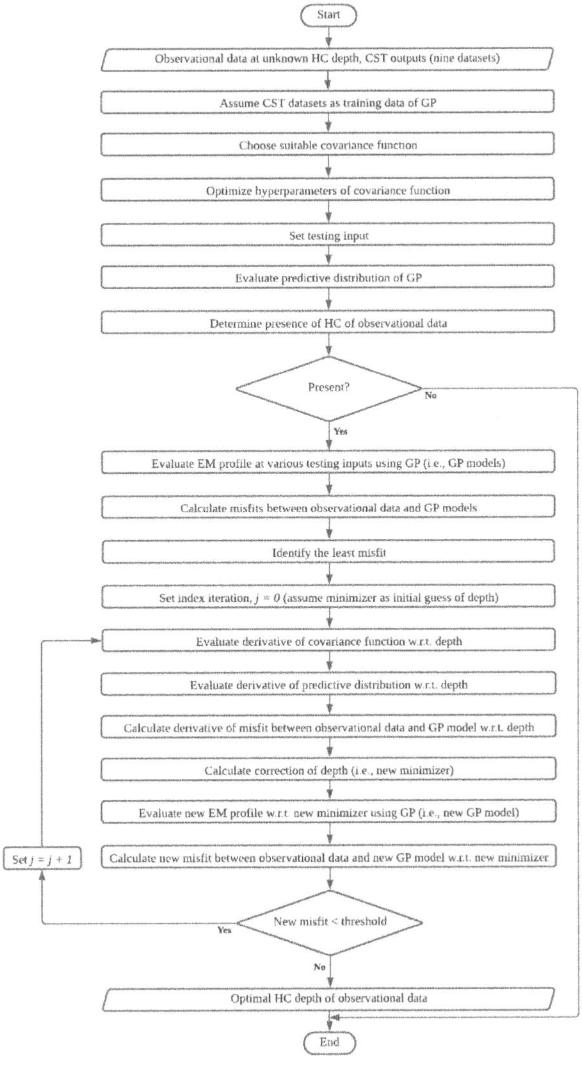

Figure 3. Methodological flow of Gaussian process (GP)-based inverse modeling. Note that all the seabed logging (SBL) models replicated by the computer simulation technology (CST) software are target models (i.e., with hydrocarbon layer). Hence, determining the presence of hydrocarbon is not discussed in this work. HC stands for hydrocarbon and w.r.t. represents with respect to.

4. Results and Discussion

All prior electromagnetic (EM) responses were generated using the computer simulation technology (CST) software. Nine seabed logging (SBL) models were parameterized with the same input properties, except for its depth of hydrocarbon, which were 200 m–1000 m with a gap of 100 m each. It must be noted that these EM responses were evaluated using the finite element (FE) method, which was used exclusively by the CST software for the low-frequency applications. In this work, only magnitude of electric (E)-field was considered as the corresponding output. Figure 4 shows the magnitude versus offset plot for all nine SBL datasets.

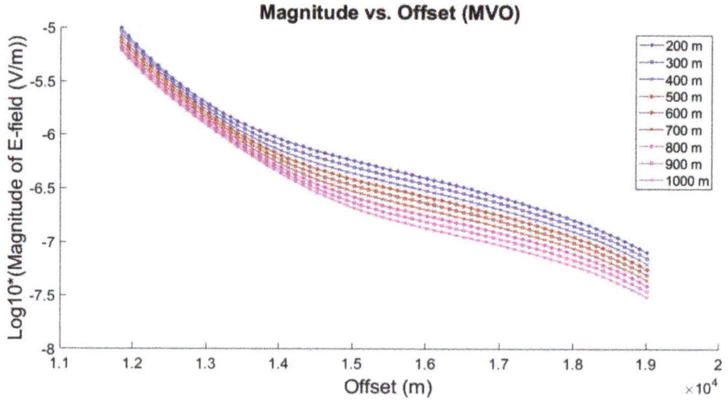

Figure 4. Magnitude of electric (E-)-field versus offset for nine different depths of hydrocarbon from 200 m to 1000 m with an increment of 100 m each.

Based on Figure 4, the x-axis represents the offset distances from approximately 11,840.80 m to 19,004.97 m and the y-axis represents the magnitude of E-field with a logarithmic scale. This figure reveals that the CST was capable of simulating the SBL application in the deep marine environment. The generated EM responses behave similarly to the real behavior of EM signals in media with various conductivities. The EM signals travel through a high-resistive body with less attenuation. Thus, a model with higher hydrocarbon depth generates a much smaller magnitude of E-field due to the factor of far EM signal propagation. Next, these EM responses were fully utilized by the GP regression in order to provide more EM profiles at "untried" depths. The contour plot of the two-dimensional (2D) GP model is depicted in Figure 5.

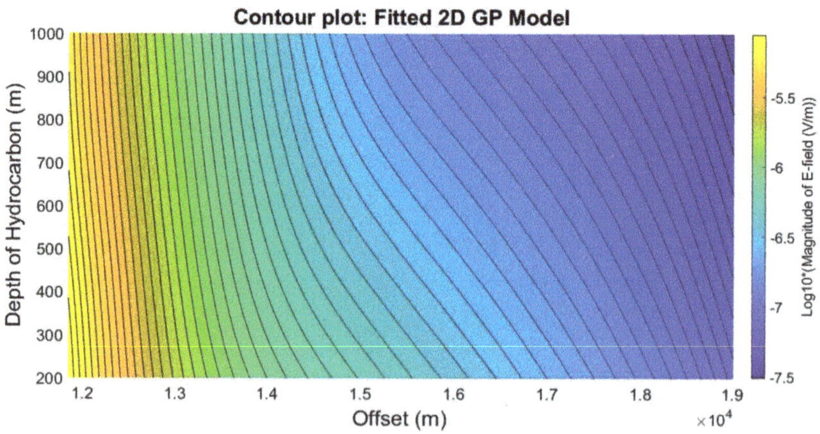

Figure 5. Two-dimensional (2D) forward Gaussian process (GP) model with 41 different electromagnetic (EM) datasets with depths from 200 m to 1000 m with an increment of 20 m.

Based on Figure 5, the x-axis is the offset distance and the y-axis is the 41 depths of hydrocarbon from 200 m to 1000 m, while the magnitudes of the E-field are represented in color codes. This forward GP model is a combination of EM profiles at "tried" and "untried" hydrocarbon depths. The purpose of this GP modeling was to evaluate as many EM responses as possible at various depths of hydrocarbon for the inversion scheme, for which generating all these EM responses using the CST software could take high computational time. Simulating and generating all these 41 datasets using the CST could require around 3 h and 30 min, whereas the GP modeling only took 45 min for the prior knowledge acquisition and less than 1 min for developing the forward GP model and evaluating 41 profiles.

In this work, three random observational datasets were generated separately using the CST software where the depths of hydrocarbon were assumed to be "unknown", i.e., 350, 650, and 950 m. Here, the goal was to find the best estimates of hydrocarbon depths of the observational datasets with the smallest MSE values. In other words, a small MSE value indicates that the EM profile evaluated by the forward GP model gives the most likely values for the observational dataset, thereby giving the best depth information. As discussed before, the step size of the optimization was determined by the learning rate. For this work, three different learning rates were used for every inversion scheme. The first scheme (1st observational dataset) used 0.051 as its learning rate, whereas the second (2nd observational dataset) and third (3rd observational dataset) schemes used 0.087 and 0.081, respectively. Details of how this work determined the suitable learning rates are provided in Appendix C. Plots of MSE versus learning rate and iteration number versus learning rate for every observational dataset are depicted in Figures 2–6 and A1. After identifying the learning rate, optimization using the gradient descent was executed. Thus, three MSE versus hydrocarbon depth plots are plotted as shown in Figures 6–8.

By referring to Figures 6–8, the x-axis is the depth of hydrocarbon and the y-axis is the misfit, which is the MSE. The blue points are the 41 MSE values between the GP model from depths 200 m to 1000 m and the observational data. The red points indicate the optimized MSEs with respect to the updated minimizers. Every optimization must satisfy the convergence condition. Thus, in order to monitor the convergence status, this work illustrates MSE versus number of iterations plots for every observational dataset, which are depicted in Figures 9–11. In these figures, the x-axis and y-axis represent the number of iterations and MSE, respectively. Based on these figures, even though the first inversion scheme needed a higher number of iterations than the second and third ones, all the iterations were smaller than the defined maximum number, which was 100. Moreover, all the MSEs converged where the optimizations stopped at the lowest MSE values. To make it easy to interpret, the main results are presented in Table 2.

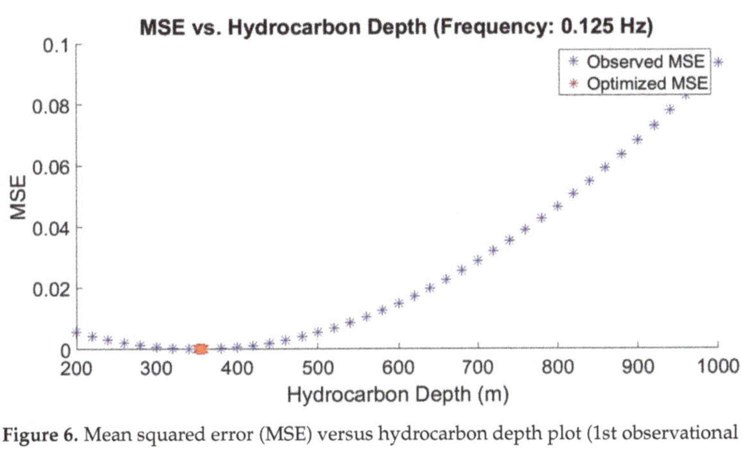

Figure 6. Mean squared error (MSE) versus hydrocarbon depth plot (1st observational dataset).

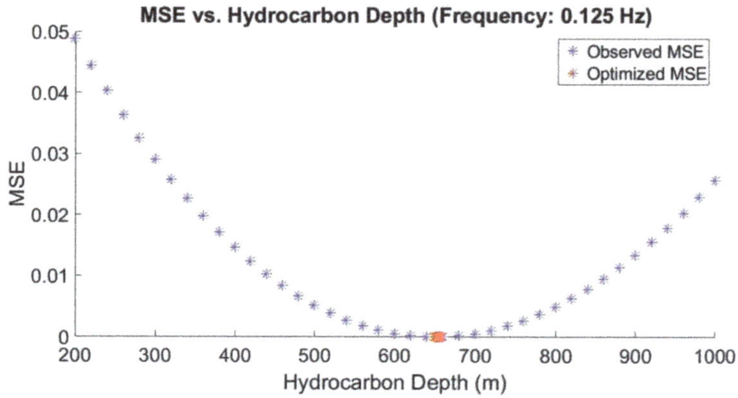

Figure 7. Mean squared error (MSE) versus hydrocarbon depth plot (2nd observational dataset).

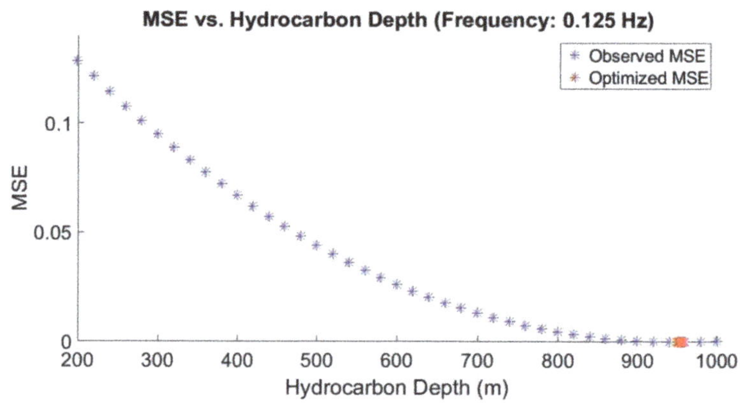

Figure 8. Mean squared error (MSE) versus hydrocarbon depth plot (3rd observational dataset).

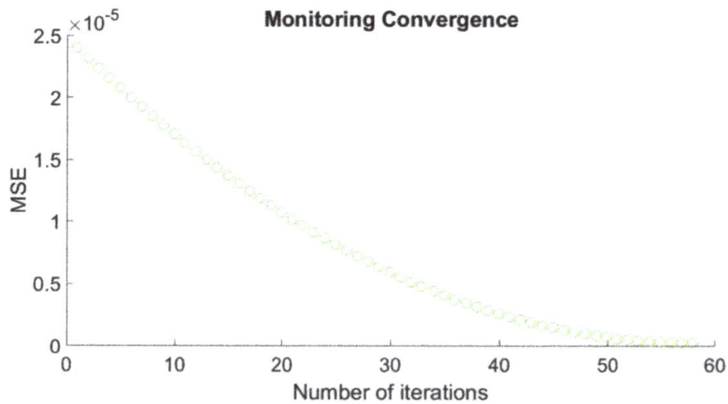

Figure 9. Convergence test (1st observational dataset).

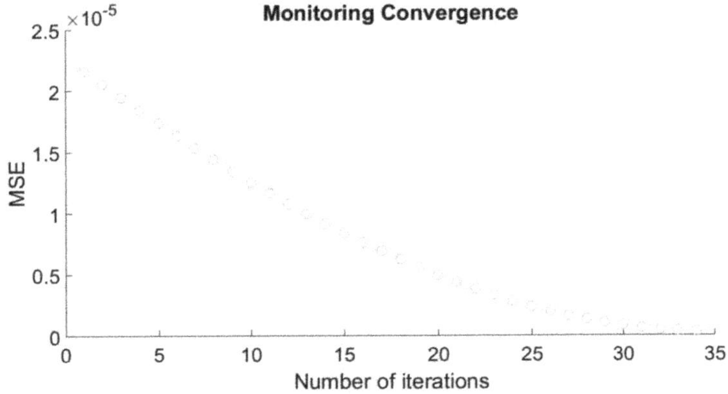

Figure 10. Convergence test (2nd observational dataset).

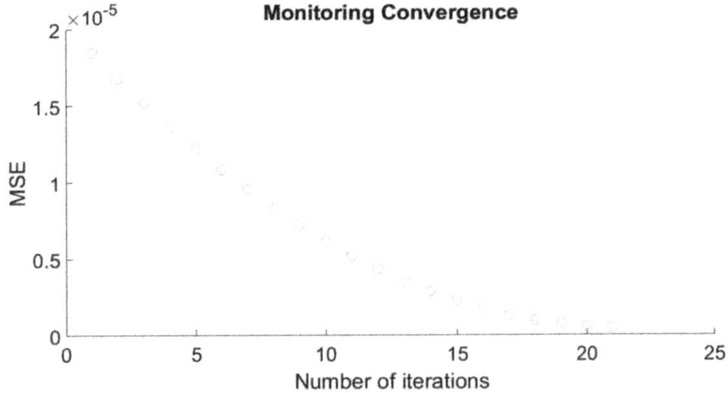

Figure 11. Convergence test (3rd observational dataset).

Table 2. The main Gaussian process (GP)-based inverse modeling results (all three observational datasets).

Dataset	Index	Iteration Number	Estimate (m)	MSE	RPC (%)
1	h_{58}	58	349.92	0.00000033	0.04977145
2	h_{34}	34	650.20	0.00000039	0.04978711
3	h_{21}	21	950.22	0.00000051	0.04991896

Based on Table 2, the first observational dataset required 58 iterations to yield the best estimate of hydrocarbon depth, which was 349.92 m. The second observational dataset needed 34 iterations to get its best estimate, which was 650.20 m. In contrast, only 21 iterations were needed by the third observational dataset to produce the estimate of 950.22 m. Moreover, only 114.047 s, 73.760 s, and 60.556 s were taken by the GP-based inversion methodology to reach the estimates of 349.92 m, 650.20 m, and 950.22 m, respectively. From the table, the calculated MSE values were also very small. This means that the deviations between its true EM values and the predicted EM profiles were very low. All the final RPCs were also lesser than 0.05 as compared to what was defined in the methodology. In addition, in the context of this work, the reliability of the GP regression in evaluating the EM profiles was investigated as well by considering data modeled by the FE method as the true EM values. Thus, MADs, MAPEs, and RMSEs between the EM responses modeled by FE and EM evaluated by the GP (at depths 200–1000 m with a gap of 100 m each) were calculated. The results are presented in Table 3. Furthermore, to find out the accuracy and performance of the GP-based inverse modeling in predicting the hydrocarbon depth, percentage errors between the true depths (i.e., computer input) and the estimates were calculated as well. The percentage errors are indicated in Table 4.

Table 3. Validation of the Gaussian process (GP) model.

Hydrocarbon Depth (m)	MAD	MAPE (%)	RMSE
200	1.39×10^{-9}	0.0849	2.63×10^{-9}
300	1.84×10^{-9}	0.1027	4.25×10^{-9}
400	1.48×10^{-9}	0.1035	3.26×10^{-9}
500	1.64×10^{-9}	0.1144	3.49×10^{-9}
600	1.68×10^{-9}	0.1168	4.28×10^{-9}
700	1.21×10^{-9}	0.1270	2.33×10^{-9}
800	1.37×10^{-9}	0.1204	3.57×10^{-9}
900	1.85×10^{-9}	0.1346	5.79×10^{-9}
1000	9.64×10^{-9}	0.1212	1.81×10^{-9}
Average	1.49×10^{-9}	0.1140	3.49×10^{-9}

Table 4. Percentage errors of hydrocarbon depth prediction.

True Depth (m)	Estimate (m)	Percentage Error (%)
350.0000	349.9172	0.0237
650.0000	650.2015	0.0310
950.0000	950.2225	0.0234

Based on Table 3, all the MADs, MAPEs, and RMSEs for depths 200 m to 1000 m with 100 m gap each were very small, which were 1.49×10^{-9} (on average), 0.114% (on average), and 3.49×10^{-9} (on average), respectively. As seen in Table 4, the percentage errors were very low as well, which was 0.0260% on average. All the results revealed that GP regression was capable of evaluating the EM profiles at various depths of hydrocarbon accurately and with low time consumption. The main goal was also achieved in that the GP-based inversion methodology can predict the depth of hydrocarbon in the SBL application efficiently.

5. Conclusions

This work proposed Gaussian process (GP)-based inverse modeling to predict the depth of hydrocarbon in the seabed logging (SBL) application. The capability of computational modeling and inversion of the SBL data with good accuracy and low computational effort is one of the main concerns in electromagnetic (EM) problems. Thus, this work used machine learning-based approaches, namely GP, as the forward modeling of EM profiles and the gradient descent (GD) method as the optimizer for finding the input parameter of the hydrocarbon (e.g., depth) in the SBL data processing. This work utilized three datasets generated through the computer simulation technology (CST) software with random "unknown" hydrocarbon depths as the observational responses. Based on the results, GP regression can provide accurate EM profiles at various depths of hydrocarbon with low computational time (45 min for the prior data acquisition) and less than 2 min for the GP-based inversion methodology. We believed that if the forward modeling performed by the GP could be done rapidly for the repeated use, the inverse modeling (by the GD) would not significantly affect the total computational time of the inversion methodology. Next, the resulting mean absolute deviations (MADs), mean absolute percentage errors (MAPEs), and root mean squared errors (RMSEs) between the GP model and the data modeled by finite element (FE) method were very small. Furthermore, the GP-based inverse modeling could also predict the depth of hydrocarbon for each observation dataset accurately. All the calculated percentage errors between the estimates (by the GP-based inversion methodology) and the true depth values (computer input) were very small as well. Therefore, it can be concluded that the GP-based inverse modeling was successfully developed to predict the depth of hydrocarbon in the SBL. This attempt could help marine EM data processing, especially in order to de-risk hydrocarbon exploration in the deep marine environment.

Author Contributions: Conceptualization, H.D. and S.C.D.; methodology, M.N.M.A. and S.C.D.; software, M.N.M.A. and H.D.; validation, S.C.D. and K.A.M.N.; writing—original draft preparation, M.N.M.A. and K.A.M.N.; writing—review and editing, M.N.M.A. and K.A.M.N.; supervision, H.D., K.A.M.N., and S.C.D.; funding acquisition, H.D. and S.C.D. All authors have read and agreed to the published version of the manuscript.

Funding: This work was funded by Yayasan Universiti Teknologi PETRONAS–Fundamental Research Grant (YUTP–FRG) with cost center of 015LC0-055.

Institutional Review Board Statement: Not applicable.

Informed Consent Statement: Not applicable.

Data Availability Statement: The data presented in this study are openly available in Mendeley Data at 10.17632/bvwfy54j2d.1. The data can be found here [https://data.mendeley.com/datasets/bvwfy54j2d/1].

Acknowledgments: We would like to express our deepest appreciation to Universiti Teknologi PETRONAS (UTP) for the tremendous support in completing this work.

Conflicts of Interest: The authors declare no conflict of interest.

Appendix A. Data Acquisition Using CST Software

Typical synthetic seabed logging (SBL) models were replicated using the computer simulation technology (CST) software. The input parameters can be seen in Table 1. The first layer was air, followed by seawater and sediment. Based on Figure A1, the isotropic hydrocarbon layer was positioned in the sediment layer at 200 m from the seafloor (for demonstration purposes, only one model is shown here). The source was placed at the center of the model and elevated at 30 m from the seafloor (coordinate: x = 10,000 m, y = 10,000 m, z = 1270 m), and inline receivers were located on the seafloor (i.e., static source–receiver separation distances). All the generated electromagnetic (EM) responses were symmetrical from the center to the left and right boundaries.

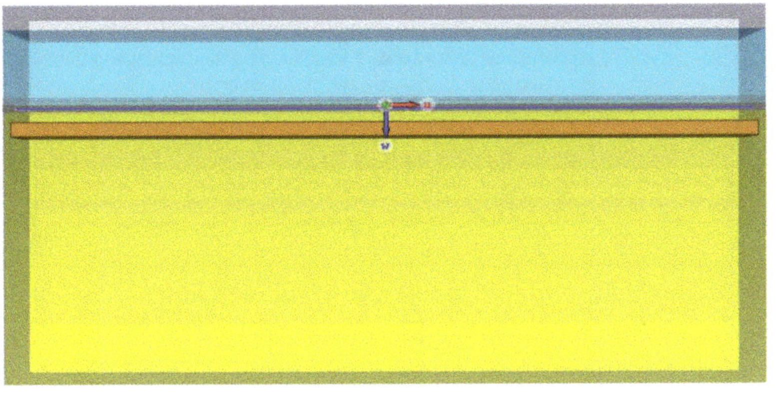

Figure A1. Setup of the seabed logging (SBL) model in the computer simulation technology (CST) software. The depth of hydrocarbon is 200 m from the seafloor.

Appendix B. Derivation of First-Partial Derivative

Equation (A1) shows the derivation of the first-partial derivative of the mean squared error (MSE) with respect to h_j.

$$
\begin{aligned}
L(h_j) &= \tfrac{1}{N}\sum_{i=1}^{N}\left[Y_{fit}(s_i,h_j) - Y_{obs}(s_i)\right]^2 \\
\tfrac{\partial}{\partial h_j}L(h_j) &= \tfrac{\partial}{\partial h_j}\left[\tfrac{1}{N}\sum_{i=1}^{N}\left[Y_{fit}(s_i,h_j) - Y_{obs}(s_i)\right]^2\right] \\
\tfrac{\partial}{\partial h_j}L(h_j) &= \tfrac{1}{N}\cdot\tfrac{\partial}{\partial h_j}\left[\sum_{i=1}^{N}\left[Y_{fit}(s_i,h_j) - Y_{obs}(s_i)\right]^2\right] \\
\tfrac{\partial}{\partial h_j}L(h_j) &= \tfrac{2}{N}\cdot\sum_{i=1}^{N}\left[Y_{fit}(s_i,h_j) - Y_{obs}(s_i)\right]\cdot\tfrac{\partial}{\partial h_j}\left[Y_{fit}(s_i,h_j) - Y_{obs}(s_i)\right] \\
\tfrac{\partial}{\partial h_j}L(h_j) &= \tfrac{2}{N}\cdot\sum_{i=1}^{N}\left[Y_{fit}(s_i,h_j) - Y_{obs}(s_i)\right]\cdot\tfrac{\partial}{\partial h_j}Y_{fit}(s_i,h_j) \\
L'(h_j) &= \tfrac{2}{N}\left(\sum_{i=1}^{N}\left[Y_{fit}(s_i,h_j) - Y_{obs}(s_i)\right]\right)\cdot Y'_{fit}(s_i,h_j)
\end{aligned} \tag{A1}
$$

Equation (A2) shows the derivation of the derivative of the squared exponential (SE) covariance function with respect to the testing point, x_*.

$$
\begin{aligned}
k(x,x_*) &= \sigma_f^2\left[exp\left(-\tfrac{|x-x_*|^2}{2\ell^2}\right)\right] \\
\tfrac{\partial}{\partial x_*}k(x,x_*) &= \tfrac{\partial}{\partial x_*}\left[\sigma_f^2\left[exp\left(-\tfrac{|x-x_*|^2}{2\ell^2}\right)\right]\right] \\
\tfrac{\partial}{\partial x_*}k(x,x_*) &= \sigma_f^2\cdot\tfrac{\partial}{\partial x_*}\left[exp\left(-\tfrac{|x-x_*|^2}{2\ell^2}\right)\right] \\
\tfrac{\partial}{\partial x_*}k(x,x_*) &= \sigma_f^2\cdot exp\left(-\tfrac{|x-x_*|^2}{2\ell^2}\right)\cdot\tfrac{\partial}{\partial x_*}\left[-\tfrac{|x-x_*|^2}{2\ell^2}\right] \\
\tfrac{\partial}{\partial x_*}k(x,x_*) &= \sigma_f^2\cdot exp\left(-\tfrac{|x-x_*|^2}{2\ell^2}\right)\cdot\left(\tfrac{-1}{2\ell^2}\right)\cdot\tfrac{\partial}{\partial x_*}|x-x_*|^2 \\
\tfrac{\partial}{\partial x_*}k(x,x_*) &= \sigma_f^2\cdot exp\left(-\tfrac{|x-x_*|^2}{2\ell^2}\right)\cdot\left(\tfrac{-1}{2\ell^2}\right)\cdot -2|x-x_*| \\
k'(x,x_*) &= \sigma_f^2\cdot\left(\tfrac{|x-x_*|}{\ell^2}\right)\cdot exp\left(-\tfrac{|x-x_*|^2}{2\ell^2}\right)
\end{aligned} \tag{A2}
$$

Appendix C. Identifying Learning Rate (α) of Gradient Descent

Geometric sequences with a common ratio of 10 were tested first in order to indicate which learning rates would give the significant optimization. From here, other values of learning rate were examined as well to find the smallest mean squared error (MSE) that could be optimized. Based on Figure 2, Figure 4, and Figure 6 3.30×10^{-7}, 3.90×10^{-7}, and 5.10×10^{-7} were the smallest optimized MSEs for all observational data. Thus, based on the MSEs, the size of the learning rates was narrowed down. Next, the number of iterations was then considered as plotted in Figure 3, Figure 5, and Figure 7. From here, learning rates that gave the smallest MSE with the lowest iteration number and followed the relative percentage change (RPC) condition were chosen for the Gaussian process (GP)-based inverse modeling.

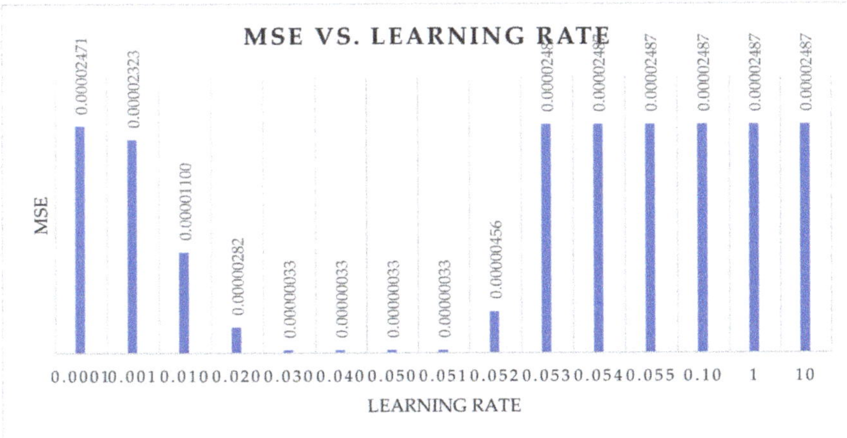

Figure 2. Bar plot of mean squared error (MSE) versus learning rate (1st observational dataset).

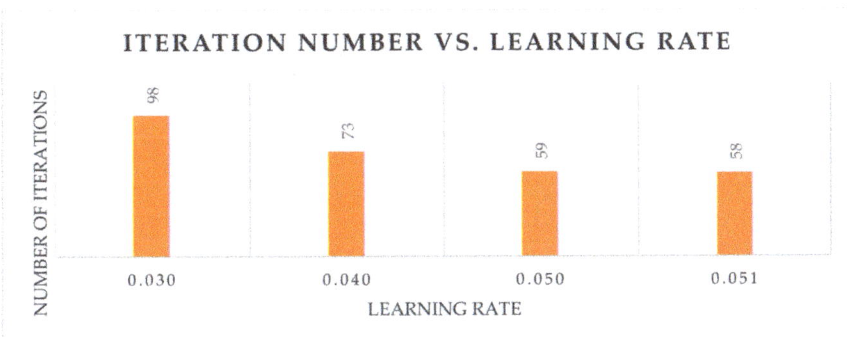

Figure 3. Bar plot of iteration number versus learning rate for the smallest mean squared error (MSE) (1st observational dataset).

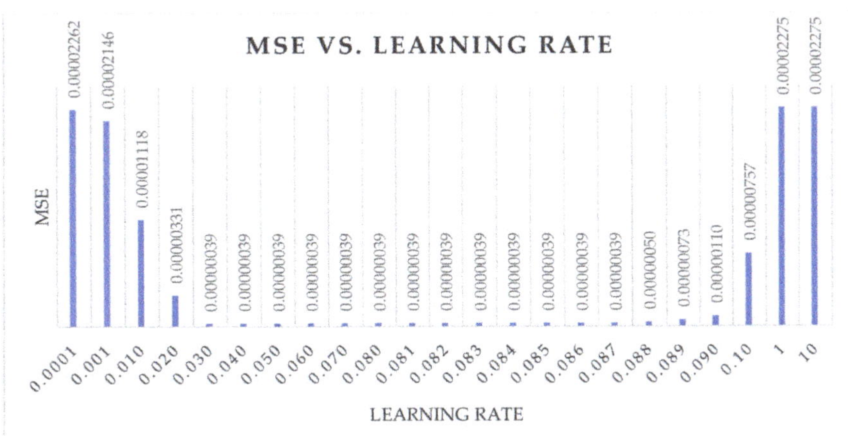

Figure 4. Bar plot of mean squared error (MSE) versus learning rate (2nd observational dataset).

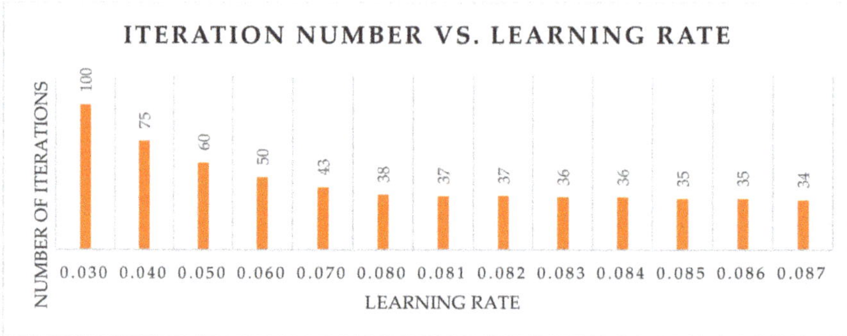

Figure 5. Bar plot of iteration number versus learning rate for the smallest mean squared error (MSE) (2nd observational dataset).

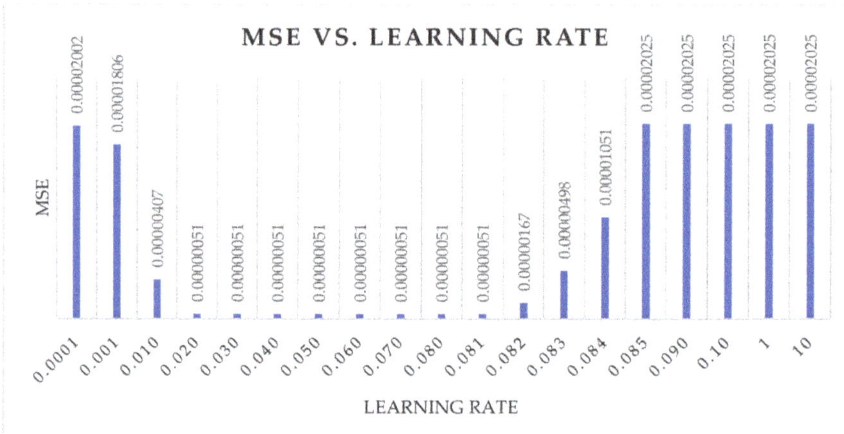

Figure 6. Bar plot of mean squared error (MSE) versus learning rate (3rd observational dataset).

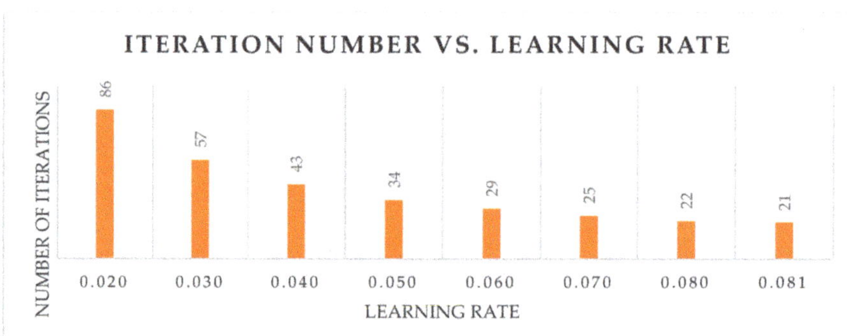

Figure 7. Bar plot of iteration number versus learning rate for the smallest mean squared error (MSE) (3rd observational dataset).

References

1. Eidesmo, T.; Ellingsrud, S.; MacGregor, L.M.; Constable, S.; Sinha, M.C.; Johansen, S.; Kong, F.N.; Westerdahl, H. Sea Bed Logging (SBL), a new method for remote and direct identification of hydrocarbon filled layers in deep water areas. *First Break* **2002**, *20*, 144–152.

2. Ellingsrud, S.; Eidesmo, T.; Johansen, S. Remote sensing of hydrocarbon layers by seabed logging: Results from a cruise offshore Angola. *Lead. Edge* **2002**, *21*, 972–982. [CrossRef]
3. Hesthammer, J.; Boulaenko, M. The offshore EM challenges. *First Break* **2005**, *23*, 59–66. [CrossRef]
4. Carazzone, J.J.; Burtz, O.M.; Green, K.E.; Pavlov, D.A. Three dimensional imaging of marine controlled source EM data. *SEG Expand. Abstr.* **2005**, *24*, 575.
5. Srnka, L.J.; Carazzone, J.J.; Ephron, M.S.; Eriksen, E.A. Remote reservoir resistivity mapping. *Lead. Edge* **2006**, *25*, 972–975. [CrossRef]
6. Constable, S.; Srnka, L.J. An introduction to marine controlled-source electromagnetic methods for hydrocarbon exploration. *Geophysics* **2007**, *72*, WA3–WA12. [CrossRef]
7. Um, E.S.; Alumbaugh, D.L. On the physics of the marine controlled-source electromagnetic method. *Geophysics* **2007**, *72*, W13–WA26. [CrossRef]
8. Andréis, D.; MacGregor, L.M. Controlled-source electromagnetic sounding in shallow water: Principles and applications. *Geophysics* **2008**, *73*, F21–F32. [CrossRef]
9. Zhdanov, M.S. Electromagnetic geophysics: Notes from the past and the road ahead. *Geophysics* **2010**, *75*, A49–A66. [CrossRef]
10. Mohd Aris, M.N.; Daud, H.; Dass, S.C.; Mohd Noh, K.A. Gaussian Process Methodology for Multi-Frequency Marine Controlled-Source Electromagnetic Profile Estimation in Isotropic Medium. *Processes* **2019**, *7*, 661. [CrossRef]
11. Puzyrev, V.; Koldan, J.; de la Puente, J.; Houzeaux, G.; Vázquez, M.; Cela, J.M. A parallel finite-element method for three-dimensional controlled-source electromagnetic forward modelling. *Geophys. J. Int.* **2013**, *193*, 678–693. [CrossRef]
12. Li, Y.; Key, K. 2D marine controlled-source electromagnetic modeling: Part 1—An adaptive finite-element algorithm. *Geophysics* **2007**, *75*, WA51–WA62. [CrossRef]
13. Bakr, S.A.; Pardo, D.; Mannseth, T. Domain decomposition Fourier finite element method for the simulation of 3D marine CSEM measurements. *J. Comput. Phys.* **2013**, *255*, 456–470. [CrossRef]
14. Daeung, Y.; Micheal, S.Z.; Hongzhu, C.; Alexander, G. A hybrid finite difference and integral equation method for modeling and inversion of marine CSEM data. In Proceedings of the SEG 2015 New Orleans Annual Meeting, New Orleans, LA, USA, 18–23 October 2015; pp. 823–827.
15. Harari, O.; Steinberg, D.M. Optimal designs for Gaussian process models | via spectral decomposition. *J. Stat. Plan. Inference* **2014**, *154*, 87–101. [CrossRef]
16. Chan, L.L.T.; Liu, Y.; Chen, J. Nonlinear system identification with selective recursive Gaussian process models. *Ind. Eng. Chem. Res.* **2013**, *52*, 18276–18286. [CrossRef]
17. Fang, D.; Zhang, X.; Yu, Q.; Jin, T.C.; Tian, L. A novel method for carbon dioxide emission forecasting based on improved Gaussian processes regression. *J. Clean. Prod.* **2018**, *173*, 143–150. [CrossRef]
18. Kong, D.; Chen, Y.; Li, N. Gaussian process regression for tool wear prediction. *Mech. Syst. Signal. Process.* **2018**, *104*, 556–574. [CrossRef]
19. Petelin, D.; Grancharova, A.; Kochigan, J. Evolving Gaussian Process models for prediction of Ozone concentration in the air. *Simul. Model. Pract. Theory* **2013**, *33*, 68–80. [CrossRef]
20. Pal, M.; Deswal, S. Modelling pile capacity using Gaussian process regression. *Comput. Geotech.* **2010**, *37*, 942–947. [CrossRef]
21. Yin, F.; Zhao, Y.; Gunnarsson, F.; Gustafsson, F. Received-Signal-Strength Threshold Optimization Using Gaussian Processes. *IEEE Trans. Signal. Process.* **2017**, *65*, 2164–2177. [CrossRef]
22. Liu, D.; Pang, J.; Zhou, J.; Peng, Y.; Pecht, M. Prognostics for state of health estimation of lithium-ion batteries based on combination Gaussian process functional regression. *Microelectron. Reliab.* **2013**, *53*, 832–839. [CrossRef]
23. Wang, H.; Gao, X.; Zhang, K.; Li, J. Fast single image super-resolution using sparse Gaussian process regression. *Signal. Process.* **2017**, *134*, 52–62. [CrossRef]
24. Yu, Z.; Shi, X.; Zhou, J.; Huang, R.; Gou, Y. Advanced Prediction of Roadway Broken Rock Zone Based on a Novel Hybrid Soft Computing Model Using Gaussian Process and Particle Swarm Optimization. *Appl. Sci.* **2020**, *10*, 6031. [CrossRef]
25. Asante-Okyere, S.; Shen, C.; Ziggah, Y.Y.; Rulegeya, M.M.; Zhu, X. Investigating the predictive performance of Gaussian process regression in evaluating reservoir porosity and permeability. *Energies* **2018**, *11*, 3261. [CrossRef]
26. Mohd Aris, M.N.; Daud, H.; Mohd Noh, K.A.; Dass, S.C.; Mukhtar, S.M. Modelling 1-D synthetic seabed logging data for thin hydrocarbon detection: An application of Gaussian process. In Proceedings of the National Symposium of Mathematical Sciences, Bangi, Malaysia, 26–27 November 2019.
27. Li, J.; He, M.; Cui, G.; Wang, X.; Wang, W.; Wang, J. A Novel Method of Seismic Signal Detection Using Waveform Features. *Appl. Sci.* **2020**, *10*, 2919. [CrossRef]
28. Mohd Aris, M.N.; Daud, H.; Mohd Noh, K.A.; Dass, S.C. Model Calibration of Stochastic Process and Computer Experiment for MVO Analysis of Multi-Low-Frequency Electromagnetic Data. *Processes* **2020**, *8*, 605. [CrossRef]
29. Goodfellow, I.; Bengio, Y.; Courville, A. *Deep Learning*; MIT Press: Cambridge, MA, USA, 2016.
30. Yoo, H.; Han, S.; Chung, K. A Frequency Pattern Mining Model Based on Deep Neural Network for Real-Time Classification of Heart Conditions. *Healthcare* **2020**, *8*, 234. [CrossRef] [PubMed]
31. Liu, S.; Takaki, Y. Optimization of Phase-Only Computer-Generated Holograms Based on the Gradient Descent Method. *Appl. Sci.* **2020**, *10*, 4283. [CrossRef]

32. Vlasenko, A.; Vlasenko, N.; Vynokurova, O.; Bodyanskiy, Y.; Peleshko, D. A Novel Ensemble Neuro-Fuzzy Model for Financial Time Series Forecasting. *Data* **2019**, *4*, 126. [CrossRef]
33. Gomez-Pulido, J.A.; Duran-Dominguez, A.; Pajuelo-Holguera, F. Optimizing Latent Factors and Collaborative Filtering for Students' Performance Prediction. *Appl. Sci.* **2020**, *10*, 5601. [CrossRef]
34. Annala, L.; Ayramo, S.; Polonen, I. Comparison of Machine Learning Methods in Stochastic Skin Optical Model Inversion. *Appl. Sci.* **2020**, *10*, 7097. [CrossRef]
35. Electromagnetic Simulation Solvers (CST Studio Suite). Available online: https://www.3ds.com/products-services/simulia/products/cst-studio-suite/solvers/ (accessed on 2 November 2020).
36. Documentation for GPML Matlab Code Version 4.2. Available online: http://www.gaussianprocess.org/gpml/code/matlab/doc/ (accessed on 9 September 2020).
37. Rasmussen, C.E.; Williams, C.K.I. *Gaussian Processes for Machine Learning*; MIT Press: Cambridge, MA, USA, 2006.

applied sciences

MDPI

Review

Advanced Development of Sensors' Roles in Maritime-Based Industry and Research: From Field Monitoring to High-Risk Phenomenon Measurement

Aditya Rio Prabowo [1,*], Tuswan Tuswan [2] and Ridwan Ridwan [1,3]

1 Department of Mechanical Engineering, Universitas Sebelas Maret, Surakarta 57126, Indonesia; ridwan97@student.uns.ac.id
2 Department of Naval Architecture, Institut Teknologi Sepuluh Nopember, Surabaya 60111, Indonesia; tuswan.18041@mhs.its.ac.id
3 Department of Mechanical Engineering, Universitas Muhammadiyah Ponorogo, Ponorogo 63471, Indonesia
* Correspondence: aditya@ft.uns.ac.id

Abstract: The development of human civilization over the last decade has reached a landmark as Industry 4.0 has been widely introduced. Several aspects of industry and manufacturing activities are changing due to the Internet of Things (IoT), location detection technologies, and advanced human–machine interfaces. To enact industrial affairs under those specifications, a sensor is required to transform physical events into numerical information. The use of sensors in marine applications also appears in research and studies, in which the sensor is used for both monitoring the phenomena of a designated subject and data acquisition. Achievements in quantifying complex phenomena in critical maritime designs are fascinating subjects to discuss regarding their development and current states, which may be reliable references for further research on developing sensors and related measurement analysis tools in marine, shipbuilding, and shipping fields. This comprehensive review covers several discussion topics, including the origins and development of sensor technology, applied sensor engineering in logistic and shipping activities, the hydrodynamic characterization of designed hulls, the monitoring of advanced machinery performance, Arctic-based field observations, the detection of vibration-based damage to offshore structures, corrosion control and monitoring, and the measurement of explosions on critical maritime infrastructures.

Keywords: sensor technology; logistic and shipping; hull hydrodynamic; machinery performance; Arctic field; vibration and corrosion; explosion loading

Citation: Prabowo, A.R.; Tuswan, T.; Ridwan, R. Advanced Development of Sensors' Roles in Maritime-Based Industry and Research: From Field Monitoring to High-Risk Phenomenon Measurement. *Appl. Sci.* **2021**, *11*, 3954. https://doi.org/10.3390/app11093954

Academic Editor: Dimitrios-Nikolaos Pagonis

Received: 4 April 2021
Accepted: 25 April 2021
Published: 27 April 2021

Publisher's Note: MDPI stays neutral with regard to jurisdictional claims in published maps and institutional affiliations.

1. Introduction

The Fourth Industrial Revolution (or Industry 4.0 [1–9]) is the ongoing automation of traditional manufacturing and industrial practices using modern smart technology. Large-scale machine-to-machine communication (M2M) [10–12] and the Internet of Things (IoT) [13–15] are being integrated for increased automation, improved communication and self-monitoring, and the production of smart machines that can analyze and diagnose issues without the need for human intervention. This phase is the latest development in a series of industrial revolutions, as summarized here and in Figure 1:

- The First Industrial Revolution [16–20] began in the 18th century through steam power and production mechanization. Compared to thread production on simple spinning wheels, the mechanized version achieved eight times the volume in the same amount of time. Its use for industrial purposes was the most remarkable breakthrough for advancing humanity. Instead of weaving looms powered by muscle, steam engines could be used for power.
- The Second Industrial Revolution [21–25], also known as the Technological Revolution, was between 1871 and 1914. It resulted from the installation of extensive railroads and telegraph networks, which allowed for faster transfer of people, ideas, and electricity.

- The Third Industrial Revolution began with the first computer era [26–30]. These early computers were often very simple, unwieldy, and incredibly large relative to the computing power they were able to provide, but they laid the groundwork for the world today, which one is hard-pressed to imagine without computer technology. Around 1970, the Third Industrial Revolution involved electronics and IT (information technology) in furthering automation in production. In the early 2000s, computational instruments grew rapidly, which were followed by the development of numerical analysis, e.g., the finite element method (FEM) [31–35] and computational fluid dynamics (CFD) [36–40]. These have been continuously deployed and improved, especially for the calculation of complex phenomena.

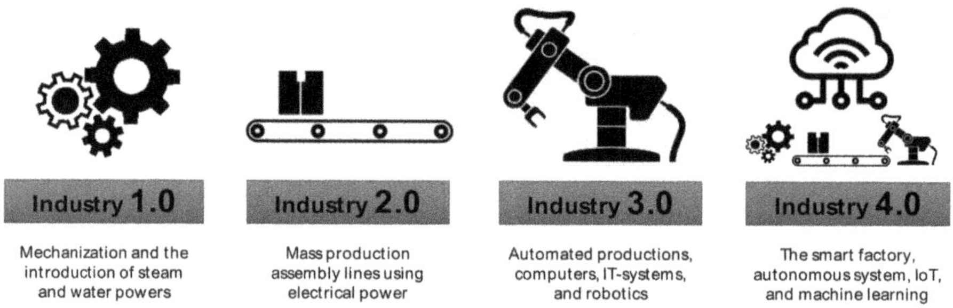

Figure 1. An outline of the industrial revolutions.

The Industrial Revolution era was expected to influence maritime-based industry and research even before it was openly discussed in the World Economic Forum in 2016. In terms of the relationship between industry and research, the maritime-based field is vast and rapidly growing due to the collaboration of these efforts. It is a complex constitution consisting of shipping and expedition, marine offshore condition monitoring, and Arctic–Antarctica exploration. Development is mandatory to maintain the sustainability of these fields as an essential contribution to the top ten global industries [41–45]. To increase the competitiveness of the maritime industry, the speed of digitalization and application of the latest technologies must increase rapidly, which will lead to many technological and regulatory challenges. The main aims of saving costs and efficiency will open up a wide range of opportunities for maritime-based industry and research 4.0. Combined with the other latest technologies, sensors will become one of the most powerful tools in Maritime Industry 4.0. The physical processes of sensors are seamlessly integrated with software and computational processes in cyber-physical systems. These systems typically involve intelligent sensing technologies at the data measurement and monitoring level of the system's architecture. In order to comprehend such vital tasks, sensors are required to provide assistance to operators, technicians, and engineers [46–50]. The characteristics of the sensor are directed to provide real-time, meaningful information to increase productivity, efficiency, and flexibility in various sectors, including commercial and academic research.

An urgent need to apply sensors in both the maritime industry and research activities, as well as a tendency to do so, has been shown since early 2010, and it is important to quantify the frequency with which they are applied to predict the trends and developments of the new decade initiated in 2021. The current development of sensor technology will provide an excellent opportunity to improve the efficiency and safety of marine structures and environments. Real-time monitoring and analysis strategies will be vital to improving commercial automation in shipping, logistics, and marine and offshore activities in Maritime 4.0. IoT devices and machine learning algorithms will contribute a certain level of intelligence for industrial control systems and answer all problems that may arise. By collecting high-quality data with reliable sensor technology, the possibility of extending the life cycle of marine structures can be improved according to the highest standards of operation and maintenance. The current sensor technology and the development of a new generation of sensors and robust networking architectures will be on the cusp of revolutionary changes in environmental monitoring and data collection in Maritime 4.0.

This work aims to present a schematic review of sensor-based technology and its applications that is thoroughly focused on maritime industry and research. Contemporary subjects, including logistics and shipping, hull hydrodynamics, machinery performance, Arctic expedition, vibration and corrosion-based damage, and explosion analysis, are discussed and summarized to provide a charted trend of sensor application as an observational and measurement instrument. Predictions related to the technology's applications in the new decade based on the state and development of the previous decade are presented.

2. Applied Sensor Engineering in Logistic and Shipping Activities

Technological and digital revolutions in the transport sector, including logistic and shipping activities of Maritime 4.0, have redefined the functional roles of shipping and ports. Ships have an important role in industrial activities as means of the transportation of passengers or goods. In 2018, approximately 11 billion tons of cargo were transported using ships [51]. Furthermore, several regulations were introduced in the 20th century to maintain passenger safety in maritime areas and strengthen ships. One of the regulations was the Safety of Life at Sea (SOLAS) convention established in 1914 [52]. Table 1 lists milestones along with regulations in international maritime safety. At present, several developments have been shown to increase the utility of the ship-using industry, including satellites, radars, and sensors. These developments can be used to determine the position of a ship and for maritime monitoring.

Table 1. Milestones in international maritime safety [53].

Year	Initiative or Regulation
1914	Safety of Life at Sea (SOLAS): Ship design and life-saving equipment
1929	First international conference to consider hull subdivision regulations
1948	The International Maritime (Consultative) Organization (IMO) is set up as a United Nations agency
1966	Load Line Convention: Maximum loading and hull strength Rules of the road The International Association of Classification Societies (IACS): Harmonization of classification rules and regulations
1969	Tonnage Convention
1972	International Convention on the International Regulations for Preventing Collisions at Sea (COLREG)
1974	IMO resolution on probabilistic analysis of hull subdivision
1973	Marine Pollution Convention (MARPOL 73)
1978	International Convention on Standards for Training, Certification and Watchkeeping for Seafarers (STCW)
1979	International Convention on Maritime Search and Rescue (SAR)
1988	The Global Maritime Distress and Safety System (GMDSS)

In 2007, Nezlin et al. [54] investigated satellite imagery as an alternative approach for assessing stormwater in the ocean region. The research examined the image quality of four types of satellite data: (1) Sea-viewing Wide Field-of-view Sensor (SeaWiFS) Level 2 normalized water-leaving radiation of 555 nm wavelength; (2) SeaWiFS Level 3 chlorophyll; (3) Advanced Very High-Resolution Radiometer (AVHRR) Level 3 sea surface temperature (SST); and (4) Moderate-Resolution Imaging Spectroradiometer (MODIS) Level 3 SST. One of the crucial conclusions of the research is that the MODIS sensors provided better coverage than SeaWiFS or AVHRR did due to better spectral, spatial, and particularly, temporal resolution. Song et al. [55] researched the detection of ships in an inland river using high-resolution optical satellite imagery. This research's background shows that it is more difficult to detect ships in an inland river due to several challenges compared to ship

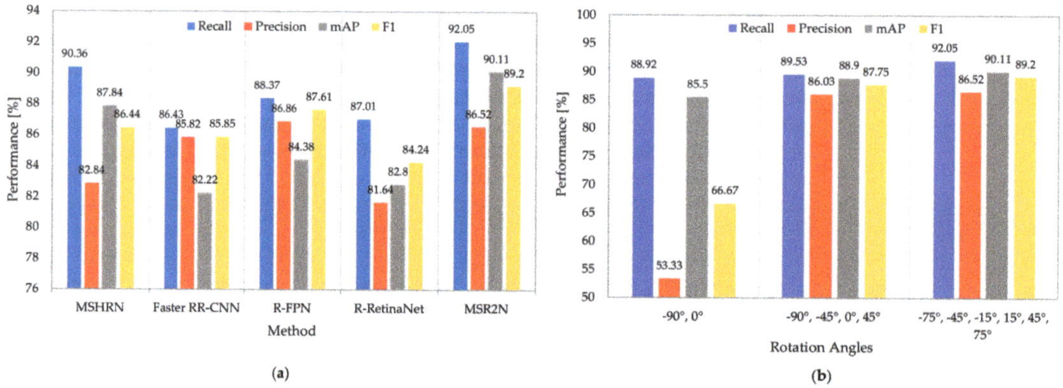

Figure 5. The experimental results of the multi-stage rotational region-based network (MSR2N): (**a**) results of different methods on the SAR ship detection dataset (SSDD); (**b**) results of rotation angles (MSR2N) (redrawn based on data in Pan et al. [63]).

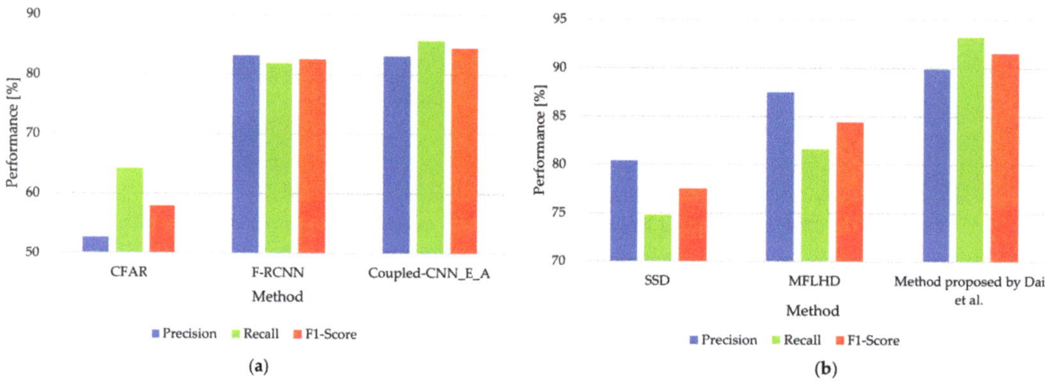

Figure 6. Comparison of the evaluated detection performance among (**a**) CFAR, F-RCNN, and Coupled-CNN E A; (**b**) SSD, MFLHD, and the method proposed by Dai et al. (redrawn based on data in Dai et al. [64] and Gui et al. [65]).

Table 2. Landmark research on sensor engineering applied in logistic and shipping activities.

Milestone	Author(s)	Observational Subject	Tools/Instrument	Important Remarks
2007	Nezlin et al. [54]	Ship-based sampling for evaluation of stormwater in the ocean region	High-resolution satellite imagery	The study shows that Moderate-Resolution Imaging Spectroradiometer (MODIS) sensors provided better coverage than Sea-viewing Wide Field-of-view Sensor (SeaWiFS) or Advanced Very High-Resolution Radiometer (AVHRR) due to greater spectral, spatial, and particularly, temporal resolution (twice a day), thereby significantly improving information about plume dynamics.

Table 2. *Cont.*

Milestone	Author(s)	Observational Subject	Tools/Instrument	Important Remarks
2017	De Maio et al. [57]	Oil spill detection in the sea territory	Multi-polarization synthetic aperture radar (SAR) images	The presence of both simulated and real data in the one-sided generalized likelihood ratio test (GLRT) confirms the practical effectiveness for oil detection.
2018	Kanjir et al. [56]	Vessel detection in the sea territory	Optical satellite imagery	The result shows that vessel monitoring from spaceborne optical images is a thriving research topic and will have excellent operational potential in the future due to a massive number of satellite data, much of it being free and open.
2018	Xie et al. [66]	Inshore ship detection	Synthetic aperture radar (SAR)	The result shows that inshore ships can be detected using SAR images with reasonable accuracy and integrity.
2019	Gui et al. [65]	Ship detection on the surface of the sea	Multilayer fusion light-head detector (MFLHD) Synthetic aperture radar (SAR)	The extensive experiments on the multilayer fusion light-head detector (MFLHD) achieved superior performance in SAR ship detection, in terms of both accuracy and speed.
2019	Song et al. [55]	Ship detection in the inland river	High-resolution optical satellite imagery	The result shows that using a mixture of multi-scale Deformable Part Models (DPMs) and a Histogram of Oriented Gradient (HOG) is effective in ship detection and performs properly in separating ships clustered together and those alongside the pier.
2020	Zou et al. [61]	Ship detection on the surface of the sea	Multiscale Wasserstein auxiliary classifier generative adversarial network (MW-ACGAN) and the Yolo v3 network Synthetic aperture radar (SAR)	The Multiscale Wasserstein Auxiliary Classifier Generative Adversarial Network (MW-ACGAN) can effectively generate realistic multiclass ship images.
2020	Dai et al. [64]	Ship detection on the surface of the sea	Synthetic aperture radar (SAR)	The novel detector based on convolution neural networks for multiscale SAR ship detection showed better performance compared to the method from the SAR ship detection dataset (SSDD).

Recently, research on performance evaluations in marine navigation systems was intensely reviewed by Fukuda et al. [67]. The global navigation satellite system (GNSS) has been widely used for ship navigation, and spoofing poses an important threat to maritime logistics. Fukuda et al. [67] conducted research on the global navigation satellite system (GNSS) by combining it with an inertial measurement unit (IMU) and a Doppler velocity log (DVL). An inertial measurement unit (IMU) can identify three-axis acceleration and three-axis angular velocity. On the other hand, a Doppler velocity log (DVL) can measure the speed of a ship. Based on this research, the method can achieve less than 1 km horizontal error in an hour. A critical research issue for ship traffic safety in narrow waters is the ship speed limit, which is one of the most common causes of maritime accidents. Wang et al. [68] presented a new methodology to quantify the shipping traffic volume in the Yangtze River under current speed limits by using the automatic identification system (AIS) big data. The steps of the proposed method are as follows: (1) cleaning the raw AIS data to filter out unavoidable errors to improve the quality of the data, (2) geocoding the AIS data, (3) calculating the ship traffic aspect, and (4) estimating the shipping speed and traffic volume to analyze the speed limit. This work provides valuable insights to maritime safety authorities regarding adjustment of ship speeds to avoid heavy traffic congestion in narrow waterways and minimize the associated maritime risks. Feng et al. [69] also applied AIS data to propose a time efficiency assessment that evaluates the amount of time each ship spends in the different areas within a port (i.e., berth, anchorage, and fairway) based on the space–time trajectories of ship movements in Shanghai Yangshan Port and Xiamen Port in China. In this study, the space–time path concept from time geography is used to generate space–time trajectories from collected AIS data for vessel traffic services (VTS). The basic idea in this proposed work assumes that a ship is considered a moving object whose location changes over time in a port. The AIS is integrated with a GPS receiver and gyrocompass, and reports specific information, including Maritime Mobile Service Identity, navigation status, turn rate, speed over ground, position coordinate, etc. The AIS trajectory-based time efficiency assessment framework is divided into five different steps: trajectory construction, geographic zone identification, status classification, time efficiency statistics, and time efficiency assessment, as illustrated in Figure 7. This study provides useful information to shipping lines and port authorities for efficient scheduling and logistic support during operation. The challenge for future research is the integration of AIS data with other relevant ship information (e.g., cargo volume and cargo type) and port operation data (e.g., operation schedule and loading/unloading capacity) that can further enhance the proposed framework to conduct more comprehensive efficiency assessments.

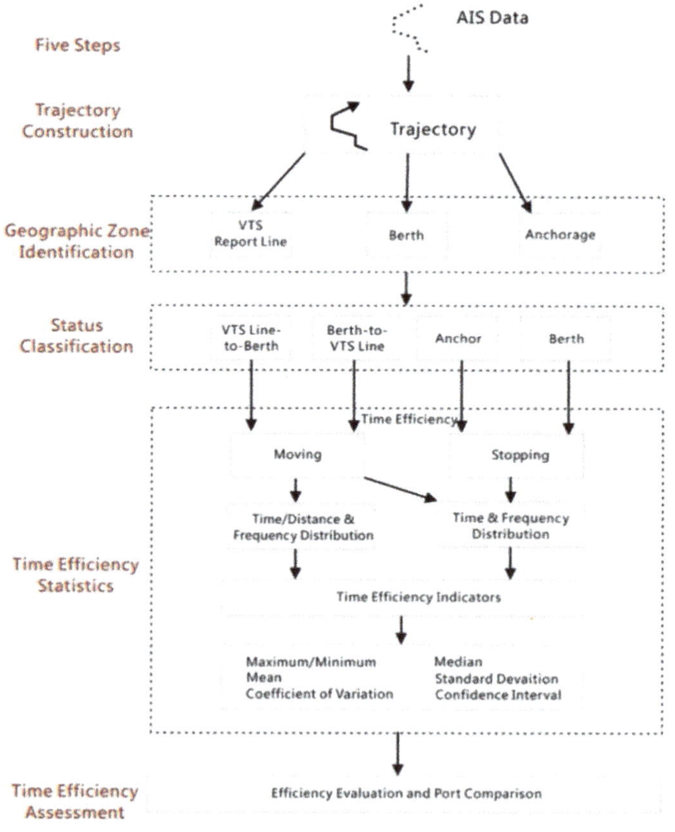

Figure 7. AIS-based time efficiency assessment framework [69].

3. Hydrodynamic Characterization of Designed Hulls

Hydrodynamic characterization is one of the most critical topics in marine structures for understanding the performance of a designed model. Hydrodynamic performance can be divided into resistance and propulsion, seakeeping and ship vibrations, and maneuvering. The measurement of hydrodynamic performance can be classified into three approaches: an empirical/statistical approach, a numerical approach (CFD based), and an experimental approach (scaled model test and full-scale trial). In this section, landmark research on sensor applications for characterizing the hydrodynamic performances of marine structures is comprehensively reviewed. Tzabiras and Kontogiannis [70] conducted lab-based resistance measurement of passenger ship models using different bows to verify the capability of the computational result. Three bulb-shape models were tested in the towing tank, where five resistance-type wave probes were mounted on the beam at five locations. The resistance was measured together with the dynamic sinkage and trim of the free-towed model. In 2011, the effect of heave and pitch motions on resistance and the ice-breaking pattern was investigated by Polach and Ehlers [71]. The towing mode test in an ice-going tanker ice basin was idealized as a spring–mass system, with a load cell as the spring element and a ship model and its mass as the mass element. Several sensors were used to measure the response. A pointer in the carriage was used to measure the broken channel, and a dynamic motion unit (DMU) at the bow measured the heave and pitch accelerations. Moreover, two cameras using the coordinate system were used to measure the cusp width.

In the following year, Lee et al. [72] conducted research on the hydrodynamic design and performance of an underwater hull-cleaning robot. Several sensors and instruments were mounted on the HCR's body to measure the hydrodynamic performance on the towing tank. A resistance dynamometer with a capacity of 3000 N was mounted to measure the drag force. The HCR's horizontal speed was measured using an electromagnetic flowmeter, and a commercial weighing scale was used to measure vertical thrust. The following year, in a collaboration project between IIHR and CNR-INSEAN, Bouscasse et al. [73] proposed a comprehensive experiment to investigate the seakeeping behavior of Delft 372 fast catamaran advancing in head waves. The catamaran model was towed through a kinematic system at the CNR-INSEAN water tank. Each demi-hull of the catamaran was connected by a transversal aluminum beam to ensure free pitch motion and restrain surge, sway, and roll movements. The position and motion of the model were identified using a Krypton optical system with three infrared LEDs. Two transducers (a Kenek probe and a Keyence ultrasound wave probe) fixed to the carriage measured the incident wave system. For wave elevation measurement, a second wave system (set of two probes) was mounted aside the hull at the LCG position. Two HYDRONICS load cells mounted between the gimbal and the model were used to measure the total resistance directly in waves. A data acquisition system with a sample rate of 300 Hz was used to record all of the physical quantities. Jang et al. [74] performed an experimental investigation in a water tunnel to measure frictional resistance reduction. Skin friction reduction using air lubrication techniques generated on the hull's lower surface at various flow rates was observed. In this work, different flow rates were controlled by a flow meter. Images captured the air layer types, and two floating-plate-type friction sensors were used to measure local friction changes.

Fossati et al. [75], under the Lecco Innovation Hub project, reported a full-scale measurement using the 10-m Sailing Yacht Lab obtained during the first sea trials carried out in offshore Colico Marina. A description of the scientific framework, measurement capabilities, and data acquisition procedure is provided. In the data collection system architecture, six load cells located near the possibly highest loads were used to estimate the overall forces and moments transmitted by the sails and rig to the hull. The proposed GPS-aided inertial navigation test system computes the yacht attitude and boat dynamic. To acquire the angular rate, acceleration, and local magnetic field, gyroscopes, accelerometers, and magnetometers were arranged on the three primary axes, respectively. A GEMAC

23554 analog inclination sensor installed beside the inertial navigation system was used to calculate the heel and trim angle. Furthermore, NMEA-GPS provided navigation data (wind speed and direction, boat speed, and depth). An ultrasonic 3D anemometer set on the yacht bow was used for the addition of wind measurement. Additionally, time of flight (TOF) technology was used to measure the 3D geometry of the sail flying shape, and MEMS sensors and pressure pads were used to estimate the sail pressure. All the obtained signals were collected and synchronized using data recording equipment.

The majority of hydrodynamic tests are conducted in a towing tank environment, where the wave conditions are different from actual sea waves. In 2016, Rajendran et al. [76] measured wave-induced motion and bending moment tested at a basin with realistic extreme seas on a cruise ship. In the experimental program, resistive wave sensors measured the wave-induced motion, and strain gauges calculated the vertical bending moment of the model. Jiao et al. [77] comprehensively proposed a comparative analysis between a small-scale laboratory seakeeping test and a large-scale seakeeping measurement test with realistic sea waves of a 72,000-ton class ship. In this research project, advanced onboard remote control and an experimental telemetry system were developed to realize seakeeping measurement at sea near the shore of the bay of Huludao, China. The project focused on measurement of wave data, navigational information, global motions and loads, accelerations, impact pressures, and visual recording (green water on deck, wave run-up, and slamming). The sensors used in the small-scale laboratory tests included wave probes, strain gauges, pressure sensors, accelerometers, rotary potentiometers, and data collectors. For the large-scale measurement systems, an anemometer, wave buoy, and tachometer measured the ocean environment. A fiber optic sensor was needed for the real-time hull stress monitoring and evaluation system. Then, sailing trace, position, speeds, and motions were calculated by a GPS/INS system. The radio control and telemetry system were developed to allow the conduction of large-scale model trials in sea conditions. As seen from the result, the proposed large-scale experimental test has tremendous significance for the future development of a large-scale seakeeping measurement technique.

In 2019, Carchen et al. [78] proposed full-scale measurement onboard The Princess Royal research vessel to estimate the effect of biofouling growth on hydrodynamic performance. An automated sensor system was placed onboard the research vessel to acquire high-quality and quantitative hydrodynamic performance data. The results prove the effectiveness of the proposed method. In 2020, onboard measurement using several sensors installed on a bulk carrier ship was also investigated in depth by Kim et al. [79]. This study proposed a data-driven approach to predict added resistance and shaft horsepower using big data from onboard sensor measurement and the NOAA database. For input features, various sensing methods and protocols were used to collect the input data. A GPS sensor was used to measure ship velocity, and a draft sensor was used to calculate the draft of the vessel. An RPM indicator and echo sounder were then used to determine the RPM and sea depth, respectively. To measure the tide, wave height, and wind vector, a Doppler sensor, accelerometer, and anemometer were installed onboard, respectively. Moreover, a shaft torque sensor using a strain gauge that converts torque to resistance/MODBUS was applied to calculate output propulsion power. The collected raw data were then processed by the data acquisition unit of a VDR system. The landmark studies on the application of sensors for hydrodynamic characterization in marine structures from the early 2010s until 2020 are summarized in Table 3.

Table 3. Landmark research on the sensing technology for hydrodynamic characterization in marine structures.

Year	Authors	Observational Subject and Purpose	Sensing Technology and Instrument	Important Remarks
2010	Tzabiras and Kontogiannis [70]	Hydrodynamic resistance of three twin-screw and low-c_B passenger ship models with different bulbous bows	Five resistance-type wave probes	Wave probe is sensitive to environmental changes, so repeated calibration is needed in each test.
2011	Polach and Ehlers [71]	Heave and pitch motions on the resistance in ice going tanker "Uikku"	Dynamic motion unit (DMU), load cell, cameras (video recording), pointer	The oscillations and vibrations possibly interfere with resistance signal and motion measurements.
2012	Lee et al. [72]	Hydrodynamic design and performance of an underwater hull cleaning robot (HCR)	Resistance dynamometer, electromagnetic flow meter, camera, weighing scale	The actual drag force is larger than that obtained by CFD. This discrepancy is assumed by a different setup between CFD and the actual test.
2013	Bouscasse et al. [73]	Seakeeping behavior of Delft 372 fast catamaran	Krypton optical system (Infrared LEDs), a finger probe and Keyence ultrasound wave probe, second wave system, HYDRONICS load cells, acquisition system	The Kenek probe is applicable for low *Fr* test and wave steepness, and the ultrasound probe is more accurate for higher *Fr* and larger wave steepness.
2014	Jang et al. [74]	Frictional resistance reduction of hull bottom of 66K DWT Supramax bulk carrier	Floating-plate-type friction sensors, camera, flow meter	The air layer is not fully developed on the floating plate. The air layer is split into left and right in front of the aft local friction sensor.
2015	Fossati et al. [75]	A full-scale test of 10-m Sailing Yacht Lab	Strain gauges based mono-axial HBM S9M load cells, a 3DM-GX3-35 GPS (AHRS and GPS receiver), GEMAC 23554 analog inclination sensor, gyroscope, accelerometer, magnetometer, ultrasonic 3D anemometer, TOF flying shape detection system, MEMS sensor, and pressure pads	A graphical user interface (GUI) is used to synchronize different types of signals and sampled signals.
2016	Rajendran et al. [76]	Vertical motions and bending moments on a cruise ship	Resistive wave sensors, strain gauge	Strain gauges are placed at the same height as tVCG to calculate the vertical bending moment.

Table 3. *Cont.*

Year	Authors	Observational Subject and Purpose	Sensing Technology and Instrument	Important Remarks
2017	Jiao et al. [77]	A large-scale seakeeping test at realistic sea conditions of 72,000-ton class ship	The sensors of small-scale laboratory test: wave probes, strain gauges, pressure sensors, accelerometers, rotary potentiometers, and data collectors The sensors for large-scale measurement: Ocean environment monitoring sensors, fiber optic sensor, GPS/INS system, radio control and telemetry system (self-propulsion system, autopilot system, radio communication system, telemetry system)	The proposed remote control and experimental telemetry system for large-scale measurement in actual sea waves were successfully developed.
2019	Carchen et al. [78]	Ship performance monitoring system on The Princess Royal vessel	DGPS and gyrocompass, Doppler speed log, electro-magnetic speed log, instrumented shafts, rudder potentiometer, wave radar, wave buoy, weather station (ultrasonic anemometer), thermosalinograph	The sampling frequency selection is crucial for acquiring the resolution of the phenomena.
2020	Kim et al. [79]	Propulsion power measurement of a 200,000-ton bulk cargo ship	GPS sensor/NMEA0183, Draft sensor (Hydrostatic level pressure transmitters), RPM indicator, Echo sounder, Doppler log, anemometer, accelerometer, shaft torque sensor (strain gauge), VDR	The shaft power measurement and recorded raw information are transformed through the data acquisition unit of a voyage data recorder (VDR).

Several researchers recently proposed a new direction for full-scale measurement using onboard sensors of ship performance on actual seas. Full-scale measurements are conducted to obtain a better understanding of ship hydrodynamics. The method can achieve a similar sailing performance to that in the real sea environment, and the response characteristics are more realistic than those in a towing tank. One comprehensive measurement system included resistance and self-propulsion tests of a 25-m-long ore carrier "CSB FORTUNE," which were conducted by Lin et al. [80] in the sea near Qingdao, China. Figure 8 illustrates the ship layout, which includes propulsion systems, sensor arrangements, and autopilot systems. A block diagram of two test systems for the actual ocean environment is shown in Figure 9. For the resistance testing procedure, the sea condition during the experiments was measured and recorded to meet the required level. Then, the model was hauled by two winches at a constant speed in the harbor, and a force balance was used to obtain data for 40 s of resistance measurement. For the self-propulsion testing procedure, the test was conducted at sea 5 km from the shore of Qingdao. The first test was performed with zero ship speed to check the propeller torque and measure the wind speed and direction. Then, the model sailed in a fixed direction at a different speed, and the thrust and torque data of the propeller were measured. It was found that the proposed procedure can be used for various applications, including as a guide for other large-scale ship performance tests. In other work using the same large-scale model, Guo et al. [81] studied the self-propulsion test for propulsion performance and the effects of complex wake field environments on propulsion performance. The GPS/INS integrated navigation system consisted of GPS antennas and an INS core, and a GNSS receiver obtained real-time motion data, including speed, track, heading, roll, pitch, and yaw rate. In the data measurement and acquisition system, the model used non-remote sensing control, which consisted of acceleration sensors, a self-developed instrument, a data collector, and a computer. Furthermore, with the measurement data of the same model as previous studies, Su et al. [82] investigated the effect of energy-saving devices' (ESDs) application on ship propulsion performance. The interaction mechanism between the hull and the ESDs, including the pre-swirl stator and rudder bulb, was studied in a large-scale ship model. Future studies should focus on the interaction between the pre-swirl stator and rudder bulb system with different types of hulls.

Ship maneuvering in waves is a hot topic for researchers. In one recent study, a novel full-scale maneuvering sea trial correction method was developed by Mei et al. [83]. A trial test of the full-scale motor vessel Yukun with a constant engine setting was conducted in the northwest of the Yellow Sea to measure the effects of wind, waves, and currents on the maneuvering motions, including surge, sway, and turning data. The measuring instruments installed onboard of Yukun are illustrated in Figure 10. A differential global positioning system (DGPS), wind measuring system, speed log, and fiber optic gyro were installed on the bridge, mast, ship bow, and gyro deck. In this work, the full-scale Yukun model's maneuvering was established, whereby zig-zag test training data were used to predict the turning circle test. The result revealed that Yukun meets the IMO standard for ship maneuverability, with the accuracy of the advance and tactical diameter reaching 93% and 95%, respectively.

In 2021, Ha et al. [84] investigated the characteristics of slamming loads on the bow of an FPSO in the ocean wave basin of the Korea Research Institute of Ships and Ocean Engineering (KRISO). In this experiment, a slamming impact load was generated with three heading angles (180°, 165°, and 150°) and measured using 15 force sensors attached to the bow. The 15 sensors were KISTLER 4576 force sensors with a sampling rate of 20 kHz. Moreover, the 6-DOF motions of the model were measured by a non-contact optical measurement system with a sampling rate of 200 Hz. At the time of the impact event, a high-speed camera captured the incoming waves at a frame rate of 5000 frames/s. The comprehensive configuration of the sensing technology used in this work is demonstrated in Figure 11. The results showed that the slamming impact loads can be classified as loads with one peak or two peaks, where the second peak is smaller due to the damping occurrence of the first slamming event. It was revealed that the 150° heading angle produced the most significant slamming impact loads in this work.

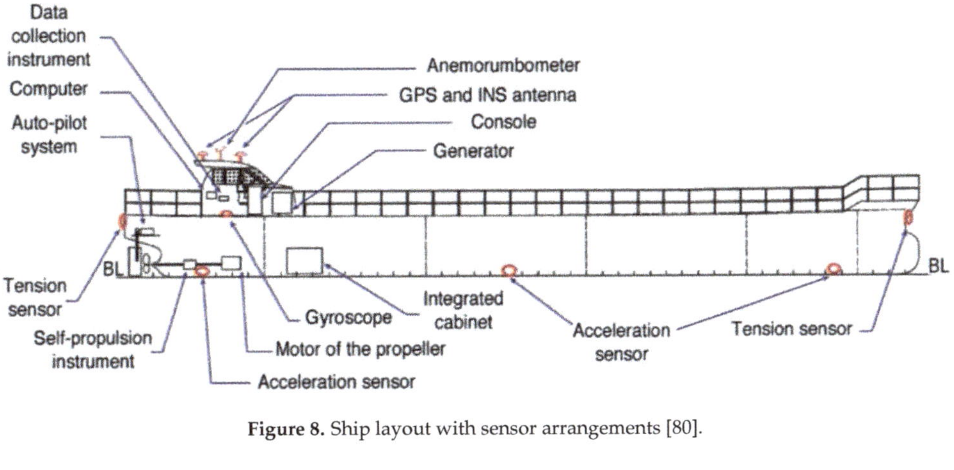

Figure 8. Ship layout with sensor arrangements [80].

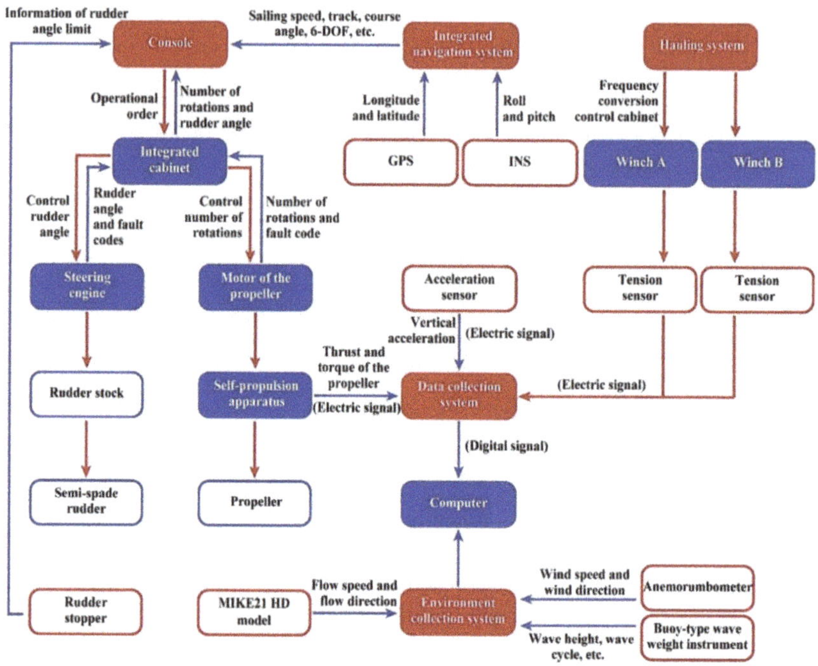

Figure 9. Block diagram of two test systems in actual seas [80].

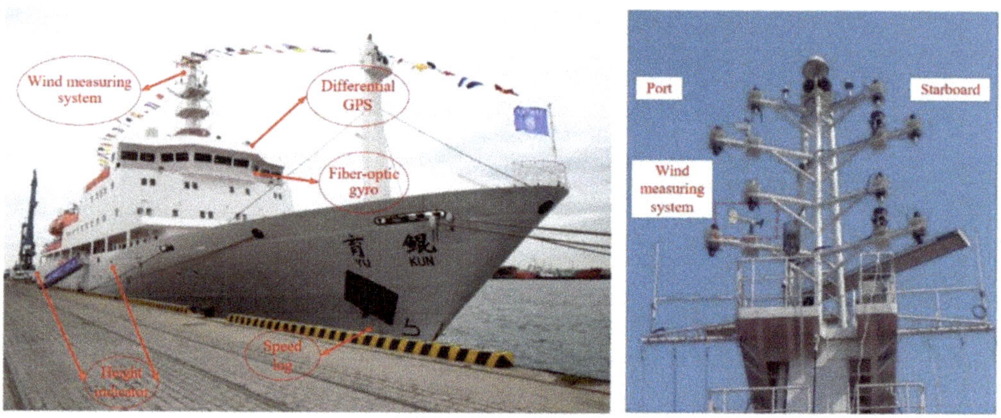

Figure 10. The measuring instruments installed onboard of Yukun [83].

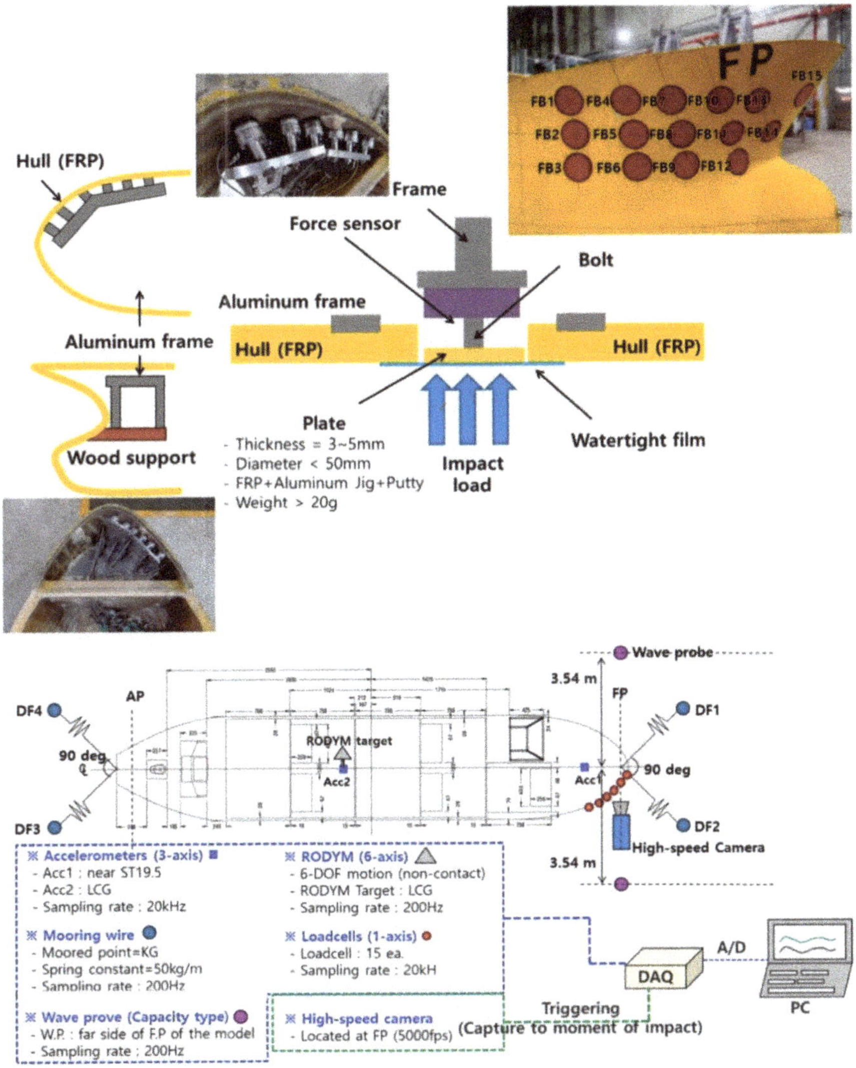

Figure 11. TENG tilt sensor experimental apparatus [84].

4. Advanced Machinery Performance Monitoring

In recent years, thousands of commercial ships have come to sail across the oceans every day. These commercial ships are operated by companies incorporated in different jurisdictions. Furthermore, these ships are built by hundreds of shipbuilding facilities and have to comply with several international, regional, and national regulations in order to fulfill the safety requirements across a worldwide network of commercial ports. One of the purposes of sailing ships is to carry passengers and goods across oceans. Moreover, the commercial shipping industry and its regulations make international trade possible, safe, and efficient.

Recently, there has been increasing interest in developing and implementing ship hull structure monitoring systems. However, ship losses are still high. The total ship loss between 2016 and 2019 is presented in Figure 12. During this period, cargo ships faced the highest loss at a total of 120 ships [85]. Furthermore, ships are extremely complex systems. Several essential components on a ship, such as the navigation system, power system, i.e., main engine, and the cargo, make this ship reliable. However, all of the very complex systems of a ship need supervision by monitoring and control systems. As can be seen in Figure 12, a total of 21 ships were lost due to machinery failure between 2016 and 2019. In this section, a review of advanced machinery performance monitoring on ships is presented. Furthermore, these technologies are expected to facilitate the maintenance, safety, and effectiveness of ship operation.

Several ship management technologies directed at enhancing the effectiveness and safety of ships have been developed and proposed by many researchers, e.g., Paik et al. [86,87], who performed research based on the employment of wireless sensor networks for full-scale in-ship application [86]. Their research examined a wireless sensor network (WSN) used to communicate in the ship environment. Two ZPAs (Zigbee Protocol Analyzers) were used as data transmission tests. The results of the wireless communication network are presented in Figure 13. As can be seen, the average delivery ratio was nearly 100% at all proposed distances. Based on the work, the wireless communication network can be used for the abnormal operation of equipment and fire prevention in dangerous regions. Paik et al. [87] also conducted follow-up research in 2009 on the characteristics of wireless sensor networks for full-scale ship application. The ship Hannara, a 3000-ton class training ship, was chosen for this study. The result showed that the characteristics of the wireless sensor network on a test bed ashore also have a successful connection.

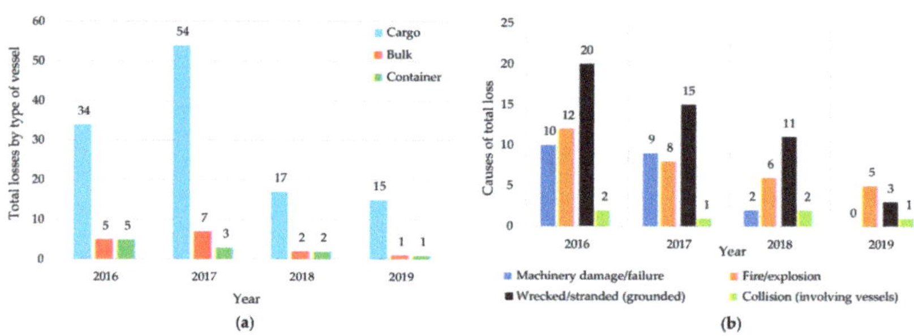

Figure 12. Total ship loss in 2016–2019. (**a**) Total loss by type of vessel and (**b**) all causes of loss (redrawn based on data in Allianz [85]).

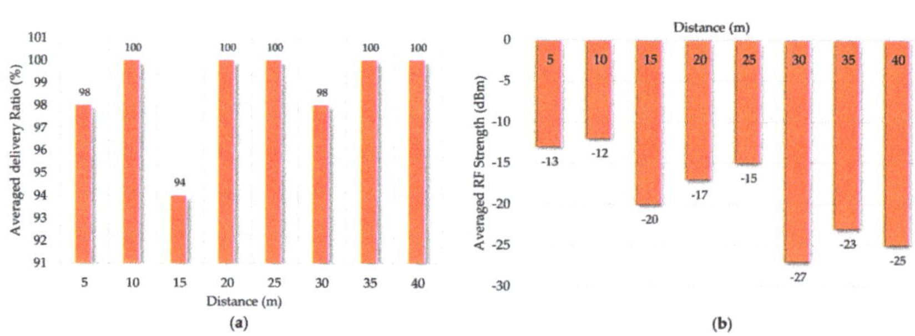

Figure 13. Result of the wireless communication tests for the long corridor in the shelter deck. (**a**) Average delivery ratio and (**b**) average radio frequency strength (redrawn based on data in Paik et al. [86]).

To this day, as the number of passengers onboard commercial vessels has grown, the concern for ship safety has risen. The safety of passengers on a large ship is the priority. As such, ships have been extensively fitted with several safety systems, i.e., for fire, evacuation, stability, and flooding control. This flooding control, specifically by breach detection, has attracted researchers such as Penttilä and Ruponen [88], who presented in the 5th International Conference on Collision and Grounding of Ships (ICCGS) in Espoo, Finland. This conference is very well known in the field of impact phenomena, i.e., in ship collision and grounding. This study is expected to provide milestone processes to improve the safety of ships. In the work, various combinations of sensor density, noise, and filter length combined with an inverse method were introduced for detecting a breach. It was also noted that flood water can penetrate through non-watertight structures. They investigated a total of 2392 cases, and the results strongly indicate that the inverse method is applicable in determining a breach if the sensor arrangement is dense enough. The method could determine the floodwater origin accurately in 71.1% of the cases. The success rate of calculating the correct breach is presented in Figure 14. As seen, the average success rate of finding the primarily flooded room was 98.6%.

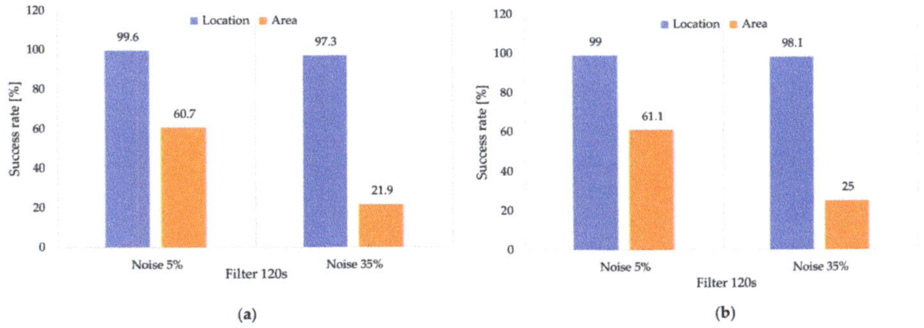

Figure 14. The success rate of calculating the correct breach. (**a**) All doors closed and (**b**) fireproof doors open (redrawn based on data in Penttilä and Ruponen [88]).

Other research has focused on advanced ship systems and ship sensors' data collection for condition monitoring. Such research was conducted by Lazakis et al. [89] and Raptodimos et al. [90]. Lazakis et al. [89] investigated the Inspection Capabilities for Enhanced Ship Safety (INCASS) framework to bring an innovative solution for inspection and maintenance of ship structures and machinery. The INCASS framework consists of advanced structural risk assessment (SRA) and machinery risk assessment (MRA) tools. The results show that the INCASS framework has a reliable performance according to ship owners and operators. Furthermore, a follow-up study was presented by Raptodimos et al. [90] in the same year of 2016, where they reported that monitoring additional systems under actual operational situations would ensure safer operation and increased operational efficiency, including improved fuel efficiency and reduced emitions. Based on the implementation, sensors installed on a specific system allow for improving measurements and data collection and can be used further in analysis and assessment to enhance ship safety, maintenance, and performance and efficiency. Furthermore, research on robotic cleaning has also been increasingly discussed in maritime industries. Noordstrand [91] introduced the Fleet Cleaner robot, developed and deployed in the Netherlands. This remotely controlled robot is used for cleaning in ships, especially hull cleaning. The landmark research on the monitoring of advanced machinery performance is summarized in Table 4.

Table 4. Research landmark of the advanced machinery performance monitoring.

Milestone	Author(s)	Observational Subject	Tools/Instrument	Important Remarks
2007	Paik et al. [86]	Communication based in the ship environment	Wireless sensor network (WSN) ZPAs (Zigbee Protocol Analyzers)	The wireless sensor network (WSN) can be used in the ship environment (in this case, Hannara ship), with an average delivery ratio of nearly 100% in all proposed distances in this case.
2009	Paik et al. [87]	Communication based in the ship environment	Wireless sensor network (WSN) ZPAs (Zigbee Protocol Analyzers)	Wireless communication that can be used regularly is desirable within a ship environment, such as for fire prevention in dangerous regions.
2010	Penttilä and Ruponen [88]	Breach and floodwater detection in the ship	Sensor density, noise, and filter length combined with an inverse method	The result shows that the method is able to determine the floodwater origin accurately with an average success rate of 98.6% in finding the primarily flooded room.
2016	Lazakis et al. [89]	Ship inspection and maintenance for ship structures and machinery	Inspection Capabilities for Enhanced Ship Safety (INCASS) Structural risk assessment (SRA) and machinery risk assessment tools (MRA)	It is noted that the development and implementation of the INCASS framework has a reliable performance according to ship owners and operators.
2016	Raptodimos et al. [90]	Ship inspection and maintenance for ship structures and machinery	Inspection Capabilities for Enhanced Ship Safety (INCASS) Structural risk assessment (SRA) and machinery risk assessment tools (MRA)	Sensors installed on a specific system allow for improving measurements and data collection and can be used further in analysis and assessment to enhance ship safety, maintenance, and performance and efficiency.
2018	Noordstrand [91]	Remotely controlled robot for ship hull cleaning	Fleet Cleaner robot	The Fleet Cleaner robot is very reliable for cleaning a ship's hull because it can be used underwater and above water at a high curvature.
2021	Drewing et al. [92]	Coking in marine diesel engine injector nozzle	Laser heads clocked at 16 MHz	Coking in a marine diesel engine injector nozzle can be diagnosed by spectral analysis of the shaft's torsional vibrations as measured by optical sensors.
2021	Kozak et al. [93]	Semiconductor power electronics devices, including thyristors	Acoustic emission sensor	The method proposed for detecting early-stage damage of the gate turn-off thyristor using an acoustic emission sensor is viable and valuable to observe changes in volume caused by abnormal conditions, such as overheating, internal dislocations, or a decrease in wafer plate pressure.

Table 4. *Cont.*

Milestone	Author(s)	Observational Subject	Tools/Instrument	Important Remarks
2021	Kyzioł et al. [94]	Material properties of polymer composites in hull boats	Acoustic emission Kolmogorov–Sinai (K–S) metric entropy	The results show that the proposed method using acoustic emission and Kolmogorov–Sinai (K–S) metric entropy can determine the mechanical properties of the transition material from the elastic to the plastic phase of polymer composites.

In recent years, research focused on sensors for monitoring advanced machinery performance has been conducted by Drewing et al. [92] and Kozak et al. [93]. Drewing et al. [92] investigated the diagnosis of the coking of a marine diesel engine injector nozzle by performing a spectral analysis of the crankshaft's torsional vibrations. Laser heads clocked at 16 MHz were used for the measurements. The result showed that the data recorded from the 16-megahertz laser heads were satisfactory to determine the diesel–electric unit's shaft vibrations. It was also noted that it is possible to diagnose coking of a ship's diesel engine injection nozzle using spectral analysis of the shaft's torsional vibrations, as measured by optical sensors. The location of the ETNP-10 laser heads on both ends of the shaft can be seen in Figure 15. Recent ships are regularly provided with converters that utilize semiconductor power electronics devices such as thyristors or power transistors. Semiconductor power electronics attracted research such as that of Kozak et al. [93], which aimed to detect early-stage damage. These semiconductor power electronics devices are mainly used in auxiliary podded drives and thrusters. Kozak et al. proposed a method for detecting early-stage damage of a gate turn-off thyristor using an acoustic emission sensor. The acoustic emission sensor and the thyristor packages are presented in Figure 16. The study showed that the acoustic emission sensor can detect waves from inside the semiconductor structure. Furthermore, changes in volume caused by abnormal conditions such as overheating, internal dislocations, or a decrease in wafer plate pressure can be observed. Polymer composites are sometimes implemented as hull materials in boats [94]. Due to their unique properties, a method was proposed for determining the mechanical properties of composite materials using acoustic emission and Kolmogorov–Sinai (K–S) metric entropy. It was noted that since the composite materials do not have a yield point, the proposed method shows that acoustic emission and Kolmogorov–Sinai (K–S) metric entropy can determine the transition of a material from elastic to plastic phase. In fact, knowing the mechanical properties of the material is extremely important when designing a structure.

(a) (b)

Figure 15. ETNP-10 laser heads installed on both ends of the shaft: (**a**) view from the engine side; (**b**) view from the generator side [92].

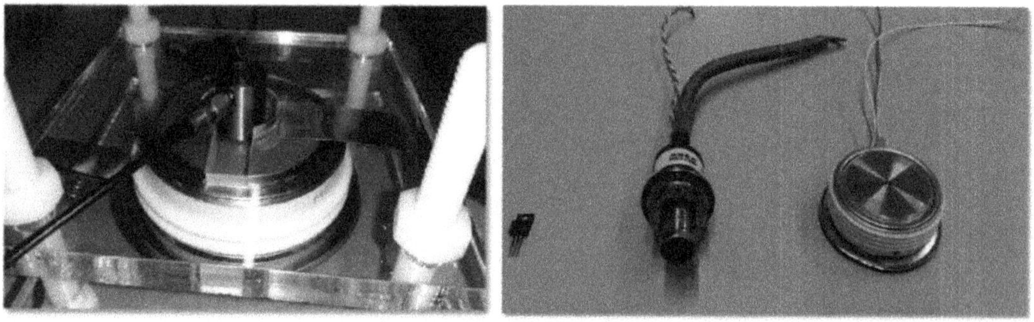

Figure 16. (**a**) The acoustic emission sensor and (**b**) the model of the thyristors in plastic (**left**), stud-mount (**middle**), and press-pack packaging (**right**) [93].

5. Arctic-Based Field Observations

The Arctic is a polar region located at the northernmost part of Earth. The Arctic region is a unique area among Earth's ecosystems. The cultures in the region and the Arctic indigenous peoples have adapted to its cold and extreme conditions. This region is also notably susceptible to climate change phenomena, which directly affect components of the environmental ecosystem. The impacts of climate change are seriously observed in the Arctic, and with more immediate and severe consequences than in most parts of the world. Field observations indicate that the Arctic is warming at almost twice the rate of the global average. Furthermore, reductions in sea ice and changes in weather are continuously visible. In the past decade, various measurements have been conducted—i.e., observation of carbon cycle and methane production, mapping of sea surface temperature, and measurement of ice extent.

In 2010, Hill and Zimmerman [95] performed a series of estimations of biomass and primary production (PP), which was measured by $H^{14}CO_3$ uptake during in situ incubation. Active remote sensing was deployed in the research by using spaceborne LIDAR to provide information on the distribution of particles in the water column. In 2012, Høyer et al. [96] deployed research to validate the performance of the Arctic satellite, which observes sea surface temperature. Several sensors are considered in this work, which concluded that the AVHRR (NAVO-GAC) and the AATSR produce satisfactory results with very small bias. Additionally, it was noted that the amount of synchronized data validation of the AMSR-E is very high. A follow-up observation of the high-Arctic wet tundra area was extended by Tagesson et al. [97]. In this research, methane (CH_4) fluxes were observed using the in situ measurement method. A laser off-axis integrated cavity output spectroscopy analyzer was deployed to measure chamber concentration. The issue of global warming in the Arctic region was intensely studied by Lund et al. [98] as the summer-time energy exchange characteristics were spotted and found to cause high sensible heat flux (H)/net radiation (*Rn*) and β and low latent heat flux (LE)/net radiation (Rn). In later years, dedicated research to observe the annual balance of CO_2 and CH_4 fluxes in the Arctic environment was performed by Goodrich et al. [99]. Sensor performance was addressed by assessing sensor combinations on fluxes with different instrumentation at the same sites. Besides atmospheric conditions, concern regarding the Arctic sea ice extent has been addressed, as a reduction in volume is expected with increasing environmental temperature. Connolly et al. [100] re-calibrated the ice dataset using satellite sensing, as before the satellite era, geopolitical considerations were also a problem for the compilers of datasets. According to this research, the sea ice trend since the 1970s has been similar to the trend of ice growth after the mid-1940s. Other research addressed aerosol process and emission modeling, in which measurement of aerosol optical depth (AOD) is the critical process. Such research was conducted by Hesaraki et al. [101] using sun photometry and sky radiometry instrumentation. The data modeling was based on the input retrieval from the three AEROCAN sites (AEROCAN is the Canadian federated sub-network of AERONET). Later, an observation on ice melting was re-conducted, but in a more specific area. Howell et al. [102] considered the Arctic region near Canada and Greenland for their work. Sentinel and synthetic aperture radar (SAR) were deployed for melt onset detection as two collaborating instruments in the observation. One of the notable remarks of the observation was that localized sea ice dynamics will need to be considered for future melt onset algorithm development from multi-sensor $\gamma_C{}^{\circ}$ SAR products. The ability of the multi-sensor products to identify the ice melting phenomenon shows that these products could be utilized for lead detection studies.

At the end of the 2010s, research addressed the development and validation of low-cost sensors for Arctic-based research. Carotenuto et al. [103] provided a study to ensure the operational performance of a low-cost tool for atmospheric and meteorological observations. For verification of the proposed tool, Vaisala HMP45AC and Campbell-Scientific EC150 were taken as reference sensors. On the other hand, Pomerleau et al. [104] constructed an instrument to enhance progressive research on measuring snowpack and ice thickness. The tool adopts the concept of the frequency-modulated continuous-wave (FMCW) radar. Even though this instrument can be attached to an autonomous vehicle, an unfavorable environment outside the radar's limits may hinder the measurement process and lead to unrealistic results, e.g., during ice and wet snow situations. Landmarks studies of sensor application in Arctic research from the early 2010s until 2020 are summarized in Table 5.

Table 5. Landmark research on the roles of sensors in Arctic studies and observations.

Milestone	Author(s)	Observational Subject	Tools/Instrument	Important Remarks
2010	Hill and Zimmerman [95]	Annual production of global carbon cycle/primary production in the Arctic Ocean	Active sensors: utilizing light detection and ranging (LIDAR)	Vertical data based on LIDAR needed to be merged with the horizontal surface gradients observed using ocean color sensors to extrapolate the narrow LIDAR beam across the study area.
2012	Høyer et al. [96]	Sea surface temperature in the Arctic Ocean	Advanced Along-Tracking Scanning Radiometer (AATSR) Advanced Very High Resolution Radiometer (AVHRR) Advanced Microwave Scanning Radiometer-EOS (AMSR-E)	The most accurate satellite observations on the subject were obtained from IR observations using either the AVHRR (NAVO-GAC) or the AATSR. The results indicate low standard deviations.
2013	Tagesson et al. [97]	Methane fluxes in the high-Arctic wet tundra area	LGR, DLT200 Fast Methane Analyzer Moderate-resolution imaging spectroradiometer (MODIS)	Studied models performed well, but the observed models involving satellite-based normalized difference water index (NDWI) reproduced the CH4 flux variability better than the models not including NDWI.

Table 5. *Cont.*

Milestone	Author(s)	Observational Subject	Tools/Instrument	Important Remarks
2014	Lund et al. [98]	Surface energy balance of Arctic ecosystems	Infrared gas analyzer LI-6262 and 7000 3D sonic anemometer Gill R2 and R3	Results indicated that H/Rn and LE/Rn increased, and b and G/Rn were decreased when subjected to the increasing active layer depth in the 2000–2010 period.
2016	Goodrich et al. [99]	CO_2 and CH_4 fluxes in the Arctic region	Metek uSonic-3 Class A (non-orthogonal ultrasonic-anemometer) CSAT3 anemometer	Application of the different eddy covariance sensor combinations indicated that seasonal and annual CO_2 and CH_4 fluxes from the different gas analyzers and anemometers were within uncertainties.
2017	Connolly et al. [100]	Ice extent on the Arctic sea	Satellite sensing (Nimbus-7, DMSP-F8, DMSP-F11, DMSP-F13, and DMSP-F17)	The recent period of Arctic sea ice retreat since the 1970s followed a period of sea ice growth after the mid-1940s. After the 1910s, data reconstructions agree with previous studies that pointed to a decreasing tendency in Arctic sea ice extent.
2017	Hesaraki et al. [101]	Aerosol optical depth (AOD)	Aerosol Robotic Network (AERONET) GEOS-Chem (GC)	The histograms of both the AERONET retrievals and the GC estimates were better represented by a lognormal distribution. AERONET and GC simulations displayed seasonal variations, reaching peaks in April/May and August/September.
2019	Howell et al. [102]	Melt onset over sea ice on the Canadian Arctic and Greenland	Sentinel-1 RADARSAT-2	Passive microwave (PMW) algorithm provides the best measurement of melt onset. However, SAR imagery can provide more robustness in estimating the subject, especially with the spatiotemporal constraints of C-band SAR satellites.
2020	Carotenuto et al. [103]	Atmospheric (temperature, humidity) sensor performance in the Arctic	AIRQino	AIRQino is proven to be more economical in terms of power consumption, and it can produce reliable data compared to other high-cost reference sensors.
2020	Pomerleau et al. [104]	Snowpack and ice thickness	Autonomous frequency-modulated continuous-wave (FMCW) radar	Ice thickness measurements provide a 2-cm accuracy above a 4-cm resolution limit. The radar can be put onboard a light remotely piloted aircraft to enable surveys on unsafe thin ice and long river transects.

Based on the review, it can be concluded that atmospheric and hydrological systems are the primary research subjects using sensor tools/instruments in the Arctic region. At the end of the decade, a sensor was used collaboratively with an autonomous vessel. As promising results were obtained, research may be continued to provide observational and measurement sensors on vessels in the current decade. This prediction is quite realistic, as Løken et al. [105] addressed their concern regarding wave measurement of the Arctic marginal ice zone (MIZ) by using a mounted sensor on ship waves. An ultrasonic gauge (UG) was taken as the setup instrument to measure ocean surface elevation. Inertial motion units (IMUs) were deployed to measure data reliability and changes in dynamic sea motion. Notable results were processed with wave spectral models, i.e., wave attenuation measurement (WAM) and ERA5 (see Figure 17). A comparative study was also conducted to analyze wave measurement from several instruments, namely, WAM and wave-in-ice instruments (WIIs), as presented in Figure 18. The results indicate that the greatest disparity (represented by the mean percentage error) in the wave height and zero-up crossing period from the WAM and WII was below 9.7%, observed at stop 1.3 with a difference of 87.8%.

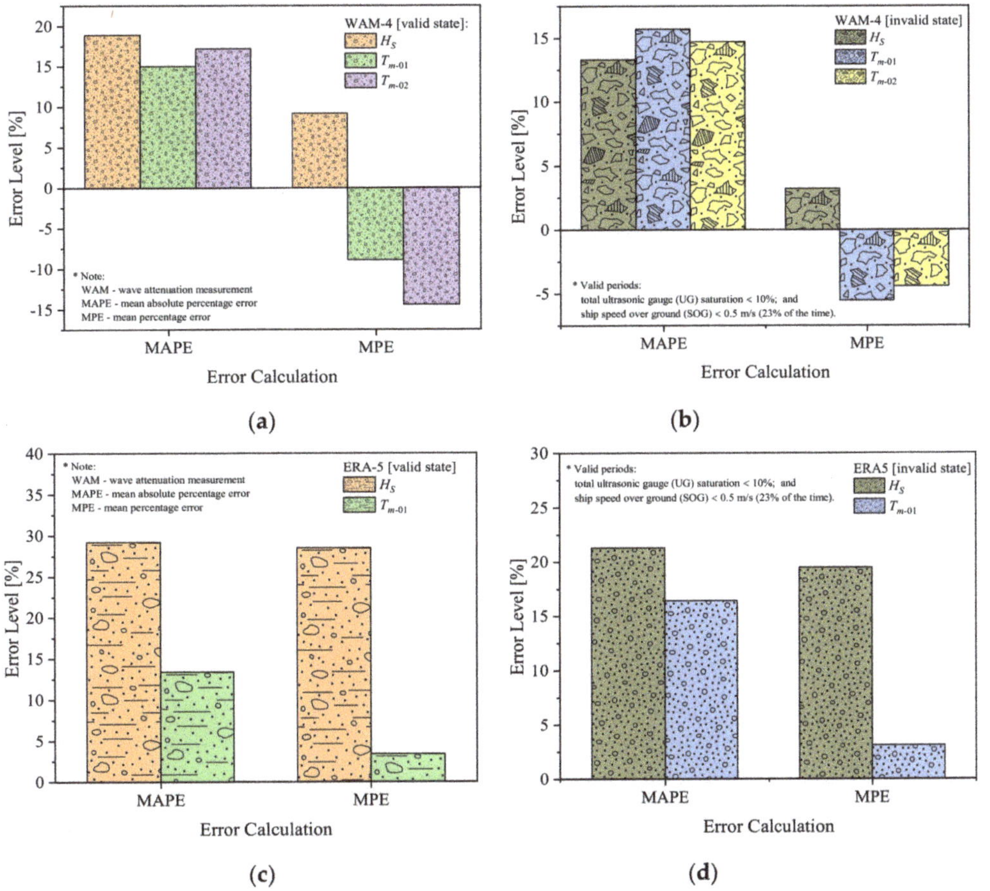

Figure 17. Comparison of the measurement data from WAM-4 (**a**,**b**) and ERA5 (**c**,**d**): the wave height (H_S) and periods ($T_{m\text{-}01}$ and $T_{m\text{-}02}$) (based on data in [105]).

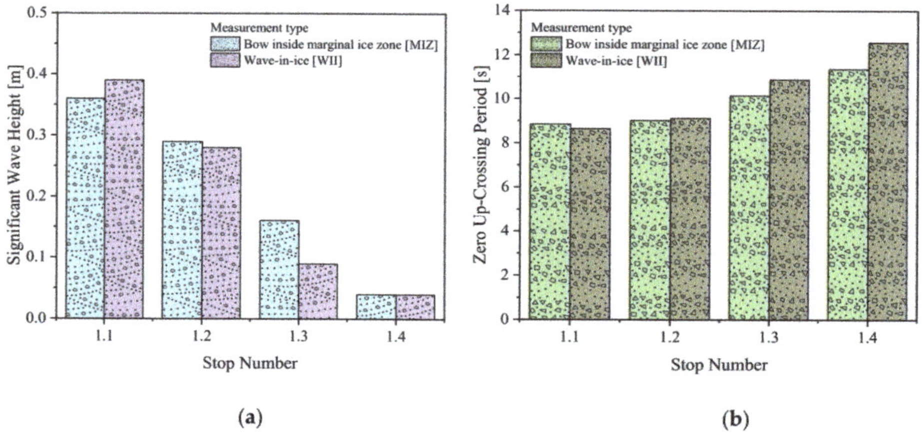

Figure 18. Measurements of the bow and WII: (**a**) wave height and (**b**) crossing period (based on data in [105]).

In other recent work, a mega project of data collection on the Arctic Ocean was initiated by researchers and consultants from Russia and United States. Nikishin et al. [106] addressed their focus to several subjects, e.g., seismic data, sub-bottom profiler data, geological sampling, borehole drilling, gravity, magnetic anomalies, and offshore geodetic data. Various equipment setups were prepared in terms of the seismic data, such as bolt APG air guns, control, and the ORCA or QINSy monitor navigation system; the DigiSHOT digital air gun controller; the ION DigiSTREAMER solid-filled seismic streamer; and the ION DigiSTREAMER integrated offshore seismic data acquisition system. Furthermore, preparation for a sub-bottom profiler survey was conducted, including the acquisition of additional data by multi-beam and single-beam echo sounders. In an effort to obtain a rock sample, the Arktika-2012 expedition was initiated, for which several sequence steps were taken, i.e., detecting the location of bedrock exposure, determining site parameters, and conducting video-photometric and sonar detection (see the summarized illustration in Figure 19). Acquisition of gravity and magnetic data was conducted later using two gravimeters, Chekan-AM and Shelf-E (manufactured by Elektobribor), which were mounted on two involved research vessels in the Arktika-2014 expedition. Additionally, the SeaPath 330 satellite integrated navigation system was used for offshore geodetic data collection. As backup secondary systems, C-Nav-2050R and C-Nav-3050 were also set to confirm the designated location reading.

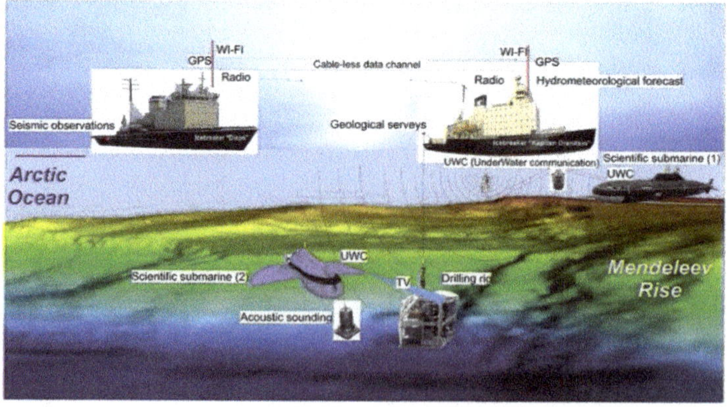

Figure 19. The technical process of rock sampling in Arctic expeditions [106].

Similarly to the discussed work, researchers from Japan and Australia performed co-research, but on a quite different subject. As presented by Kodaira et al. [107], the consortium conducted a field survey during the R/V Mirai Arctic expedition in 2019 to observe on-ice wind waves. The specific location was also designated, i.e., the western part of the Arctic Ocean. In the expedition, near-surface atmospheric and oceanographic variables were measured using the Shipboard Oceanographic and Atmospheric Radiation (SOAR) measurement system. Specifically, the air temperature and wind characteristics were recorded using an anemometer (Model 05106). To provide atmospheric data, including land and oceanic climate per hour, ERA5 (atmospheric reanalysis introduced by European Centre for Medium-Range Weather Forecasts—ECMWF) was deployed. Several buoys, namely, spotter buoys (low-cost, compact, and solar-powered buoy developed by Sofar Technologies, Inc. (San Francisco, CA, USA), were deployed in a designated location (see Figure 20) to measure the spectral distinction between the wave characteristics in the non-ice ocean and those in the sea ice-covered area. Satellite instrumentation was also considered in this work. Synthetic aperture radar (SAR) was used to provide different images of the sea ice-covered ocean, such as drifting of sea ice, and ice classification. Finally, passive microwave measurement from the satellite was performed to obtain information regarding the sea ice concentration (SIC) in the marginal ice zone (MIZ). The images from the SAR data acquisition were compared to the retrieved SIC data to quantify wave prediction uncertainties. In this step, the SIC dataset was created by JAXA using the Advanced Microwave Scanning Radiometer 2 (AMSR2).

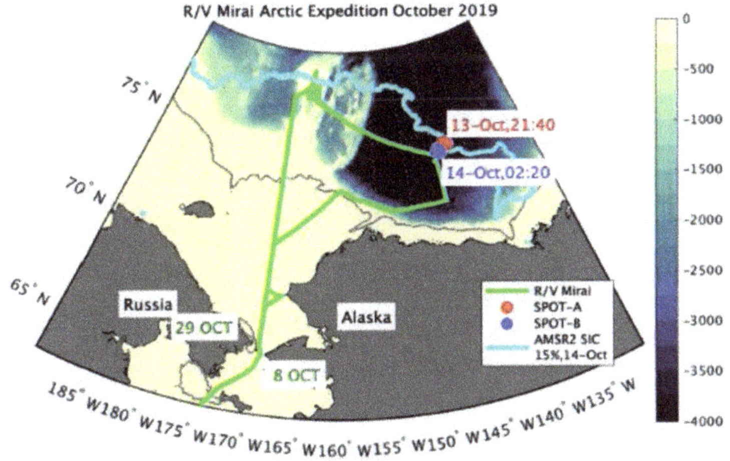

Figure 20. Positions of the deployed "spotter" buoys to measure wave characteristics [107].

6. Vibration-Based Damage Detection on Offshore Structures

Offshore platforms are enormous, complex structures that cover a large span of the ocean for various functions and are repeatedly endangered by harsh environmental loads. Fixed offshore platforms including jacket platforms, wind turbines (WTs), marine current turbines (CTs), etc., are the most common types, which suffer severe unforeseen damage during their service life. Therefore, permanent, continuous, and automatic monitoring of the platforms' dynamic behavior needs to be developed to evaluate their safety and durability for a long-term operation.

Most recently developed offshore platforms are designed to be self-operating and self-producing (unmanned conditions). They have been synchronously equipped with structural health monitoring (SHM) systems to investigate their structural status and assess their functional fitness and performance levels. SHM techniques consist of either global or local damage detection. Local damage techniques, as the most used measuring techniques today, involve non-destructive testing (NDT) methods. The most frequently used NDT tests are visual, magnetic field, acoustic, eddy current, strain measurement, etc. However, detecting damage using local damage inspection is not possible when the structure is enormous and partially submerged in the water. SHM using vibration-based damage detection techniques can identify damage that affects the overall structure and examine changes in global structural vibration characteristics. Damage identification in vibration-based damage techniques is classified into four levels: (i) Level 1, determination of damage occurrence; (ii) Level 2, detection of damage location; (iii) Level 3, measurement of damage severity; and (iv) Level 4, prediction of the remaining service life of the structure.

In recent work, considerable new developments in vibration-based damage detection have emerged, particularly the application of advanced sensing technology, signal processing techniques, machine learning (ML), and artificial intelligence (AI). This chapter will review the development of vibration-based damage detection techniques in offshore platforms between 2010 and 2021, classified by various application areas, types of damage, damage tools and instruments, and damage detection strategies. The recent developments and challenges are also summarized to help researchers and practitioners to develop more reliable, practical methods for developing offshore structures in the future.

In 2010, Elshafey et al. [108] performed an in-air laboratory experimental test to detect damage occurrence and location in a four-legged scaled jacket platform using two different random loadings. Accelerometers and strain gauges were used to measure the responses. The combination of the random decrement technique and a neural network can be applied to detect damage magnitude and location. However, it is not recommended that the methodology replaces the traditional inspection method. Damage detection in another offshore structure was performed by Razi and Taheri [109] in 2014. The propagating notch of the offshore submerged pipe's girth weld was detected using numerical simulation and experiment-based tests. An experiment using a piezoelectric sensor was conducted using impact and chirp excitations to verify the correctness of the finite element model. The approach selected empirical mode decomposition (EMD) to create energy-based damage indices (EMD_EDIs). The results confirmed that the proposed method can identify the existence and progression of the propagating notch.

In the same year, Liu et al. [110] introduced a new modal strain energy change ratio (MSEC) damage indicator for reducing the effects of damage accumulated before the first measurement in an offshore scaled structure. The method is essential because such platforms have seldom been tested for damage detection during their previous lifetime. An experiment-based dynamic test in a water tank using triaxial capacitive accelerometers was conducted to evaluate the proposed method's efficiency. It was found that the method can accurately localize the damage between two adjacent measurements, even with spatially incomplete data. Research on damage detection based on laboratory tests considering semi-real conditions with different boundary conditions is scarce. Asgarian et al. [111] proposed an experimental test of a scaled six-legged jacket-type offshore platform in hinge-based conditions and pile-supported conditions, performed using a small and limited number of sensors to detect the damage location. The rate of signal energy (RSE) using wavelet packet transform to detect the damage location was introduced. In this work, the dynamic responses of the real offshore platform from forced vibration tests could be used as inputs. Furthermore, at present, two crucial challenges in damage detection for situ platforms are spatial incompleteness of data and noisy data measurement.

In 2016, Hosseinlou and Mojtahedi [112] developed experimental and numerical-based damage detection in a fixed offshore platform using a combination of the Cross Model Cross Mode (CMCM) method and the Pseudo Simplified (PS) model. The major problems in detecting damage in an offshore platform, such as the process being time-consuming and expensive and involving a high computational burden, are solved using this methodology. Furthermore, ocean wind turbines have attracted increasing interest due to their potential energy resources. However, several technological issues, such as damage detection, still require further development for their successful application. Opoka et al. [113] addressed crack detection and localization based on strain measurements in an offshore tripod-supported ocean wind turbine structure using experimental and numerical tests. An FBG sensor was used in the experimental test to obtain strain measurement data. The proposed approach can be applied in frames and truss structures. Such research was also performed by Mieloszyk and Ostachowicz [114]. In this work, an experimental laboratory investigation in a water basin using FBG sensors installed on the underwater level was conducted to monitor crack occurrence and location in an offshore wind turbine support structure. The structure was excited by both artificial irregular waves and wind conditions. The result verified the convenience of FBG strain sensors permanently placed on the underwater element of a wind turbine tripod.

To establish an automated methodology for crack propagation detection in operational conditions of a wind turbine in the Polish Baltic sea, Luczak et al. [115] compared the result of acceleration-based damage detection between experimental modal analysis (EMA) in dry conditions and operational modal analysis (OMA) in a towing tank with different real operational conditions. Piezoelectric tri-axial accelerometers were deployed to measure the acceleration signal. The developed method revealed consistent results with either the EMA or OMA method, and its applicability in different operational conditions. In real applications, when monitoring a jacket platform, it is necessary to consider unstable wave excitations that influence the structural response. To eliminate the influence of environmental effects, Li et al. [116] proposed combining the cross-correlation function and principal component analysis (PCA) method using both experimental and numerical tests. The result showed that the proposed method has good accuracy in different damage locations and severities under random wave excitations. However, complicated noises and operational effects still existed in the practical test. Fathi et al. [117] proposed a Bayesian approach for damage severity assessment for the first time in offshore structures. A 2D fixed offshore platform with three damage scenarios was developed, and an experimental test using uni-axial accelerometers was carried out to validate the method. The proposed method was more efficient under high-level noisy and incomplete data and more accurate than previous methods for damage detection in offshore platforms. A review of the research on vibration-based damage detection methods on offshore structures from the early 2010s until 2020 is outlined in Table 6.

Table 6. Landmark research on the roles of sensors in vibration-based damage detection on offshore structures.

Authors	Application and Damage Scenario	SHM Tools/Instruments	Damage Detection Strategy	Important Remarks
Elshafey et al. (2010) [108]	A four-legged scaled jacket platform with four braced floors and one non-braced floor Damage is assumed by cutting one section of the member with four damaged locations	4 wired accelerometers and eight strain gages, hydraulic actuator, computer (DAQ system and Labview), and 407-controller	In-air damage detection (Level 2) experiment using random excitation (white noise and Pierson–Moskowitz spectra) was conducted. Sensors were used to measure free decay responses and extract them using random decrement. The free decay and its time derivative were used as inputs for a neural network for damage detection (Level 2).	The success level of the method depended on the damage location. It is more sensitive to predicting damage in diagonal members than in horizontal members.

Table 6. *Cont.*

Authors	Application and Damage Scenario	SHM Tools/Instruments	Damage Detection Strategy	Important Remarks
Raji and Taheri (2014) [109]	Offshore submerged pipeline Damage in the pipe's girth weld against a propagating notch from 1 mm in depth to a depth of 4 mm	8 flexible piezoelectric actuator/sensors (pa16n), pneumatic hammer, piezoelectric transducers, signal generator (33210A), power amplifier (790 series)	Damage detection (Level 2) of numerical simulations using ABAQUS and experimental test in the laboratory tank was compared using two excitations (impact and chirp excitation). The experimental test verified the integrity of the numerical model.	The chirp excitation method was more effective than the impact excitation for damage detection in the submerged pipes. SHM of the submerged pipe's girth weld could be efficiently accomplished with a minimum of two transducers—one acting as a sensor and the other as an actuator—bonded on each side of the girth weld.
Liu et al. (2014) [110]	A four-legged offshore structure with steel tubular members Damage modeled by removing the bolt of the flanges in two locations	12 triaxial capacitive accelerometers (4803A-0002), PL64-DCB8, integrated measurement and control cooperation, data acquisition system (NI-9215)	New proposed method of damage localization (Level 2) was developed to identify the damage between two adjacent measurements using modal strain energy change ratio (MSEC). The first scenario assumes that damage is modeled, and the first measurement is conducted. Then, the damage is modeled in the two other locations, and the second lot of measurement data are collected.	The new proposed method can be used to accurately localize damage that occurs in the time interval of two adjacent measurements, even in spatially incomplete situations.
Asgarian et al. (2015) [111]	A scaled six-legged jacket-type offshore platform in the Persian Gulf 11 damage scenarios in hinge-based condition and 7 damage scenarios in pile-supported condition	8 accelerometers (2 three-dimensional, 2 two-dimensional, and 4 one-dimensional accelerometers), eccentric mass shaker, actuator system, 4-channel TMR data logger	Experimental dynamic test was analyzed using the rate of signal energy (RSE) using wavelet packet transform to detect damage location (Level 2). Two sets of signals from the intact and damaged conditions are required to detect damage. RSE is calculated for each node where sensors exist, and high RSE represents damage location.	Only sensors on damaged members have desirable RSE, and the other sensors have lower values of RSE. RSE has desirable values for sensors near damage location and non-significant values for other sensors.

Table 6. *Cont.*

Authors	Application and Damage Scenario	SHM Tools/Instruments	Damage Detection Strategy	Important Remarks
Hosseinlou and Mojtahedi (2016) [112]	1:100 scaled fixed jacket offshore platform (SPD9 in the Persian Gulf) with four legs constructed by ABS tubes Damage severity (3 damage scenarios) defined as the percentage of stiffness loss of the FE model	2 light uni-axial accelerometers (4508 BandK), load cell, power amplifier (model 2706), electrodynamic exciter (type 4809) with a force sensor (AC20, APTech), PULSE software processing package	An experimental test is performed to provide validation of the FE result. Then, the FE-based updating method is developed, and sensitivity analysis is used as the basis to validate of Pseudo Simplified (PS) model. The cross model cross mode (CMCM) method and Pseudo Simplified (PS) model technique are proposed to detect damage severity (Level 3) in using spatially incomplete and with noise in modal data.	The proposed method can be applied for an offshore platform in service to save both time and cost. The result indicates that the improved CMCM method with the PS model technique is capable of damage diagnosis with limited, spatially incomplete modal data.
Opoka et al. (2016) [113]	Scaled offshore wind turbine with tripod support structure Crack is assumed by dismantled flange located in one of the upper braces of the tripod model	8 optical three-element delta rosettes with fiber Bragg grating (FBG) sensors, impact hammer SmartScan, interrogator smart fibers for data acquisition	Both experimental and numerical vibration tests are conducted to detect and localize damage (Level 2) using the root mean square deviation (RMSD) estimator as a damage indicator.	The FBG sensors can measure strains along the grating length, and at least two sensors directed along directions are needed to increase confidence result in the specified location. The effectiveness of localization depends on the chosen new frequency peak used in the analysis.
Mieloszyk and Ostachowicz (2017) [114]	Offshore wind turbine with tripod support structure made of aluminum alloy (EN AW6060T66) Pipes Crack in the circumferential direction of the brace	15 fiber Bragg grating (FBG) single-axis sensors (Micron Optic, os3120), 1 temperature compensation sensor (Micron Optics, os4100), interrogator (Micron Optic si425-500), water basin, wave generator, rotor	The damage detection (Level 2) is analyzed by an experimental laboratory test in a water basin that excites both artificial irregular waves and wind using FBG strain sensors. Strain measurement from FGB sensor is used for the frequency domain decomposition (FDD) method to detect and localize damage using different damage indexes.	The damage is most visible on sensors from the S1 set than on the others. Comparing strain values for every sensor to detect damage is possible for the same conditions only. The damage index can develop an automatic calculation process and apply it for unmanned SHM.

Table 6. *Cont.*

Authors	Application and Damage Scenario	SHM Tools/Instruments	Damage Detection Strategy	Important Remarks
Luczak et al. (2019) [115]	Scaled model of a tripod-type supporting structure of an offshore wind turbine in the Polish Baltic sea made of cylindrical aluminum tubes Five circumferential crack scenarios in the upper central joint of the tripod support structure	5 piezoelectric tri-axial accelerometers, transducer, electrodynamic shaker, hammer, Ship Design and Research left Towing Tank in Gdansk (Poland)	Vibration-based damage detection (Level 1) of the tripod model is analyzed by comparing acceleration signal between experimental modal analysis (EMA) in air using a shaker as input and operational modal analysis (OMA) in a towing tank with different operational conditions simulating real-life operation.	The low-frequency excitation cannot provide a suitable energy level at higher frequencies, with consequent difficulty in the post-processing of the acquired data. The sensor should be installed below the sea surface at a considerable distance from the acquisition equipment placed on the upper part of the turbine with consequent signal amplification.
Li et al. (2020) [116]	NB35-2 WHPB jacket oil platform Member loss, stiffness reduction, and supporting capacity reduction in the piled foundation in various locations	Servo electric cylinder, force sensor, 18 three-directional AKE398B-08 accelerometers (capacity ±8 g), data acquisition system 500 Hz	Damage location and severity (Level 3) measurement using a combination of the cross-correlation function and PCA-based outlier algorithm.	The accelerometers in the experimental test were installed on the deck, with the positions being consistent with the measuring points set in the numerical simulations.
Fathi et al. (2020) [117]	A scaled 2D fixed offshore platform in Bohai sea Damage scenarios modeled by removing the branches	15 uni-axial accelerometers, electrodynamic exciter (type 4809), force sensor (AC20, APTech), power amplifier (model 2706)	Damage detection (Level 3) using a new Bayesian model updating framework is proposed with incomplete frequency response function (FRF) data.	The proposed damage detection method is efficient under noisy high-level data; the accuracy decreases as the number of measured DOFs is reduced.

The past work mainly presented lab-based vibration measurements of structures with known structural and external load parameters. However, for a long-served offshore platform, the structural stiffness degrades, and the mass and load parameters change from the original values; therefore, it will be difficult to measure the structure's comprehensive modal parameters. Therefore, a more suitable model updating method needs to be developed to solve offshore platforms' unknown parameters. Liu et al. [118] conducted comprehensive measurement of vibration induced by working equipment on a long-served eight-legged offshore platform (W12-1) to analyze the vibration characteristics and simulate the offshore responses. In situ measurement was carried out on the decks and columns of the platform using several cases, with 51 measured points. The layout and the in situ measurement points are shown in Figure 21. In terms of vibration signal acquisition, the data acquisition device SDY2400A was used for real-time acquisition of signals. CA-YD-108 acceleration sensors were placed at all measurement points to collect acceleration data in the site test. The 3D vibration signals from sensors were collected on the compressor and column, and only the vertical vibration signals of the points on the deck were collected. In this work, a new updating method for the finite element (FE) model was developed to determine the unknown structural parameters of offshore platforms. A comparison of the vibration displacement amplitude at the main frequency of 16.5 Hz obtained from the FEM simulation and the site measurement is shown in Figure 22. As shown, each case's simulated values are in good agreement with the measured values, thereby indicating that the modified model is reliable and the updating method for determining unknown parameters in a complex offshore structure is reliable.

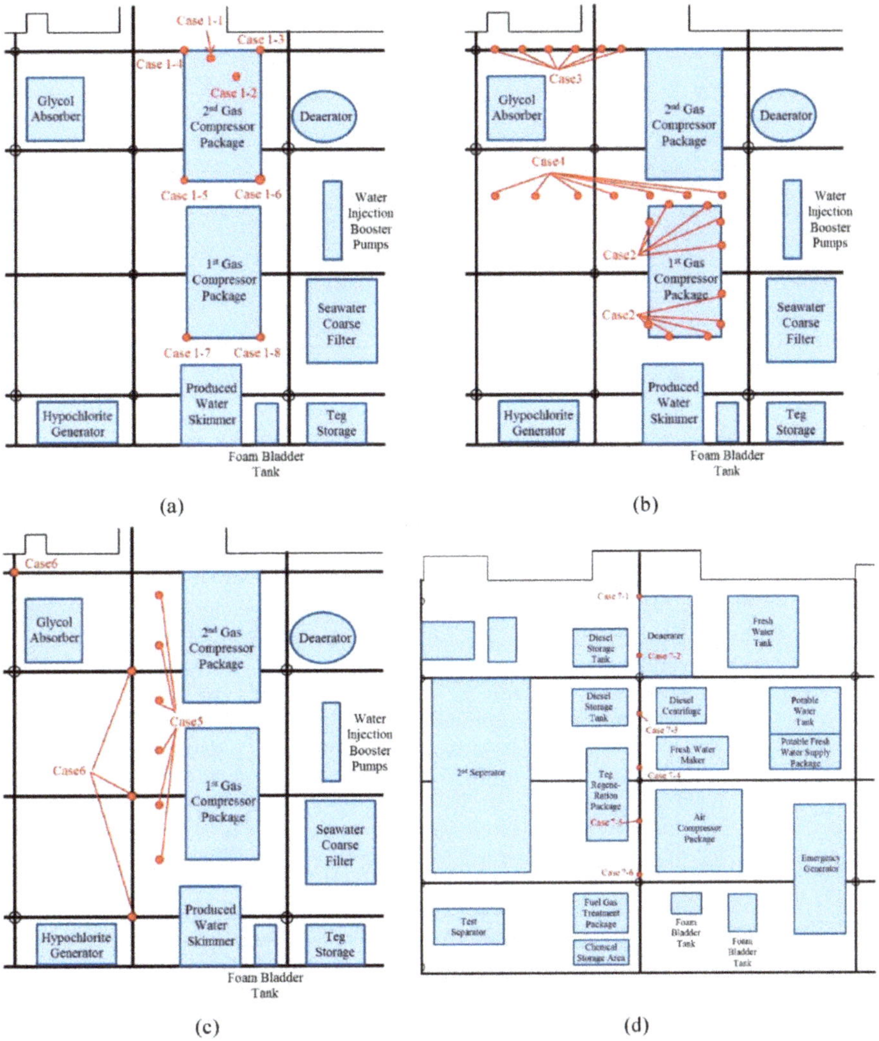

Figure 21. The layout and the in situ measurement points: (**a**) case 1; (**b**) cases 2–4; (**c**) cases 5 and 6; (**d**) case 7 [118].

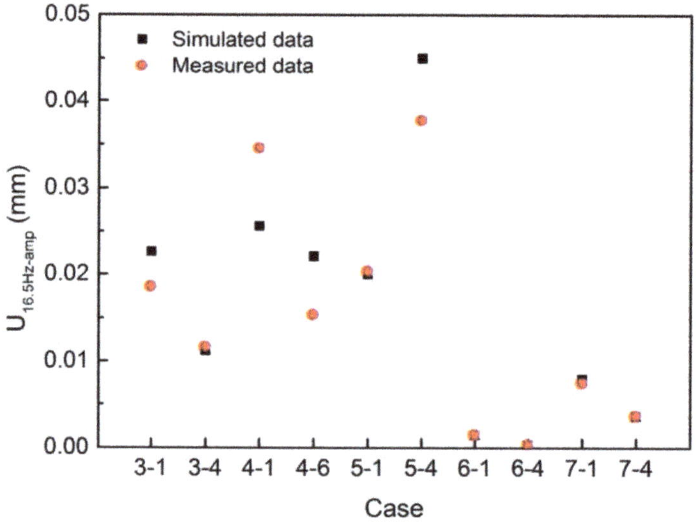

Figure 22. A comparison of the vibration displacement amplitude at the main frequency of 16.5 Hz [118].

In a recent development, machine learning (ML) application for vibration-based damage detection in a wind turbine has been highlighted. Vidal et al. [119] introduced a structural damage methodology in a jacket-type offshore wind turbine using a vibration response-only case. This methodology stems from the fact that in the offshore structure, excitation cannot be imposed and is often not measurable. In this work, two different machine learning algorithms, namely, k nearest neighbor (k-NN) and quadratic kernel support vector machine (SVM), are investigated. The proposed damage detection and localization method at four different crack locations was conducted in a small-scale laboratory structure excited by a steady-state wind load. A general overview of the experimental test is illustrated in Figure 23. Firstly, Gaussian white noise was applied to simulate the wind load, given by a functional generator (GW INSTEKAF-2005 model). The signal was amplified and passed to the inertial shaker (model GW-IV47). The response was monitored by triaxial accelerometers (model 356A17, PCB Piezotronics), which were linked to a data acquisition system (cDAQ-9188 chassis and six NI-9234 modules from National Instruments). Furthermore, the raw data were pre-processed using group reshape and column scaling, and principal component analysis (PCA) was applied as a feature selection technique. After that, the two different machine learning algorithms were compared. As seen from Tables 7 and 8, the damage detection and localization method using the quadratic SVM classifier obtained a better result than that from the k-NN algorithm, achieving a higher result with high prediction speed and a short training time. In future work, the proposed method should be explored in a more realistic environment.

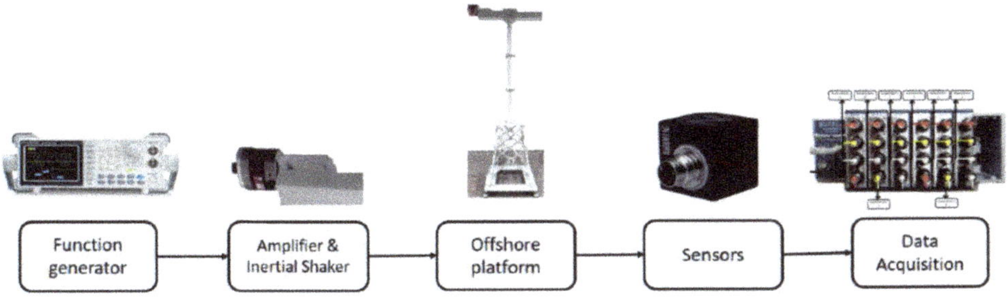

Figure 23. Overview of the experimental vibration response-only test [119].

Table 7. A confusion matrix for the k-NN algorithm [119].

	0	1	2	3	4
0	>99%		<1%		
1	<1%	98%	2%		
2	7%		93%		<1%
3				75%	25%
4	1%		<1%		99%

Table 8. A confusion matrix for the SVM model [119].

	0	1	2	3	4
0	>99%	<1%	<1%		
1	<1%	99%	<1%		
2		<1%	99%		
3				>99%	<1%
4					100%

Similar work to that of Vidal et al. [119] was performed by Puruncajas et al. [120], who investigated vibration response-only damage detection in a similar structure using different machine learning applications. A vibration response-only methodology using an accelerometer and a convolutional neural network (CNN) was proposed and tested on an experimental laboratory setup. A laboratory experiment test was conducted on a steel jacket-type offshore wind turbine under different damage scenarios to obtain accelerometer data; see Figure 24a. Several SHM tools and instruments were used in the experimental test. A white noise signal was generated by employing a function generator (GWINSTEK AF-2005 model), and the signal was amplified to a modal shaker (GW-IV47) that induces vibration. Eight triaxial accelerometers (PCB R Piezotronics, model 356A17), as illustrated in Figure 24b, were set and connected to six National Instruments^{TM} cartridges (NI 9234 model) to gather raw data. After raw time series data were obtained, the data were pre-processed (signal-to-image conversion) with sensors to obtain a dataset of 24-channel gray-scale images. A 24-channel-input CNN was designed and trained for classification of the different structural states. The proposed SHM strategy using a deep CNN with data augmentation demonstrated outstanding performance with all considered metrics, with an overall accuracy of 99.90% (see Table 9). In future research, the proposed SHM strategy of experimental laboratory testing in a water tank needs to be validated using a more realistic environment.

Table 9. Comparison of accuracy, validation error, training error, and training time when using data augmentation or when using the original dataset [120].

	Accuracy (%)	Validation Error	Training Error	Training Time	# of Images
Without data augmentation	93.81	0.1692	0.1167	11 min	6400
With data augmentation	99.90	0.0044	0.0026	1196 min	1,612,800

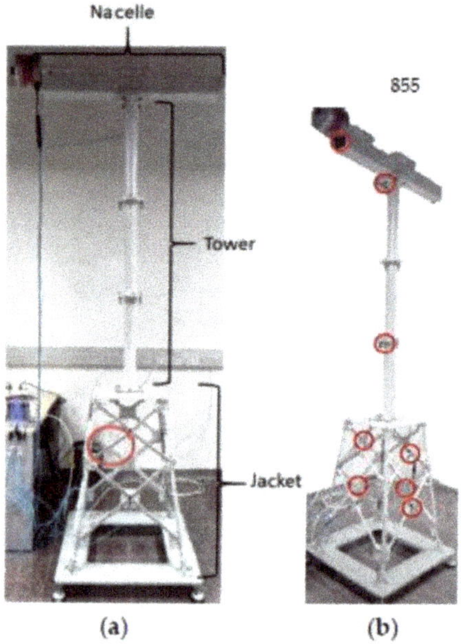

Figure 24. (**a**) An experimental setup with the damage location shown (red circle). (**b**) Locations of accelerometers [120].

Furthermore, one of the difficulties in long-term continuous monitoring of offshore platforms is the data acquisition requiring a high sampling rate and lots of computational and spatial resources. To solve this problem, the effect of sampling rate on structural damage identification in offshore structures, an FPSO soft yoke single-point mooring structure was investigated by Tang et al. [121]. This study compared the acceleration and posture sensors commonly used in offshore monitoring. In this work, random decrements and their spectrum characteristics were extracted from acceleration and displacement data under different sampling rates. The results revealed that data collected at a low sampling rate can also identify the parameter changes from structural damage.

7. Corrosion Control and Monitoring

Corrosion, one type of age-related damage experienced by most marine structures, is caused by long-term chemical and electrochemical reactions due to the harsh marine environment. Marine structures with a long-term operating life, such as a ship, offshore facility, submarine pipeline, wind turbine, etc., are vulnerable to this common type of damage. It may lead to failures of mechanical systems and thereby reduce their safety level and structural integrity. Therefore, continuous corrosion control and monitoring of marine structures are essential for ensuring their safe operation. A corrosion control and protection system can delay and mitigate corrosion initiation in marine structures and allow them to reach the expected service life. The most common corrosion control technologies used in marine structures are coating (PA), corrosion allowance (CA), and galvanic anode cathodic protection (GACP). Corrosion monitoring techniques can be divided into two major categories: offline monitoring and online monitoring. Offline monitoring measures the corrosion status within a specific time interval, while online measurement involves continuous real-time monitoring of the corrosion status of marine structures. Offline monitoring methods, including magnetic leakage and ultrasonic methods, are usually used. In contrast, electrochemical measurement, resistance probes, underwater robot probes, and imaging technology are being used in recent developments.

This section presents a comprehensive review of the development of corrosion monitoring in the marine environment over the past decade. The monitoring scopes, measurement methods and principles, and related advanced sensing technology and instruments will be highlighted. In 2010, Chen et al. [122] conducted a laboratory test to evaluate corrosion on a submerged plate using the Lamb wave-based technique and probability-based diagnostic imaging using time-of-flight (ToF). A pair of transducers were compiled at the fluid–solid interface, acting as the wave transmitter and receiver to generate Lamb wave signals. The signals were then processed with a signal digitizer (Agilent® E1438A). It was found that corrosion can be identified precisely with the support of a probability-based diagnostic imaging approach. In 2013, two sensors (Anode-Ladder-System and Corrowatch System) were developed to monitor the corrosion state in the steel rebar of coastal concrete by Xu et al. [123]. The sensor design was based on a three-electrode electrochemical test architecture, where the sensors' performance was evaluated using a dry–wet cycle test on the upper reinforcement of the concrete block. In this work, Q235 steel bars were used as working electrodes, titanium bars were used as reference electrodes, and stainless steel bars were used as counter electrodes. The investigation revealed that corrosion sensors can efficiently monitor chloride penetration into concrete.

In the same year, Mu et al. [124] conducted laboratory-based corrosion behavior monitoring of steel structures in a marine tidal zone using electrochemical impedance spectroscopy (EIS). Comb-shaped electrodes were used for the electrochemical monitor, and a ZKW881 rapid corrosion tester was used to monitor frequency impedance. In this work, four different tidal positions representing the high tidal zone, the middle tidal zone, the central zone between the middle tidal zone and low tidal zone, and the low tidal zone were assumed. The results show that the corrosion behavior of steel in immersion and wet stages is different. The work showed that the fastest corrosion rate is attributed to the longest wet time in a tidal cycle. In 2015, lab-based experimental corrosion rate monitoring of two steel rebars in wet conditions using a TFBG corrosion sensor was performed by Islam et al. [125]. During the test, two steel rebars were immersed in a container filled with water containing NaCl. A TFBG sensor was attached to the rebar that was electrically connected to an anode DC power supply. The measurement was performed by comparing the spectral response of cladding resonances and the SRI change during the corrosion process. Corrosion monitoring in different material types using other methods was presented by Nie et al. [126]. The crevice corrosion monitoring relevance of a copper alloy marine structure based on electrochemical sensing using a boron-doped diamond electrode (BDD) was investigated. The electrochemical behavior on the BDD electrode was investigated in NaCl solution (equal to average seawater salinity). The proposed method is promising for monitoring the corrosion of copper alloys in marine environments. Ships are periodically situated in a dockyard and inspected by well-trained surveyors and class societies to assess their physical and structural performance. The expenses required for onboard inspection can be significant, so introducing new technological tools is a must.

Ortiz et al. [127], under the EU-FP7 project MINOAS in 2016, developed a micro-aerial vehicle (MAC) with different locomotion capabilities to collect images focusing on a remote area that the surveyor has difficulty accessing. A novel method for detecting coating breakdown/corrosion (CBC) on a complete ship hull was performed by classifying image pixels (color and texture information) using a three-layer feed-forward artificial neural network. The MAC comprises a flight management unit (FMU) for platform stabilization, an inertial measuring unit (IMU) as a height and distance sensor, embedded PC and control software, and a camera set. The detection method using this novel technology was reported to successfully classify images between coating breakdown and corrosion. In the following year, sensing technology using an electrochemical probe was used to monitor atmospheric corrosion by Dahai et al. [128]. Monitoring of the 60-day corrosion form in the extreme atmospheric conditions in the Zhoushan offshore environment was conducted by comparing the electrochemical noise (EN) data of two different steel plates. The EN data were analyzed by wavelet transform to identify corrosion and transform it into an energy distribution plot (EDP). It was discovered that noise resistance is well correlated with weight loss data. Furthermore, the concentration of salt as a critical indicator for diagnosing the corrosion occurrence of an offshore concrete structure was investigated by Luo et al. [129]. In this work, a 48-h experimental test for salinity detection was performed using a low-cost salinity sensor with four sensor probes multiplexed into one system. The experimental findings indicate that the proposed sensor has higher sensitivity with a lower cost compared to previous work.

Two years later, the application of artificial intelligence (AI) technology for corrosion monitoring in naval architecture and ocean engineering was investigated in depth by Yao et al. [130]. Applying deep learning to ship corrosion monitoring reduces workforce and financial resources. The corrosion of a ship hull plate was detected and monitored using a convolutional neural network (CNN) model. A total of 30,000 corrosion image datasets were trained using AlexNet as the HCDR network, and a classifier model to recognize the position and location of the corrosion was developed with an overlap-scanning sliding window algorithm. The proposed CNN model could detect various types of superficial structural damage, such as delamination, voids, and spalling and corrosion damage. In 2020, a comparison of the monitoring of long-term corrosion of carbon and stainless steel reinforcements exposed to the actual marine environment was intensely studied by Gartner et al. [131]. In the monitoring strategy, electrical resistance (ER) probes and coupled multi-electrodes (CMEs) were placed at five different exposure zones of these concrete columns for 52 months. The ER probes and CMEs were virtually connected through zero resistance ammeters (ZRAs). The ZRAs were then connected to a 16-bit analog-to-digital (A/D) converter. The observations showed that ER probes are able to detect corrosion initiation and propagation and detect localized corrosion. At the same time, CMEs can measure the spatio-temporal evolution of the corrosion process (microcell corrosion). The landmark research on sensor technology application for monitoring corrosion in the marine environment over the last decade is summarized in Table 10.

Table 10. Landmark research on the sensor technology applications for monitoring corrosion in various marine structures.

Authors	Observational Platforms	Tools/Instruments	Important Remark
Chen et al. (2010) [122]	Quasi-circular zone chemical corrosion on the submerged aluminum plate	Lamb-wave-based identification (pulse-echo measurement scheme) with the assistance of probability-based diagnostic imaging technique	Rectification and compensation are required for medium coupling when applying Lamb wave-based corrosion monitoring with coupled media.
Xu et al. (2013) [123]	Steel rebar of coastal concrete	Anode–ladder system and Corrowatch System	Monitoring test corrosion state is developed based on electrochemical polarization dynamics. The ohm-drop effect can be ignored to obtain a more accurate result.
Mu et al. (2014) [124]	Steel structures (Q235B) in the marine tidal zone	Electrochemical impedance spectroscopy (EIS)	Simulating the marine tidal corrosion under laboratory test is feasible, and the comb-shaped double electrode probe is suitable for in situ EIS monitoring.

Table 10. *Cont.*

Authors	Observational Platforms	Tools/Instruments	Important Remark
Islam et al. (2015) [125]	Steel reinforcing bar of the marine concrete structure	Tilted fiber Bragg grating (TFBG) sensors	Higher-order cladding resonances exhibit relatively higher sensitivity to corrosion times than the lower-order cladding resonance existing near the Bragg wavelength.
Nie et al. (2016) [126]	Copper alloy marine structures	Oxygen-terminated Boron-doped diamond (BDD) disk by Diamond Detectors Ltd.	The monitoring by quantification of copper ions using the DPV technique results in an excellent linear relationship between peak current density and copper ion concentration.
Ortiz et al. (2016) [127]	Coating breakdown/corrosion (CBC) in vessel structure (bulk carrier)	Semi-autonomous micro-aerial vehicles (MACs)—INCASS aerial platform Features: Laser scanner Hokuyo UST-20LX, downward-looking LIDAR-Lite laser range finder, cameras and video footage, LED, Intel NUC D54250WYB embedded PC	Future enhancement steps should comprise the fusion of the laser scanner with optical flow sensors to enlarge the inspection area.
Dahai et al. (2017) [128]	Atmospheric corrosion of T91 steel and Q235B exposed to Zhoushan offshore environment	Electrochemical noise (EN) is composed of a compact (cRIO) module, two zero resistance ammeter (ZRA) modules, a lithium-ion battery, DC power supply, and two atmospheric-corrosion-monitoring probes	The application of atmospheric probes in the offshore environment for a long time may not be suitable due to the formed rust layer.
Luo et al. (2017) [129]	Offshore concrete structure	Optical fiber-based salinity sensor based on multimode plastic optical fiber (POF) as a sensor probe	More improvements are needed to reach the industrial expectation, including replacing the multi-axis stage to realize automatic survey, making the sensor probe and light source movable, and re-designing the sensor probe to attain simultaneous calculation.
Yao et al. (2019) [130]	Ship hull structural plate	Convolutional neural network (CNN) model using image data (camera)	The CNN model training requires massive image data with different categories and conditions.
Gartner et al. (2020) [131]	Concrete columns with carbon and stainless steel reinforcements exposed to the north Adriatic coast environment	Electrical resistance (ER) probes and coupled multi-electrodes (CMEs)	The infrequent sampling rate of measurement by ER and CMEs is more problematic due to the severe changing exposure conditions. Continuous monitoring with wireless data acquisition can be implemented.

Recently, Ahuir-Torres et al. [132] conducted a benchmark autonomous detection and monitoring study of offshore wind turbines using real-time remote sensing technology. Under a project funded by Innovate UK, they invented smart wireless miniaturized sensors utilizing the Internet of Things (IoT) and integrated with satellite and terrestrial communication networks, named the iWindCr project. The arrangement plans of the iWindCr WSN system and the user interface prototype of iWindCr software are comprehensively illustrated in Figures 25 and 26, respectively. To detect and monitor the progress of corrosion, two passive electrochemical sensors (open circuit potential (OCP) and zero resistance ammeter (ZRA)) are merged into the sensor interface of the iWindCr system to monitor physical and electrochemical changes remotely and quantify the electrode potential and current. In this work, the corrosion parameter output and threshold value data of the foundation, tower, and gearbox of an offshore wind turbine were collected. The database can be used as a benchmark and guideline by end-users to implement corrosion monitoring in other complex structures. The proposed benchmark study points out that the technology should be applicable in different industrial sectors, such as marine, oil and gas, automobiles, etc.

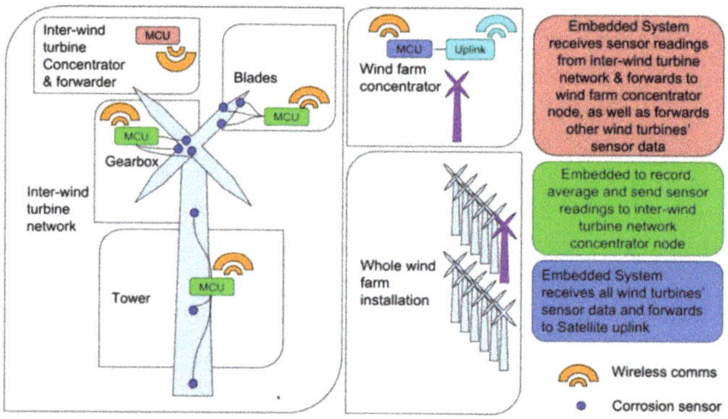

Figure 25. Arrangement plan of the iWindCr WSN system [132].

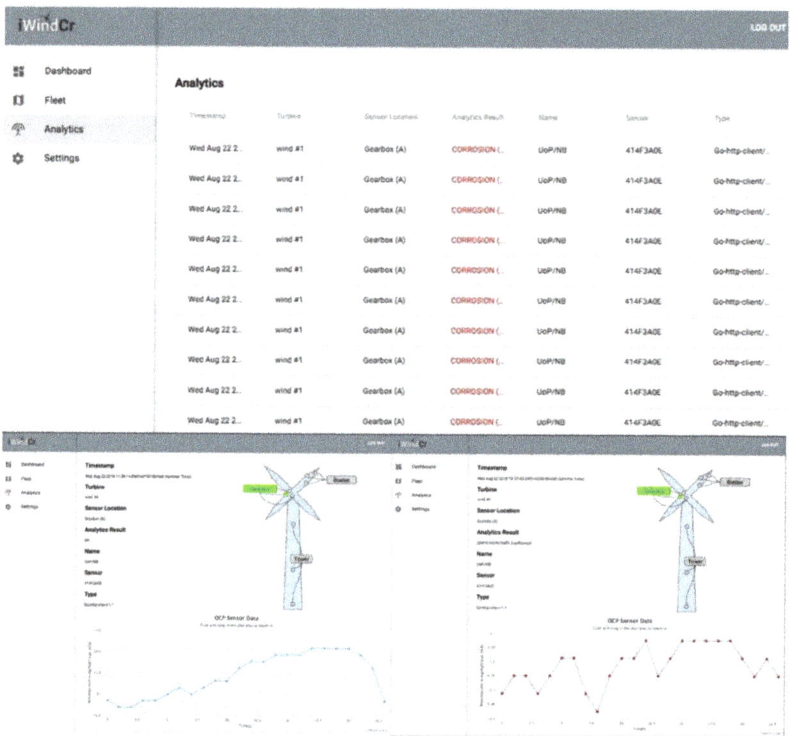

Figure 26. The user interface of iWindCr software [132].

In another recent study, Ha et al. [133] investigated the monitoring of corrosion in structural members of a steel bridge truss in a real marine environment representing a severe corrosion environment due to high levels of airborne chlorides and humidity. In this case, steel plates and atmospheric corrosion monitoring (ACM) sensors were installed in different locations to measure the corrosion depth and galvanic corrosion current. The plates and sensors were installed in several places, including on the diagonal, horizontal, vertical, and portal members. In the experimental setup, the monitoring plates were placed on horizontal or diagonal truss members. The ACM sensors were placed on members with a lead-acid battery and a data logger to record the measured corrosion current data. Furthermore, a thermo-hygrometer was also used to recognize atmospheric corrosion environments during monitoring. The corrosion monitoring data were measured after 6 and 12 months of exposure to evaluate the corrosion damage and the corrosion grade. Based on the results, it seems possible to estimate corrosion using monitoring steel plates and ACM sensors. The horizontal members experienced a more extensive corrosion depth and the highest galvanic corrosion current compared to diagonal, vertical, and portal bracing members after the 12-month exposure period. These results, especially the most critical result considering the extreme corrosion rate in the horizontal members, show that maintenance is necessary to monitor the local corrosion condition of the structures.

In Maritime 4.0, onboard inspection can be extensively assisted by a robotic platform. Traditional onboard inspection using human surveyors was confirmed to be relatively expensive, time consuming, and high risk. Considering these problems, Poggi et al. [134] launched the adoption of robotics and autonomous systems (RAS)-aided inspections under the ROBINS project, funded by the Horizon 2020 EU Research and Innovation program. The aim was to fill the gap between traditional onboard assessments and the potential application of robotic technology. In general, an overview of RAS-aided inspection platforms with different capabilities involved in the ROBINS project is highlighted to assist the whole range of inspection needs (see Figure 27).

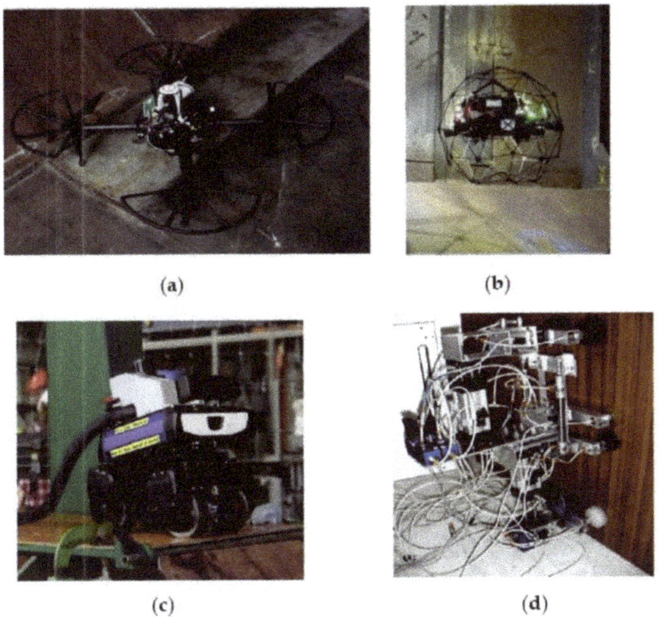

Figure 27. RAS platforms: (**a**) UIB drone, (**b**) FLY drone, (**c**) GEIR crawler, and (**d**) zoomorphic robot [134].

Two different aerial drones (UIB drone and FLY drone) are dedicated to flying in wide, open spaces and cluttered spaces for autonomous information obtention (image and video streaming). Different sensing devices, including 2D and 3D laser scanners, vision-based navigation, depth, and RGB-D cameras, are reported to offer an overview of RAS technologies. Moreover, the most common solution for close-up surveys is represented by the GEIR crawler, shown in Figure 27c. The crawler can run on the magnetic wheel directly with the structure to carry out thickness gauging and non-destructive testing. The main issue is the limited obstacle handling ability when inspecting some parts of a ship.

Another RAS platform developed by the ROBINS project to perform autonomous direct inspection using direct contact with the structure is the zoomorphic robot (Figure 27d). This portable and user-friendly platform is controlled by actuator–sensor interface technology, and each unit is equipped with a video camera and an ultrasonic thickness measurement tool. For measuring the RAS inspection abilities, a testing facility module for general purpose inspections onboard a bulk carrier has been developed (see Figure 28). The possibility of changing the environment characteristics (lightening, humidity, dirt, etc.) to simulate different conditions was also considered. The assessment is carried out by comparing the capabilities of general RAS–human inspection in orienteering, localization, and identification of actual defects found in coating conditions, such as pitting corrosion, corrosion, cracks, distortions, mechanical damage, etc. It was noted in the trials that RAS-aided inspection acquires more extensive quantitative data while minimizing intrinsic uncertainties of the collected data. However, the use of RAS-aided inspection results is considered as an assisting tool of the attending surveyor. It means that the final decision about the applicability of remote techniques is determined by surveyor experience.

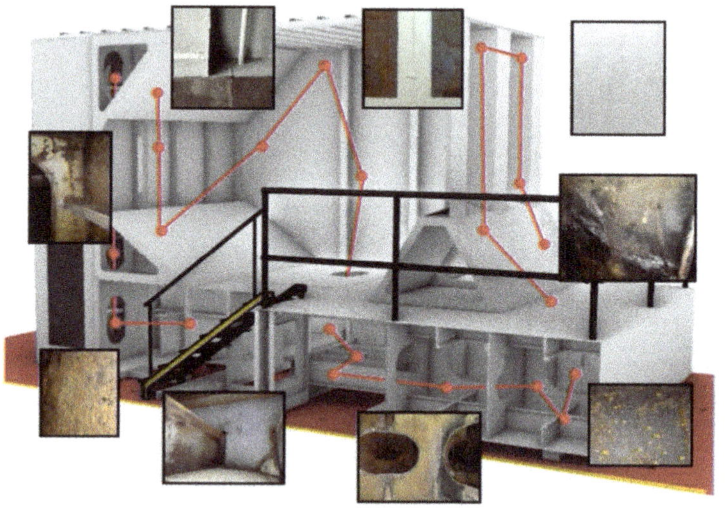

Figure 28. The construction module of a testing facility [134].

In the same year, Abdulkader et al. [135], under the National Robotics Program funding, developed a novel autonomous metal thickness inspection robot for ship hull inspection during the dry-docking condition. The proposed climbing robot, named Sparrow, is capable of conducting multiple inspection tasks on a ship hull and navigating autonomously on the vertical metal surface. According to its description, the robot platform consists of the Sparrow bot with a thickness gauge, an app with control and a video feed, storage, and the communication unit. The HMI App screen layout is split into the upper part showing the video feed from the camera and the lower part for the locomotion control buttons and autonomous buttons. An overview of the Sparrow robotic system is illustrated in Figure 29. The hardware architecture of the Sparrow robot (see Figure 30) comprises six functional units, including a Jetson Nano kit as the central computing unit; a sensor unit comprising a VectorNav IMU, a wireless 1080 p camera, and Marvelmind beacons; a thickness measuring unit; a locomotive Unit (Arduino Mega 2560 and motors); and a communication unit comprising a Bluetooth adapter. For thickness measurement, a 38DL PLUS Ultrasonic Thickness Gauge from Olympus is attached to the center part of the robot. The gauge is controlled using a linear solenoid, which makes the gauge touch the ground and obtain the thickness measurement. A series of experimental tests, including a slippage test and a thickness measurement test under various scenarios, were conducted to validate and verify the robot's parameters. It can be concluded from the test results that the proposed robot could successfully measure the metal plate's thickness in all considered scenarios. Future developments should aim at improving navigation on the curved structure and applying machine learning to classify the metal surface corrosion level.

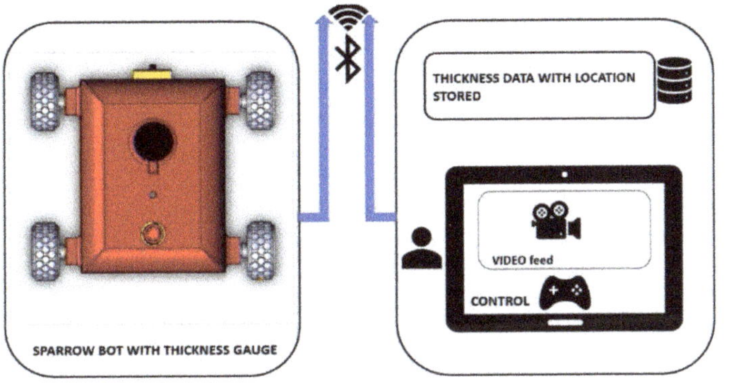

Figure 29. An overview of Sparrow, a robotic system [135].

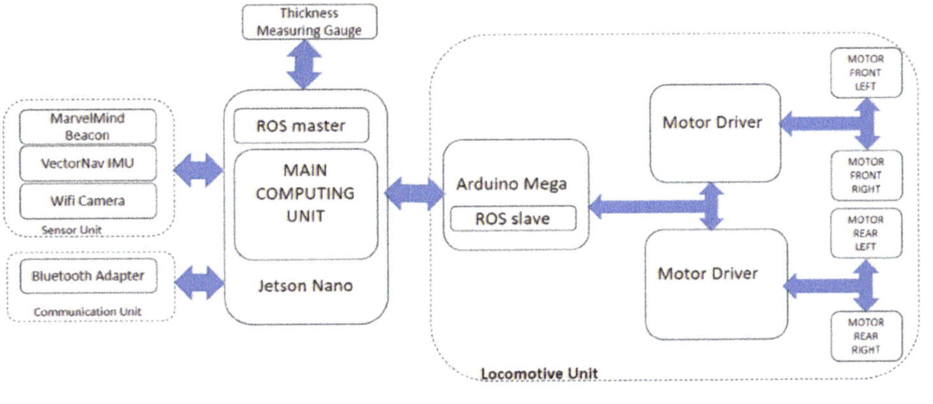

Figure 30. The hardware architecture of Sparrow, an autonomous robot [135].

8. Impact and Shock Measurements on Critical Structures

Structures in the marine environment are subject to various loadings during their building and manufacture processes due to the harsh environment, which contributes to a number of loading forms, such as nonlinear wave patterns and crack-induced corrosion. Besides these, impact phenomena also threaten the structures. The events can be triggered in almost unlimited scenarios, i.e., dust and methane explosions in a mining site, underwater explosion, and shock due to a torpedo attack in warfare state, until collision and grounding hull interaction with obstructions. Such conditions will cause severe damage to the structures and lead to immense casualties. Considering the critical influences delivered by explosions, shocks, and collision and grounding, a series of research studies have been dedicated to quantifying structural responses and loading behavior in the form of experimental tests and trials. In these, sensors are deployed to sense and record the occurred response and behavior from both the target structure and the induced impact phenomenon.

In 2010, Liu et al. [136] conducted an experiment in coal dust using a large-scale tube as the medium for explosion set-up. The values measured by 16 pressure sensors, which were mounted on the wall of the tube (along the axial direction) in the experiment, were later recorded by a data acquisition (DAQ) system with a sampling frequency of 1 MHz. Primarily, pressure sensors were used to measure the pressure histories at different points along the tube during the explosion experiment. According to the recorded measurements, propagation of the pressure wave occurs in the range of 370–420 m/s, with the maximum overpressure of the explosion being between 50 and 72 kPa. A year later, similar work was performed, but with a different explosion medium. Methane–air mixtures (with methane purity of 99.995%) were considered another possible ignition cause in a coal mine [137]. In realizing the explosion, an explosion vessel was deployed that was mounted to a pressure sensor to record the pressure variation that occurred when the ignition grew. A camera was installed on the experiment site to observe the flame structure and fireball growth. Based on the experimental observation, it was concluded that the explosion of the mixture was divided into two stages, namely, the pressure rise stage (in this state, the released energy level from the combustion reaction is higher than the heat loss to the environment) and the pressure decline stage (the reaction energy is less than the energy loss to the environment). Besides mining sites, impact and shock phenomena in sub-zero regions, i.e., the Arctic, have also received attention. The opening of the Northern Sea Route (NSR) has led to ships from Asia passing through this region to reach the European and American continents since the mid-2000s. The motive behind this was found to be the reduced sailing time, lower fuel consumption, and lower risk of pirating activities compared to the Southern Sea Route (SSR) through the Gibraltar strait. Scholars from Finland tried to quantify ice-crushing pressure on a ship hull during interaction with ice [138]. Instead of a piezoelectric pressure sensor, the experimental research used I-Scan 210 tactile sensors, in which sensing elements were installed in 44 by 44 grids with a sample rate of 50 Hz along the ship hull (in bow shoulder, midship, aft shoulder, and aft locations for the general cargo ship; and bow, bow shoulder, midship, and aft shoulder for the ice-breaker ship). Evaluation of the test results indicated that the number of contact elements is set as the number of adjacent elements with a pressure higher than the threshold (in this case, 20 kPa). This threshold was used because at a pressure level lower than this, the obtained measurement signals would be mixed with large scattering due to the various noise sources in the experimental environment.

Related to explosion loading, structural development is also accounted for in terms of sensors and measurement. Naval vessels, which are subject to explosion risk during the war state, demand more structural safety. One study improved the naval hull with a multi-layer projective structure to withstand blast and fragment/debris loadings [139]. To measure its crashworthiness, dynamic strain gauges were placed on the outer and inner plates of the liquid cabin, while accelerometers were set on the longitudinal defensive bulkhead to record impulse acceleration during the blast ignition. The experiment results concluded that the perforations due to the fragments from the ends were mainly concentrated on a square plate (side length l = 600 mm), although the travel distance of these fragments in the air reached almost 2.5 m. Additionally, the function of the liquid cabin was noted, which was installed in the second layer of the multi-layer projective structure. This part was designed to halt high-speed fragments from causing critical damage to the inner structure. Impact loadings that possibly occur are not always in the form of high-speed contact. In the case of ship collision and hard grounding, the contact is mostly classified as low-velocity impact. A series of structural trials were conducted by Pierre et al. [140] by considering the development of ship structure by arranging sandwich beams. Two cores were deployed, namely, a corrugated type and a y-frame type, as specimens that are subjects of projectile impact in the forms of quasi-static and drop-weight loading as representatives of collision and grounding. The measurements displayed different tendencies in terms of the high initial load, which was not shown by the quasi-static test. To investigate this phenomenon, an additional test was conducted, in which the initial hypothesis stating that the measured initial peak loads during the drop-weight test are highly affected by the mass of the roller and load cell was confirmed, which is the loading applied to the target beam.

The development of the structural type subjected to low-velocity impact continued in 2016 with a study in which, instead of modifying the beam geometry, a steel–concrete–polymer (SCP) composite was used to arrange a barrier to defend the internal tank construction [141]. Considering that anisotropic material was used, material characterization was performed, including the corrosion resistance sensitivity, as the application was aimed at maritime-based vessel operation. In this stage, the test was conducted according to ASTME2661/E2661M-10, which was followed by an idealized collision resistance test using a sphere with a diameter of Ø 300 mm as the indenter. In this experiment, the deformation of the construction was recorded by displacement gauges installed at the representative location points. In 2017, Moon et al. [142] re-performed the explosion test presented by the previous scholars but in a new way, i.e., underwater conditions, which was later known as an underwater explosion (UNDEX) experiment. In this work, three underwater pressure sensors (at radius r 2 m, 4 m, and 6 m from the explosive source (TNT)) and three data acquisition (DAQ) systems were deployed to measure the generated reflected wave pressure from the wall of the water tank and pressure from bubbles. In this case, an underwater shock wave was superimposed on the hydrostatic pressure. Based on the assessment of the recorded pressure history, the shock wave at a fixed location begins with an instantaneous pressure level (it reaches the peak point), which is followed by a decline. Generally, simplified estimation is represented by an exponential function. Similar to the explosion concept, shock is considered necessary in the structural performance of ship construction. The quantification of vibration, therefore, is required to be discussed in the assessment report. The research developed in [143] is a follow-up of the pioneering works in the U.S. Navy ship DDG-53, which is deployed in ship shock trials; in this work, measured data were later processed to find other vital coefficients, i.e., total damped ship vibration. The logarithmic decay rate was used to find the coefficient by fitting the peak strain values. After that, empirical mode decomposition (EMD) was used to decompose the measured data $X(t)$ as a function of time t into intrinsic mode functions (IMFs). In this solution, the high frequency of the IMFs represents noise modes that are to be eliminated to acquire data $\widetilde{y}(t)$ without noise modes. Finally, the least-squares complex exponential (LSCE) method was employed to extract modal parameters effectively.

Towards the end of the 2010s, intense underwater explosion was taken as an experimental reference. In [144], explosion was combined with the concept of material structure improvement to enhance structural performance against torpedos and mines. A polyurea coating was applied on ship hull steel to improve its underwater shock resistance, and the specimens were installed in an experimental rig. A Photron-Fastcam-SA1 high-speed camera was placed to measure the speed of the flyer plate before hitting the piston. In terms of pressure measurement, a piezoelectric pressure sensor produced by PCB Company was applied in the test rig. Later, the measured signal was acquired by a high-speed data acquisition (DAQ) instrument built by the German HBS. In the final stage, the digital imaging correlation (DIC) method was used to record the real-time displacement strain during the high-speed dynamic deformation and present the deformation process and strain field. This work offers a more complete experimental rig and setting (using piezoelectric pressure sensors), including sensors and measurement instruments, compared to the previously reviewed work in earlier years. In 2020, trials of underwater shock loading on cylindrical shells were conducted by [145]. The experimental subject in this work was very specific and quite different to that in other work, which mostly uses general plate geometry to represent thin-walled structures (except work that considers structural development). The test specimen in this work was set as a representative of the submarine pressure hull, submarine cable, torpedo, and tower structures for wind power generation. Furthermore, a different sensor type, namely, a fiber Bragg grating (FBG) sensor calibrated by a four-point bending test, was employed in this work to observe the strain of underwater structures. Other vital findings of the respective reviewed research studies are summarized in Table 11.

Table 11. Summarized findings of the sensor-based research on measuring impact and shock.

Milestone	Author(s)	Observational Subject	Tools/Instrument	Important Remarks
2010	Liu et al. [136]	Coal dust explosion in mine sites	Kistler piezoelectric pressure sensor Experimental tube	Combustion and explosion of fine particle coal dust with Ø 45–70 μm was observed faster than that of coarse particle with Ø 70–105 μm.
2011	Bai et al. [137]	The explosion of methane–air mixtures	Pressure sensor High-speed camera Explosion vessel	Methane/air mixture could be ignited in the vessel by 40-J electric spark when the methane concentration was approximately 5–13%.
2012	Kujala and Arughadhoss [138]	Steel hull–ice collision	I-Scan 210 tactile sensors	Pressure measurements conducted by the tactile sensor in the deployed model scale provide realistic distributions for the ice-induced loads. The condition is concluded from a level ice environment.
2014	Kong et al. [139]	Naval vessel's multi-layer protective structure under blast and fragment loadings	Sensors: Strain gauge Accelerometer	Stiffened plates were found to be easily penetrated by fragments. It is different compared to the thick structure, to which the fragments only caused localized damage.
2015	Pierre et al. [140]	Sandwich beams under low-velocity impact (representation of ship collision and grounding)	Phantom V12 camera Force sensors	Post-peak responses for the drop-weight and quasi-static measurements were very similar, which indicates that the beam response is not sensitive to the material strain rate or inertial effects for impact velocity of approx. 5 m/s.
2016	Niklas and Kozak [141]	Steel–concrete–polymer (SCP) composite barrier under corrosion and impact	AMSYS system VS30-V VS75-V sensors Displacement gauges	SCP structures were still in a tight condition until undergoing 5 times deeper penetration compared to the initial state.
2017	Moon et al. [142]	Ship structure subjected to the underwater explosion (UNDEX)	Underwater pressure sensor W138A10 Data acquisition (DAQ) systems: NI 9234 NI-PXIe-4497 NI PXIe-5160	Analog–digital converter (ADC) resolution may be low in the condition where the sampling rate is very high. On the other hand, even though the sampling rate is low, the effect on the measurement result will be small, with ADC resolution in a high state.

Table 11. *Cont.*

Milestone	Author(s)	Observational Subject	Tools/Instrument	Important Remarks
2018	Jun et al. [143]	Naval structure subjected to shock trial	Shock sensor Least-squares complex exponential (LSCE) method	The values of α presented in the research are less than the proposed value for DDG-53, while the values of β in this work are greater than the reference. α and β are the Rayleigh coefficients.
2019	Li et al. [144]	Polyurea-coated metal plate subjected to underwater shock wave	Piezoelectric pressure sensor Photron-Fastcam-SA1 high-speed camera Genesis data acquisition (DAQ) systems	Polyuria coating had a significant effect on the plastic deformation of the coated plate subjected to the applied loading in the trials. The recorded critical peak was approx. 87 MPa.
2020	Hsu et al. [145]	Cylindrical shell subjected to underwater shock loading (representative of marine structures)	Fiber Bragg grating (FBG) sensor Data acquisition (DAQ) systems	FBG is concluded to have excellent repeatability and linearity characteristics in recording the strain loading cycles. To avoid sensor deformation due to spontaneous loading, the sensor design may be developed with a buffer mechanism to reduce the shock effect in measurement.

In 2021, research on impacts and sensors was marked by an experimental study on aluminum plates [146]. The novelty of this work is in the specimen used, which underwent pre-cracked treatment. The primary purpose of this crack was to quantify slight defects in the dynamic response and damage degree. The launchings system embedded the experimental installation to accelerate the steel flyer plate to achieve a high speed before it flew out from the launching tube and impacted a steel piston in the starting side of the underwater shock tube. In this work, the velocity of the flyer plate was measured by a laser velocity measurement instrument, the pressure of the underwater shock was recorded by a high-range dynamic pressure sensor, and deformation of the target plate was assessed by two high-speed cameras. In terms of the deflection growth due to the crack extension of the target plate, as set according to Table 12, the final bulge height and crack length of the target plate with a small initial crack size were smaller than in the plate with the larger one under the same loading. All the measurement data related to bulge height and crack length are presented in comparative bar charts in Figures 31 and 32. Another result measured is dynamic velocity, and acceleration reflects the characteristic of bending stiffness of the target plates. In the same loading condition, specimen plates with a smaller crack size had a lower velocity and acceleration during the deformation process.

Table 12. Summarized findings of the sensor-based research on measuring impact and shock [146].

Notation	Plate Thickness (mm)	Loading Diameter (mm)	Crack Type	Crack Length (mm)
T-6	0.5	66	Single slim crack ('−')	6
T-9	0.5	66	Single slim crack ('−')	9
T+6	0.5	66	Double cross crack ('+')	6
T+9	0.5	66	Double cross crack ('+')	9

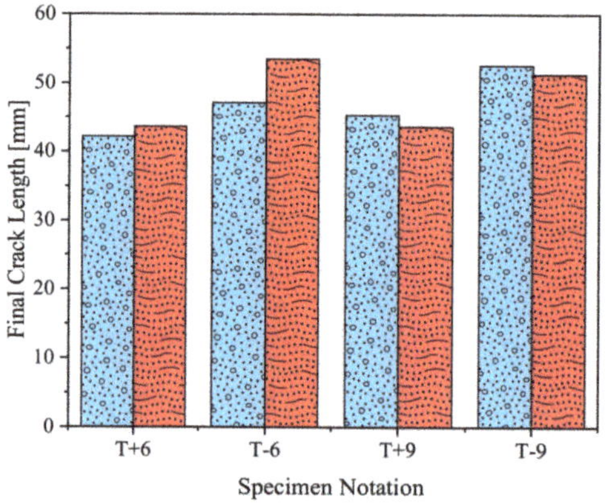

Figure 31. Measurement results of a specimen after an explosion: final crack length [146].

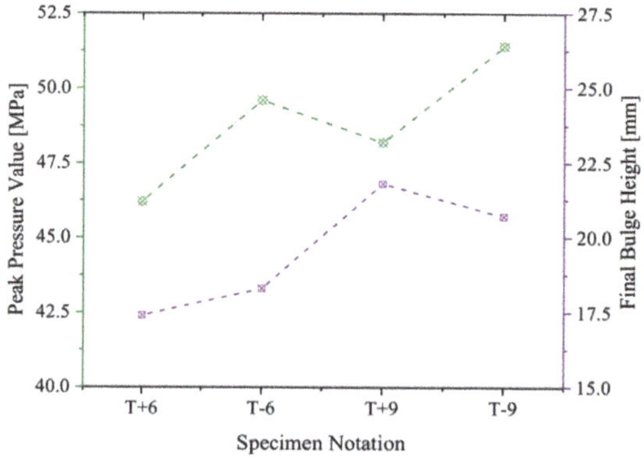

Figure 32. Recorded results of an explosion test: peak pressure value (left side) and final bulge height (right side) [146].

Another work in measurement was conducted by Mieloszyk et al. [147], who considers relatively new specimens, i.e., fiber-reinforced polymer (FRP), instead of metallic-based materials. Fiber Bragg grating (FBG) sensors (with a gauge length equal to 10 mm and without coating) were used for performing structural health monitoring (SHM), including spectra analysis and strain measurement. In order to calculate the quality of the sensors, spectra (in stages before and after embedding and loading are conducted) were determined using the Pearson correlation coefficient, as presented in Table 13. The notation set in the table is defined as follows: *F*—free sensor; E_S, E_F—start and final states of the embedding process on the boat, respectively; *R*—the finished hull of the boat; B_{10h}, B_{80h}—boat 10 h and 80 h after the sea trials, respectively. In the next experiment, strain changes (see Figure 33) were summarized based on four FBG sensors installed in the boat hull structures, which was divided into two research group stages:

- The first group (E_S and E_F) was defined as the state related to the construction process while the hull was still in form with a boat shape, and the part with embedded FBG sensors was manufactured.
- The second group (*R*, B_{10h}, and B_{80h}) was arranged to represent a finished boat mounted in a boat trailer.

Table 13. Calculated Pearson's correlation coefficients [147].

Stage	Sensors			
	S_1	S_2	S_3	S_4
Auto-correlation				
F	0.9963	0.9983	0.9997	0.9787
Cross-correlation				
E_S	0.9949	0.9943	0.9904	0.9700
E_F	0.9559	0.9854	0.9575	0.9528
R	0.6219	0.8783	0.5520	0.2515
B_{10h}	0.6950	0.8872	0.5617	0.2310
B_{80h}	0.8341	0.8612	0.5290	0.2604

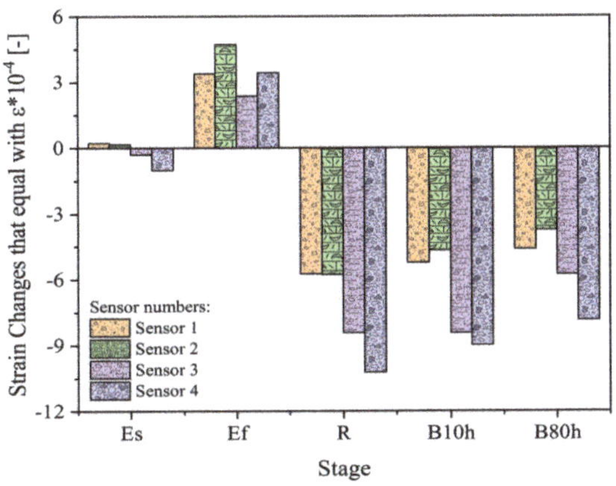

Figure 33. Measured strain changes by FBG sensors according to the defined research stages [147].

Besides measuring structural performance, recent work on ship collision was dedicated to improving a genetic algorithm (GA) to execute collision avoidance [148]. The algorithm core was designed according to international regulations for preventing collisions at sea (COLREGs)—i.e., Rule 13—Overtaking (The overtaking vessel should keep out of the way of the vessel being overtaken); Rule 14—Head-on situation (When two power-driven vessels are meeting on reciprocal or nearly reciprocal courses so as to involve risk of collision each shall alter her course to starboard so that each shall pass on the port side of the other); Rule 15—When two power-driven vessels are crossing so as to involve risk of collision, the vessel which has the other on her own starboard side shall keep out of the way and shall, if the circumstances of the case admit, avoid crossing ahead of the other vessel); Rule 16—Action by give-way vessel (Every vessel which is directed to keep out of the way of another vessel shall, so far as possible, take early and substantial action to keep well clear); and Rule 17—Action by stand-on vessel (Where one of two vessels is to keep out of the way the other shall keep her course and speed). The concept was then implemented in a multiple unmanned surface vehicles (USVs) system and sensor. In its operation, the sensor calculated the motion parameters and collision risk based on the defined variables in the system, such as USV velocity [$V_m(V_{mx}, V_{my})$], target velocity [$V_n(V_{nx}, V_{ny})$], relative velocity [$V_r(V_{rx}, V_{ry})$], target coordinate [$N(x_n, y_n)$], USV heading [β_m], target heading [β_n], relative heading [β_r], the relative distance between the USV and target [R_m], relative orientation [γ_m], and the relative direction of between the USV and target [α_m]. The characteristics of the improved GA compared to the conventional one are presented in

Table 14, which confirm the superiority of the improved version in terms of the processing time and results in reliability. Considering the variety of vessel speeds encountered when operating the sensor, a more specific comparative study (see Figures 34–36) was conducted, whose results indicated that the slower the vessel is, the longer the avoidance duration is (discrepancy of 4–8 s between the conventional and the improved versions) and the higher the survivability rate is (the improved version is upgraded by 6–15% compared to the conventional version).

Table 14. Comparison of the observed parameters based on the GA types [148].

Parameters	Conventional GA	Improved GA	Gap Value (%)
Numbers of iteration (-)	250	100	60 (reduced)
Standard deviation (-)	1.876	0.266	85.82 (reduced)
Computation time (ms)	160	90	43.75 (reduced)
Success rate (%)	80	95	15.79 (enlarged)

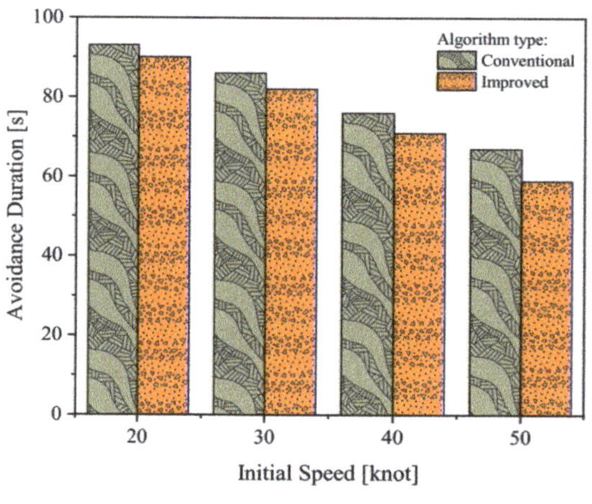

Figure 34. Comparison of the avoidance durations between two GA types [148].

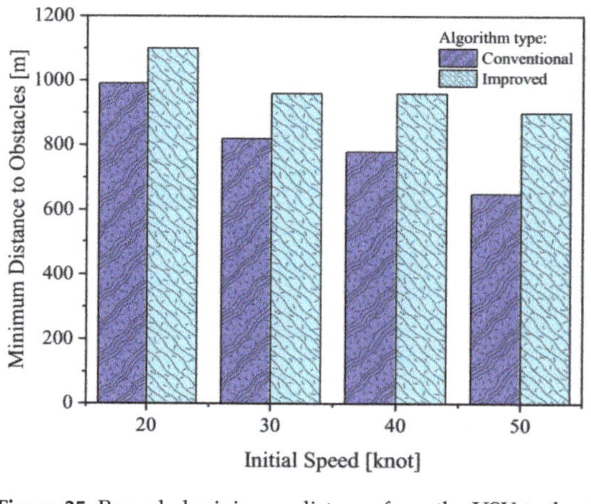

Figure 35. Recorded minimum distance from the USV to the obstacles based on GA type [148].

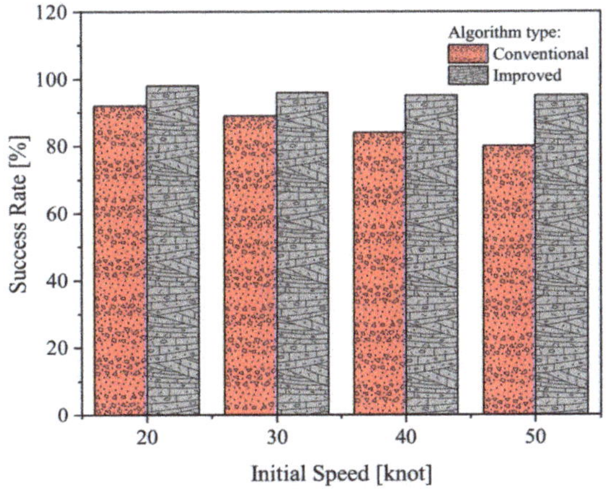

Figure 36. Influences of the deployed GA types on the success rate [148].

9. Concluding Remarks

Reviews on several aspects of sensor-based instruments and apparatus were performed in this work, including offshore engineering, ocean state, marine structures, and polar territory. According to the comprehensive review of the mentioned areas, sensors serve as a vital tool for measuring physical phenomena and observing environmental conditions. Several essential remarks are summarized as follows:

1. In terms of logistic and shipping activities, sensors are heavily involved to ensure that the pathway and sailing route of a ship/vessel are correct and clear to avoid collision or other accidents. The frequently used tools include high-resolution satellite imagery, enhanced spatio-spectral techniques, multi-polarization synthetic aperture radar (SAR), multilayer fusion light-head detectors (MFLHDs), and multiscale Wasserstein auxiliary classifier and generative adversarial networks (MW-ACGANs). These were then varied into global navigation satellite systems (GNSSs), inertial measurement units (IMU)s, Doppler velocity log (DVL), spectral entropy (SE), and C0 complexity in early 2021.
2. The hydrodynamic characteristics of marine design and structure are critical aspects that need to be accurately estimated. A sensor-based instrument is recorded for six degrees of freedom, resistance/drag, and propulsion performance. As concluded from the review, the subject of investigation is not only conventional ships, but also vessels with special purposes, i.e., cruise ship, sailing yacht, and underwater cleaning robot. It was also noted in certain works that sensors are deployed for measuring noise in the underwater territory. This contributes to an effort to quantify the effect of noise radiation on mammals' communication waves.
3. Regarding ship/vessel design, energy-saving characteristics have been assessed on machinery monitoring systems. Methodologies for designing sensors and measuring this phenomenon include first-principle safety analysis, hazard and operability studies (HAZOP), fault tree analysis (FTA), wireless sensor network (WSN), structural risk assessment (SRA), machinery risk assessment (MRA) tools, acoustic emission sensors, and laser heads clocked at 16 MHz. The evaluation of vibration interference in recorded motion data in the ice field territory is the interdisciplinary research of this field (hydrodynamic) and the polar science field.
4. Besides hull performance, machinery safety must also be evaluated from a sustainability point of view to either avoid loss in fuel consumption or environmental pollution. The summarized research indicates that machinery safety does not solely concern machine performance, but also concerns refraining the subject from undergoing physical damage (e.g., collision damage, hull joint failure, and unpredicted behavior of

anisotropic material). Therefore, the collision avoidance concept has been necessarily developed based on the beam search algorithm (BSA), followed by the deployment of laser vibrometry and a displacement laser sensor to quantify defects on the hull joint. In the case of anisotropic materials, a study was performed on polymer composites using acoustic emission Kolmogorov–Sinai (K–S) metric entropy.

5. Exploring the Arctic territory is the most addressed concern in terms of polar science and arctic engineering using sensor-based apparatus. For the past decade, works in this field have been dominated by the dedication to observing environmental conditions (of the ocean, atmospheric, and tundra areas), primarily when climate change arises due to an excessive amount of fossil fuel consumption. In recent years, the study has extended to evaluate problems such as the ice melting phenomenon. This work has been carried out especially near Canada and Greenland. Furthermore, autonomous frequency-modulated continuous-wave (FMCW) radar has been used to measure states of snowpack and ice levels.
6. Concern regarding offshore structure risk was summarized by assessing sensors' roles in damage detection, especially against vibration loadings. A fundamental conclusion regarding damage detection is that the sensor-based apparatus set consisting of four wired accelerometers and eight strain gauges, a hydraulic actuator, a computer (DAQ system and Labview), and a 407-controller is more reliable in estimating the damage of diagonal members based on an experiment on a scaled jacket platform. In 2021, a study of damage detection using CA-YD-108 acceleration sensors found good results from the scaled-based model and the in situ measurement.
7. Besides vibration, corroded steel and other marine structure parts are considered the origin of further catastrophic damage and loss. The demand to estimate this phenomenon has been fulfilled by a series of studies using various sensor instruments. On offshore platforms, optical fiber-based salinity sensors based on multimode plastic optical fiber (POF) have been used. In additional to commercial sensors (including image data harvesting using a camera for observation), developed ones, such as the tilted fiber Bragg grating (TFBG) concept, are being adopted to design customized sensors.
8. Impact and shock are addressed as important design parameters in any marine structure in which numerous scenarios may occur and inflict direct damage to the structure. In the studies, force pressure sensors, displacement gauges, high-speed cameras (e.g., Photron-Fastcam-SA1 series), accelerometers, and DAQ systems are the principal sensor-based tools used to measure design performance against impact and shock (widely known as crashworthy design). It is noted that the current technology is good enough to obtain reliable measurement of data of these phenomena. Nevertheless, an argument is raised regarding an area for improvement: measurement sensors are susceptible to impact and shock, which means that the accuracy of measurements can sometimes be questionable. Besides impact and shock, structure debris may also disturb the data recording. Thus, there is a challenge to design a buffer sensor that is strong enough to withstand such loadings.

Author Contributions: Conceptualization, A.R.P.; methodology, A.R.P.; software, A.R.P., T.T., and R.R.; validation, A.R.P., T.T., and R.R.; formal analysis, A.R.P., T.T., and R.R.; investigation, A.R.P., T.T., and R.R.; resources, A.R.P.; data curation, A.R.P., T.T., and R.R.; writing—original draft preparation, A.R.P., T.T., and R.R.; writing—review and editing, A.R.P., T.T., and R.R.; visualization, A.R.P., T.T., and R.R.; supervision, A.R.P.; project administration, A.R.P.; funding acquisition, A.R.P. All authors have read and agreed to the published version of the manuscript.

Funding: This work was supported by the RKAT PTNBH Universitas Sebelas Maret, Surakarta, under the scheme "Penelitian Kolaborasi Internasional UNS" (KI-UNS) 2021, with grant/contract number 260/UN27.22/HK.07.00/2021. The support is gratefully acknowledged by the authors.

Institutional Review Board Statement: Not applicable.

Informed Consent Statement: Not applicable.

Data Availability Statement: Not applicable.

Conflicts of Interest: The authors declare no conflict of interest.

References

1. Li, X.; Nosheen, S.; Haq, N.U.; Gao, X. Value creation during fourth industrial revolution: Use of intellectual capital by most innovative companies of the world. *Technol. Forecast. Soc. Chang.* **2021**, *163*, 120479. [CrossRef]
2. Neto, R.D.C.S.; Maia, J.S.; Neiva, S.D.S.; Scalia, M.D.; Guerra, J.B.S.O.D.A. The fourth industrial revolution and the coronavirus: A new era catalyzed by a virus. *Res. Glob.* **2020**, *2*, 100024.
3. Cowie, P.; Townsend, L.; Salemink, K. Smart rural futures: Will rural areas be left behind in the 4th industrial revolution? *J. Rural Stud.* **2020**, *79*, 169–176. [CrossRef]
4. Ćwiklicki, M.; Klich, J.; Chen, J. The adaptiveness of the healthcare system to the fourth industrial revolution: A preliminary analysis. *Futures* **2020**, *122*, 102602. [CrossRef]
5. Krafft, M.; Sajtos, L.; Haenlein, M. Challenges and Opportunities for Marketing Scholars in Times of the Fourth Industrial Revolution. *J. Interact. Mark.* **2020**, *51*, 1–8. [CrossRef]
6. Ebekozien, A.; Aigbavboa, C. COVID-19 recovery for the Nigerian construction sites: The role of the fourth industrial revolution technologies. *Sustain. Cities Soc.* **2021**, *69*, 102803. [CrossRef]
7. Dogaru, L. The Main Goals of the Fourth Industrial Revolution. Renewable Energy Perspectives. *Proc. Manuf.* **2020**, *46*, 397–401. [CrossRef]
8. Min, J.; Kim, Y.; Lee, S.; Jang, T.W.; Kim, I.; Song, J. The Fourth Industrial Revolution and Its Impact on Occupational Health and Safety, Worker's Compensation and Labor Conditions. *Saf. Health Work* **2019**, *10*, 400–408. [CrossRef]
9. Zunino, C.; Valenzano, A.; Obermaisser, R.; Petersen, S. Factory Communications at the Dawn of the Fourth Industrial Revolution. *Comput. Stand. Interface* **2020**, *71*, 103433. [CrossRef]
10. Fitzgerald, E.; Piór, M.; Tomaszewski, A. Network lifetime maximization in wireless mesh networks for machine-to-machine communication. *Ad Hoc Netw.* **2019**, *95*, 101987. [CrossRef]
11. Montori, F.; Bedogni, L.; Di Felice, M.; Bononi, L. Machine-to-machine wireless communication technologies for the Internet of Things: Taxonomy, comparison and open issues. *Pervasive Mob. Comput.* **2018**, *50*, 56–81. [CrossRef]
12. Bruns, R.; Dunkel, J.; Masbruch, H.; Stipkovic, S. Intelligent M2M: Complex event processing for machine-to-machine communication. *Expert Syst. Appl.* **2015**, *42*, 1235–1246. [CrossRef]
13. Xu, B.; Li, C. Influencing factors of college students' entrepreneurial ecosystem based on the internet of things and embedded systems. *Microprocess. Microsyst.* **2021**, *81*, 103694. [CrossRef]
14. Li, X.; Lu, Y.; Fu, X.; Qi, Y. Building the Internet of Things platform for smart maternal healthcare services with wearable devices and cloud computing. *Future Gener. Comput. Syst.* **2021**, *118*, 282–296. [CrossRef]
15. Malik, P.K.; Sharma, R.; Singh, R.; Gehlot, A.; Satapathy, S.C.; Alnumay, W.S.; Pelusi, D.; Ghosh, U.; Nayak, J. Industrial Internet of Things and its Applications in Industry 4.0: State of The Art. *Comput. Commun.* **2021**, *166*, 125–139. [CrossRef]
16. Trew, A. Spatial takeoff in the first industrial revolution. *Rev. Econ. Dyn.* **2014**, *17*, 707–725. [CrossRef]
17. Bruland, K.; Smith, K. Assessing the role of steam power in the first industrial revolution: The early work of Nick von Tunzelmann. *Res. Policy* **2013**, *42*, 1716–1723. [CrossRef]
18. Bottomley, S. In Patenting in England, Scotland and Ireland during the Industrial Revolution, 1700–1852. *Explor. Econ. Hist.* **2014**, *54*, 48–63. [CrossRef]
19. Jones, E. Space, sound and sedition on the Royal Naval ship, 1756–1815. *J. Hist. Geogr.* **2020**, *70*, 65–73. [CrossRef]
20. Hirschman, C.; Mogford, E. Immigration and the American industrial revolution from 1880 to 1920. *Soc. Sci. Res.* **2009**, *38*, 897–920. [CrossRef]
21. Iyer, A. Moving from Industry 2.0 to Industry 4.0: A case study from India on leapfrogging in smart manufacturing. *Proc. Manuf.* **2018**, *21*, 663–670. [CrossRef]
22. Kennedy, C. The energy embodied in the first and second industrial revolution. *J. Ind. Ecol.* **2020**, *24*, 887–898. [CrossRef]
23. Dillistone, F.W. Britain and the Second Industrial Revolution. *Theol. Today* **1956**, *13*, 11–17. [CrossRef]
24. Tweedale, G. Marketing in the Second Industrial Revolution: A Case Study of the Ferranti Computer Group, 1949–1963. *Bus. Hist.* **1992**, *34*, 96–127. [CrossRef]
25. Florida, R. The new industrial revolution. *Futures* **1991**, *23*, 559–576. [CrossRef]
26. Guerado, E. Scientific societies and the third industrial revolution—The future role of the OTC. *Injury* **2017**, *48*, S1–S4. [CrossRef]
27. Fitzsimmons, J. Information technology and the third industrial revolution. *Electron. Libr.* **1994**, *12*, 295–297. [CrossRef]
28. Glasnovic, Z.; Margeta, K.; Premec, K. Could Key Engine, as a new open-source for RES technology development, start the third industrial revolution? *Renew. Sustain. Energy Rev.* **2016**, *57*, 1194–1209. [CrossRef]
29. Finkelstein, J.; Newman, D. The Third industrial revolution: A special challenge to managers. *Organ. Dyn.* **1984**, *23*, 53–65. [CrossRef]
30. Rifkin, J. How the Third Industrial Revolution Will Create a Green Economy. *New Perspect. Quat.* **2016**, *33*, 6–10. [CrossRef]
31. Huda, N.; Prabowo, A.R. Investigation of Optimum Ply Angle using Finite Element (FE) Approach: References for Technical Application on the Composite Navigational Buoys. *Proc. Struct. Integr.* **2020**, *27*, 140–146. [CrossRef]

32. Ridwan, R.; Prabowo, A.R.; Muhayat, N.; Putranto, T.; Sohn, J.M. Tensile analysis and assessment of carbon and alloy steels using FE approach as an idealization of material fractures under collision and grounding. *Curved Layer. Struct.* **2020**, *7*, 188–198. [CrossRef]
33. Prabowo, A.R.; Cao, B.; Sohn, J.M.; Bae, D.M. Crashworthiness assessment of thin-walled double bottom tanker: Influences of seabed to structural damage and damage-energy formulae for grounding damage calculations. *J. Ocean Eng. Sci.* **2020**, *5*, 387–400. [CrossRef]
34. Ikhsan, I.; Triyono, J.; Prabowo, A.R.; Sohn, J.M. Investigation of Meshing Strategy on Mechanical Behaviour of Hip Stem Implant Design Using FEA. *Open Eng.* **2020**, *10*, 769–775. [CrossRef]
35. Prabowo, A.R.; Bae, D.M.; Sohn, J.M.; Zakki, A.F.; Cao, B. Effects of the rebounding of a striking ship on structural crashworthiness during ship-ship collision. *Thin Walled Struct.* **2017**, *115*, 225–239. [CrossRef]
36. Yusvika, M.; Prabowo, A.R.; Tjahjana, D.D.D.P.; Sohn, J.M. Cavitation Prediction of Ship Propeller Based on Temperature and Fluid Properties of Water. *J. Mar. Sci. Eng.* **2020**, *8*, 465. [CrossRef]
37. Bahatmaka, A.; Kim, D.J.; Prabowo, A.R.; Zaw, M.T. Investigation on the performance of the traditional Indonesian fishing vessel. *MATEC Web. Conf.* **2018**, *159*, 02056. [CrossRef]
38. Nubli, H.; Prabowo, A.R.; Sohn, J.M. Fire Phenomenon of Natural Gas Leak Accidents on the LNG-Fueled Ship Using Computational Fluid Dynamic. In Proceedings of the 39th International Conference on Offshore Mechanics and Arctic Engineering, Online, 3–7 August 2020; American Society of Mechanical Engineers: New York, NY, USA, 2020; Volume 84324, p. V02AT02A066.
39. Prabowoputra, D.M.; Prabowo, A.R.; Hadi, S.; Sohn, J.M. Assessment of turbine stages and blade numbers on modified 3D Savonius hydrokinetic turbine performance using CFD analysis. *Multidiscip. Model. Mater. Struct.* **2021**, *17*, 253–272. [CrossRef]
40. Bahatmaka, A.; Kim, D.J.; Prabowo, A.R. Numerical Investigation against Laboratory Experiment: An Overview of Damage and Wind Loads on Structural Design. *Proc. Struct. Integr.* **2020**, *27*, 6–13. [CrossRef]
41. Schwab, K. The Fourth Industrial Revolution: What It Means, How to Respond. Available online: https://www.weforum.org/agenda/2016/01/the-fourth-industrial-revolution-what-it-means-and-how-to-respond/ (accessed on 3 March 2021).
42. Fotaine, E.; Orsero, P.; Ledoux, A.; Nerzic, R.; Prevosto, M.; Quiniou, V. Reliability analysis and Response Based Design of a moored FPSO in West Africa. *Struct. Saf.* **2013**, *41*, 82–96. [CrossRef]
43. Alexandridis, G.; Kavussanos, M.G.; Kim, C.Y.; Tsouknidis, D.A.; Visvikis, I.D. A survey of shipping finance research: Setting the future research agenda. *Transp. Rep.* **2018**, *115*, 164–212. [CrossRef]
44. Nwafor, N.A.; Walker, T.R. Rethinking marine insurance and plastic pollution: Food for thought. *Resour. Conserv. Recycl.* **2020**, *161*, 104950. [CrossRef]
45. Prabowo, A.R.; Laksono, F.B.; Sohn, J.M. Investigation of structural performance subjected to impact loading using finite element approach: Case of ship-container collision. *Curved Layer Struct.* **2020**, *7*, 17–28. [CrossRef]
46. Finetti, E.S.; Bonet, C.A.; Lapeñ, O.L.; Areny, R.P. Cost-effective autonomous sensor for the long-term monitoring of water electrical conductivity of crop fields. *Comput. Electron. Agric.* **2019**, *165*, 104940.
47. Apicella, V.; Caponero, M.A.; Davino, D.; Visone, C. A magnetostrictive biased magnetic field sensor with geometrically controlled full-scale range. *Sens. Actuators A: Phys.* **2018**, *280*, 475–483. [CrossRef]
48. Kim, S.H.; Umar, A.; Hwang, S.W. Rose-like CuO nanostructures for highly sensitive glucose chemical sensor application. *Ceram. Int.* **2015**, *41*, 9468–9475. [CrossRef]
49. Luyckx, G.; Voet, E.; Lammens, N.; Degrieck, J. Strain Measurements of Composite Laminates with Embedded Fibre Bragg Gratings: Criticism and Opportunities for Research. *Sensors* **2011**, *11*, 384–408. [CrossRef]
50. Pandey, S.K.; Kim, K.H.; Lee, S.H. Use of a Dynamic Enclosure Approach to Test the Accuracy of the NDIR Sensor: Evaluation Based on the CO_2 Equilibration Pattern. *Sensors* **2007**, *7*, 3459–3471. [CrossRef]
51. UNCTAD. *Review of Maritime Transport 2017*; United Nations Publication: Geneva, Switzerland, 2017.
52. IMO. *International Convention for the Safety of Life at Sea (SOLAS)*; IMO Publications: London, UK, 2010.
53. Vassalos, D. Shaping Ship Safety: The Face of the Future. *Mar. Technol.* **1999**, *36*, 61–76.
54. Nezlin, N.P.; Weisberg, S.B.; Diehl, D.W. Relative availability of satellite imagery and ship-based sampling for assessment of stormwater runoff plumes in coastal southern California. *Estuar. Coast. Shelf Sci.* **2007**, *71*, 250–258. [CrossRef]
55. Song, P.; Qi, L.; Qian, X.; Lu, X. Detection of ships in inland river using high-resolution optical satellite imagery based on mixture of deformable part models. *J. Parallel Distrib. Comput.* **2019**, *132*, 1–7. [CrossRef]
56. Kanjir, U.; Greidanus, H.; Oštir, K. Vessel detection and classification from spaceborne optical images: A literature survey. *Remote Sens. Environ.* **2018**, *207*, 1–26. [CrossRef] [PubMed]
57. De Maio, A.; Orlando, D.; Pallotta, L.; Clemente, C. A Multifamily GLRT for Oil Spill Detection. *IEEE. Geosci. Remote Sens. Lett.* **2017**, *55*, 63–79. [CrossRef]
58. Zhang, Y.; Li, Q.Z.; Zang, F.N. Ship detection for visual maritime surveillance from non-stationary platforms. *Ocean Eng.* **2017**, *141*, 53–63. [CrossRef]
59. Liao, M.S.; Tang, J.; Wang, T.; Balz, T.; Zhang, L. Landslide monitoring with high-resolution SAR data in the Three Gorges region. *Sci. China Earth Sci.* **2012**, *55*, 590–601. [CrossRef]
60. Liu, L.; Gao, Y.; Wang, F.; Liu, X. Real-Time Optronic Beamformer on Receive in Phased Array Radar. *IEEE. Geosci. Remote Sens. Lett.* **2018**, *16*, 387–391. [CrossRef]

61. Zou, L.; Zhang, H.; Wang, C.; Wu, F.; Gu, F. Mw-acgan: Generating multiscale high-resolution SAR images for ship detection. *Sensors* **2020**, *20*, 6673. [CrossRef]
62. Gunnarsson, B. Recent ship traffic and developing shipping trends on the Northern Sea Route—Policy implications for future arctic shipping. *Mar. Policy* **2021**, *124*, 104369. [CrossRef]
63. Pan, Z.; Yang, R.; Zhang, Z. Msr2n: Multi-stage rotational region based network for arbitrary-oriented ship detection in sar images. *Sensors* **2020**, *20*, 2340. [CrossRef] [PubMed]
64. Dai, W.; Mao, Y.; Yuan, R.; Liu, Y.; Pu, X.; Li, C. A novel detector based on convolution neural networks for multiscale sar ship detection in complex background. *Sensors* **2020**, *20*, 2547. [CrossRef] [PubMed]
65. Gui, Y.; Li, X.; Xue, L. A multilayer fusion light-head detector for SAR ship detection. *Sensors* **2019**, *19*, 1124. [CrossRef] [PubMed]
66. Xie, T.; Zhang, W.; Yang, L.; Wang, Q.; Huang, J.; Yuan, N. Inshore ship detection based on level set method and visual saliency for SAR images. *Sensors* **2018**, *18*, 3877. [CrossRef]
67. Fukuda, G.; Hatta, D.; Guo, X.; Kubo, N. Performance evaluation of IMU and DVL integration in marine navigation. *Sensors* **2021**, *21*, 1056. [CrossRef] [PubMed]
68. Wang, L.; Li, Y.; Wan, Z.; Yang, Z.; Wang, T.; Guan, K.; Fu, L. Use of AIS data for performance evaluation of ship traffic with speed control. *Ocean Eng.* **2020**, *204*, 107259. [CrossRef]
69. Feng, M.; Shaw, S.-L.; Peng, G.; Fang, Z. Time efficiency assessment of ship movements in maritime ports: A case study of two ports based on AIS data. *J. Transp. Geogr.* **2020**, *86*, 102741. [CrossRef]
70. Tzabiras, G.; Kontogiannis, K. An Integrated Method for Predicting the Hydrodynamic Resistance of Low-CB Ships. *CAD Comput. Aided Des.* **2010**, *42*, 985–1000. [CrossRef]
71. Von Bock und Polach, R.; Ehlers, S. Heave and Pitch Motions of a Ship in Model Ice: An Experimental Study on Ship Resistance and Ice Breaking Pattern. *Cold Reg. Sci. Technol.* **2011**, *68*, 49–59. [CrossRef]
72. Lee, M.H.; Park, Y.D.; Park, H.G.; Park, W.C.; Hong, S.; Lee, K.S.; Chun, H.H. Hydrodynamic Design of an Underwater Hull Cleaning Robot and Its Evaluation. *Int. J. Nav. Archit. Ocean Eng.* **2012**, *4*, 335–352. [CrossRef]
73. Bouscasse, B.; Broglia, R.; Stern, F. Experimental Investigation of a Fast Catamaran in Head Waves. *Ocean Eng.* **2013**, *72*, 318–330. [CrossRef]
74. Jang, J.; Choi, S.H.; Ahn, S.M.; Kim, B.; Seo, J.S. Experimental Investigation of Frictional Resistance Reduction with Air Layer on the Hull Bottom of a Ship. *Int. J. Nav. Archit. Ocean Eng.* **2014**, *6*, 363–379. [CrossRef]
75. Fossati, F.; Bayati, I.; Orlandini, F.; Muggiasca, S.; Vandone, A.; Mainetti, G.; Sala, R.; Bertorello, C.; Begovic, E. A Novel Full Scale Laboratory for Yacht Engineering Research. *Ocean Eng.* **2015**, *104*, 219–237. [CrossRef]
76. Rajendran, S.; Fonseca, N.; Soares, C.G. Prediction of Extreme Motions and Vertical Bending Moments on a Cruise Ship and Comparison with Experimental Data. *Ocean Eng.* **2016**, *127*, 368–386. [CrossRef]
77. Jiao, J.; Ren, H.; Adenya, C.A.; Chen, C. Development of a Shipboard Remote Control and Telemetry Experimental System for Large-Scale Model's Motions and Loads Measurement in Realistic Sea Waves. *Sensors* **2017**, *17*, 2485. [CrossRef] [PubMed]
78. Carchen, A.; Atlar, M.; Turkmen, S.; Pazouki, K.; Murphy, A.J. Ship Performance Monitoring Dedicated to Biofouling Analysis: Development on a Small Size Research Catamaran. *Appl. Ocean Res.* **2019**, *89*, 224–236. [CrossRef]
79. Kim, D.; Lee, S.; Lee, J. Data-Driven Prediction of Vessel Propulsion Power Using Support Vector Regression with Onboard Measurement and Ocean Data. *Sensors* **2020**, *20*, 1588. [CrossRef]
80. Lin, J.F.; Zhao, D.G.; Guo, C.Y.; Su, Y.M.; Zhong, X.H. Comprehensive Test System for Ship-Model Resistance and Propulsion Performance in Actual Seas. *Ocean Eng.* **2020**, *197*, 106915. [CrossRef]
81. Guo, C.Y.; Zhong, X.H.; Zhao, D.G.; Wang, C.; Lin, J.F.; Song, K.W. Propulsion performance of large-scale ship model in real sea environment. *Ocean Eng.* **2020**, *210*, 107440. [CrossRef]
82. Su, Y.M.; Lin, J.F.; Zhao, D.G.; Guo, C.Y.; Guo, H. Influence of a pre-swirl stator and rudder bulb system on the propulsion performance of a large-scale ship model. *Ocean Eng.* **2020**, *218*, 108189. [CrossRef]
83. Mei, B.; Sun, L.; Shi, G. Full-Scale Maneuvering Trials Correction and Motion Modelling Based on Actual Sea and Weather Conditions. *Sensors* **2020**, *20*, 3963. [CrossRef] [PubMed]
84. Ha, Y.J.; Kim, K.H.; Nam, B.W.; Hong, S.Y.; Kim, H. Experimental Study for Characteristics of Slamming Loads on Bow of a Ship-Type FPSO under Breaking and Irregular Wave Conditions. *Ocean Eng.* **2021**, *224*, 108738. [CrossRef]
85. Allianz. *Safety and Shipping Review 2020*; Allianz Global Corporate & Specialty: Munich, Germany, 2020.
86. Paik, B.G.; Cho, S.R.; Park, B.J.; Lee, D.; Yun, J.H.; Bae, B.D. Employment of Wireless Sensor Networks for Full-Scale Ship Application. *IFIP Int. Fed. Inf. Process.* **2007**, *4808*, 113–122.
87. Paik, B.G.; Cho, S.R.; Park, B.J.; Lee, D.; Bae, B.D.; Yun, J.H. Characteristics of wireless sensor network for full-scale ship application. *J. Mar. Sci. Technol.* **2009**, *14*, 115–126. [CrossRef]
88. Penttilä, P.; Ruponen, P. Use of Level Sensors in Breach Estimation for a Damaged Ship. In Proceedings of the 5th International Conference on Collision and Grounding of Ships, Espoo, Finland, 14–16 June 2010; pp. 80–87.
89. Lazakis, I.; Dikis, K.; Michala, A.L.; Theotokatos, G. Advanced Ship Systems Condition Monitoring for Enhanced Inspection, Maintenance and Decision Making in Ship Operations. *Transp. Res. Procedia* **2016**, *14*, 1679–1688. [CrossRef]
90. Raptodimos, Y.; Lazakis, I.; Theotokatos, G.; Varelas, T.; Drikos, L. Ship sensors data collection and analysis for condition monitoring of ship structures and machinery systems. In Proceedings of the Smart Ship Technology, London, UK, 26–27 January 2016.

91. Noordstrand, A. Experience with Robotic Underwater Hull Cleaning in Dutch Ports. In Proceedings of the 3rd Hull Performance & Insight Conference, Redworth, UK, 12–14 March 2018; pp. 4–9.
92. Drewing, S.; Witkowski, K. Spectral Analysis of Torsional Vibrations Measured by Optical Sensors, as a Method for Diagnosing Injector Nozzle Coking in Marine Diesel Engines. *Sensors* **2021**, *21*, 775. [CrossRef] [PubMed]
93. Kozak, M.; Bejger, A.; Tomczak, A. Identification of Gate Turn-Off Thyristor Switching Patterns Using Acoustic Emission Sensors. *Sensors* **2021**, *21*, 70. [CrossRef] [PubMed]
94. Kyzioł, L.; Panasiuk, K.; Hajdukiewicz, G.; Dudzik, K. Acoustic emission and k-s metric entropy as methods for determining mechanical properties of composite materials. *Sensors* **2021**, *21*, 145. [CrossRef] [PubMed]
95. Hill, V.J.; Zimmerman, R.C. Estimates of primary production by remote sensing in the Arctic Ocean: Assessment of accuracy with passive and active sensors. *Deep Sea Res. I* **2010**, *157*, 1243–1254. [CrossRef]
96. Høyer, J.L.; Karagali, I.; Dybkjær, G.; Tonboe, R. Multi sensor validation and error characteristics of Arctic satellite sea surface temperature observations. *Remote Sens. Environ.* **2012**, *121*, 335–346. [CrossRef]
97. Tagesson, T.; Mastepanov, M.; Mölder, M.; Tamstorf, M.P.; Eklundh, L.; Smith, B.; Sigsgaard, C.; Lund, M.; Ekberg, A.; Falik, J.M.; et al. Modelling of growing season methane fluxes in a high-Arctic wet tundra ecosystem 1997–2010 using in situ and high-resolution satellite data. *Tellus B Chem. Phys. Meteorol.* **2013**, *65*, 19722. [CrossRef]
98. Lund, M.; Hansen, B.U.; Pedersen, S.H.; Stiegler, C.; Tamstorf, M.P. Characteristics of summer-time energy exchange in a high Arctic tundra heath 2000–2010. *Tellus B Chem. Phys. Meteorol.* **2014**, *66*, 21631. [CrossRef]
99. Goodrich, J.P.; Oechel, W.C.; Gioli, B.; Moreaux, V.; Murphy, P.C.; Burba, G.; Zona, D. Impact of different eddy covariance sensors, site set-up, and maintenance on the annual balance of CO_2 and CH_4 in the harsh Arctic environment. *Agric. For. Meteorol.* **2016**, *228–229*, 239–251. [CrossRef]
100. Connolly, R.; Connolly, M.; Soon, W. Re-calibration of Arctic sea ice extent datasets using Arctic surface air temperature records. *Hydrol. Sci. J.* **2017**, *62*, 1317–1340. [CrossRef]
101. Hesaraki, S.; O'Neil, N.T.; Lesins, G.; Saha, A.; Martin, R.V.; Fioletov, V.E.; Baibakov, K.; Abboud, I. Comparisons of a Chemical Transport Model with a Four-Year (April to September) Analysis of Fine and Coarse-Mode Aerosol Optical Depth Retrievals Over the Canadian Arctic. *Atmos. Ocean* **2017**, *55*, 213–229. [CrossRef]
102. Howell, S.E.I.; Small, D.; Rohner, C.; Mahmud, M.S.; Yackel, J.J.; Brady, M. Estimating melt onset over Arctic sea ice from time series multi-sensor Sentinel-1 and RADARSAT-2 backscatter. *Remote Sens. Environ.* **2019**, *229*, 48–59. [CrossRef]
103. Carotenuto, F.; Brilli, L.; Gioli, B.; Gualtieri, G.; Vagnoli, C.; Mazzola, M.; Viola, A.P.; Vitale, V.; Severi, M.; Traversi, R.; et al. Long-Term Performance Assessment of Low-Cost Atmospheric Sensors in the Arctic Environment. *Sensors* **2020**, *20*, 1919. [CrossRef]
104. Pomerleau, P.; Royer, A.; Langlois, A.; Cliché, P.; Courtemanche, B.; Madore, J.B.; Picard, G.; Lefebvre, É. Low Cost and Compact FMCW 24 GHz Radar Applications for Snowpack and Ice Thickness Measurements. *Sensors* **2020**, *20*, 3909. [CrossRef]
105. Løken, T.K.; Rabault, J.; Jensen, A.; Sutherland, G.; Christensen, K.H.; Müller, M. Wave measurements from ship mounted sensors in the Arctic marginal ice zone. *Cold Reg. Sci. Technol.* **2021**, *182*, 103207. [CrossRef]
106. Nikishin, A.M.; Petrov, E.I.; Cloetingh, S.; Korniychuk, A.V.; Morozov, A.F.; Petrov, O.V.; Poselov, V.A.; Beziazykov, A.V.; Skolotnev, S.G.; Malyshev, N.A.; et al. Arctic Ocean Mega Project: Paper 1—Data collection. *Earth Sci. Rev.* **2021**, in press.
107. Kodaira, T.; Waseda, T.; Nose, T.; Sato, K.; Inoue, J.; Voermans, J.; Babanin, A. Observation of on-ice wind waves under grease ice in the western Arctic Ocean. *Polar Sci.* **2021**, in press. [CrossRef]
108. Elshafey, A.A.; Haddara, M.R.; Marzouk, H. Damage detection in offshore structures using neural networks. *Mar. Struct.* **2010**, *23*, 131–145. [CrossRef]
109. Razi, P.; Taheri, F. A Vibration-Based Strategy for Health Monitoring of Offshore Pipelines' Girth-Welds. *Sensors* **2014**, *14*, 17174–17191. [CrossRef] [PubMed]
110. Liu, F.; Lu, H.; Wang, W.; Yang, D. Periodic Damage Evaluation of Aging Offshore Jackets Based on Continuous Dynamic Test. *J. Mar. Sci. Technol.* **2014**, *22*, 732–738.
111. Asgarian, B.; Aghaeidoost, V.; Shokrgozar, H.R. Damage detection of jacket type offshore platforms using rate of signal energy using wavelet packet transform. *Mar. Struct.* **2016**, *45*, 1–21. [CrossRef]
112. Hosseinlou, F.; Mojtahedi, A. Developing a robust simplified method for structural integrity monitoring of offshore jacket-type platform using recorded dynamic responses. *Appl. Ocean Res.* **2016**, *56*, 107–118. [CrossRef]
113. Opoka, S.; Soman, R.; Mieloszyk, M.; Ostachowicz, W. Damage detection and localization method based on a frequency spectrum change in a scaled tripod model with strain rosettes. *Mar. Struct.* **2016**, *49*, 163–179. [CrossRef]
114. Mieloszyk, M.; Ostachowicz, W. An application of Structural Health Monitoring system based on FBG sensors to offshore wind turbine support structure model. *Mar. Struct.* **2017**, *51*, 65–86. [CrossRef]
115. Luczak, M.M.; Telega, J.; Zagato, N.; Mucchi, E. On the damage detection of a laboratory scale model of a tripod supporting structure by vibration-based methods. *Mar. Struct.* **2019**, *64*, 146–160. [CrossRef]
116. Li, W.; Huang, Y. A Method for Damage Detection of a Jacket Platform under Random Wave Excitations Using Cross Correlation Analysis and PCA-Based Method. *Ocean Eng.* **2020**, *214*, 107734. [CrossRef]
117. Fathi, A.; Esfandiari, A.; Fadavie, M.; Mojtahedi, A. Damage detection in an offshore platform using incomplete noisy FRF data by a novel Bayesian model updating method. *Ocean Eng.* **2020**, *217*, 108023. [CrossRef]

118. Liu, Y.; Lu, Z.; Yan, X.; Liu, Z.; Tang, L. Measurement and modelling of the vibration induced by working equipment on an offshore platform. *Ocean Eng.* **2021**, *219*, 108354. [CrossRef]
119. Vidal, Y.; Aquino, G.; Pozo, F.; Gutiérrez-Arias, J.E.M. Structural Health Monitoring for Jacket-Type Offshore Wind Turbines; Experimental Proof of Concept. *Sensors* **2020**, *20*, 1835. [CrossRef]
120. Puruncajas, B.; Vidal, Y.; Tutivén, C. Vibration-Response-Only Structural Health Monitoring for Offshore Wind Turbine Jacket Foundations via Convolutional Neural Networks. *Sensors* **2020**, *20*, 3429. [CrossRef]
121. Tang, D.; Chen, J.; Wu, W.; Jin, L.; Yue, Q.; Xie, B.; Wang, S.; Feng, J. Research on sampling rate selection of sensors in offshore platform shm based on vibration. *Appl. Ocean Res.* **2020**, *101*, 102192. [CrossRef]
122. Chen, J.; Su, Z.; Cheng, L. Identification of corrosion damage in submerged structures using fundamental anti-symmetric Lamb waves. *Smart Mater. Struct.* **2010**, *19*, 1–12. [CrossRef]
123. Xu, C.; Li, Z.; Jin, W. A New Corrosion Sensor to Determine the Start and Development of Embedded Rebar Corrosion Process at Coastal Concrete. *Sensors* **2013**, *13*, 13258–13275. [CrossRef] [PubMed]
124. Mu, X.; Wei, J.; Dong, J.; Ke, W. In Situ Corrosion Monitoring of Mild Steel in a Simulated Tidal Zone without Marine Fouling Attachment by Electrochemical Impedance Spectroscopy. *J. Mater. Sci. Technol.* **2014**, *30*, 1043–1050. [CrossRef]
125. Islam, M.R.; Bagherifaez, M.; Ali, M.M.; Chai, H.K.; Lim, K.; Ahmad, H. Tilted Fiber Bragg Grating Sensors for Reinforcement Corrosion Measurement in Marine Concrete Structure. *IEEE Trans. Instrum. Meas.* **2015**, *64*, 3510–3516. [CrossRef]
126. Nie, M.; Neodo, S.; Wharton, J.A.; Cranny, A.; Harris, N.R.; Wood, R.J.K.; Stokes, K.R. Electrochemical detection of cupric ions with boron-doped diamond electrode for marine corrosion monitoring. *Electrochim. Acta* **2016**, *202*, 345–356. [CrossRef]
127. Ortiz, A.; Bonnin-Pascual, F.; Garcia-Fidalgo, E.; Company-Corcoles, J.P. Vision-Based Corrosion Detection Assisted by a Micro-Aerial Vehicle in a Vessel Inspection Application. *Sensors* **2016**, *16*, 2118. [CrossRef]
128. Dahai, X.; Shizhe, S.; Weixian, J.; Jian, L.; Zhiming, G.; Jihui, W.; Wenbin, H. Atmospheric Corrosion Monitoring of Field-exposed Q235B and T91 Steels in Zhoushan Offshore Environment Using Electrochemical Probes. *J. Wuhan Univ. Technol. Mater. Sci. Ed.* **2017**, *32*, 1433–1440.
129. Luo, D.; Li, P.; Yue, Y.; Ma, J.; Yang, H. In-Fiber Optic Salinity Sensing: A Potential Application for Offshore Concrete Structure Protection. *Sensors* **2017**, *17*, 962. [CrossRef]
130. Yao, Y.; Yang, Y.; Wang, Y.; Zhao, X. Artificial intelligence-based hull structural plate corrosion damage detection and recognition using convolutional neural network. *Appl. Ocean Res.* **2019**, *90*, 101823. [CrossRef]
131. Gartner, N.; Kosec, T.; Legat, A. Monitoring the Corrosion of Steel in Concrete Exposed to a Marine Environment. *Materials* **2020**, *13*, 407. [CrossRef]
132. Ahuir-Torres, J.I.; Bausch, N.; Farrar, A.; Webb, S.; Simandjuntak, S.; Nash, A.; Thomas, B.; Muna, J.; Jonsson, C.; Mathew, D. Benchmarking parameters for remote electrochemical corrosion detection and monitoring of offshore wind turbine structures. *Wind Energy* **2019**, *22*, 857–876. [CrossRef]
133. Ha, M.-G.; Jeon, S.H.; Jeong, Y.-S.; Mha, H.-S.; Ahn, J.-H. Corrosion Environment Monitoring of Local Structural Members of a Steel Truss Bridge under a Marine Environment. *Int. J. Steel Struct.* **2021**, *21*, 167–177. [CrossRef]
134. Poggi, L.; Gaggero, T.; Gaiotti, M.; Ravina, E.; Rizzo, C.M. Recent Developments in Remote Inspections of Ship Structures. *Int. J. Nav. Archit. Ocean Eng.* **2020**, *12*, 881–891. [CrossRef]
135. Abdulkader, R.E.; Veerajagadheswar, P.; Lin, N.H.; Kumaran, S.; Vishaal, S.R.; Mohan, R.E. Sparrow: A Magnetic Climbing Robot for Autonomous Thickness Measurement in Ship Hull Maintenance. *J. Mar. Sci. Eng.* **2020**, *8*, 469. [CrossRef]
136. Liu, Q.; Bai, C.; Li, X.; Jiang, L.; Dai, W. Coal dust/air explosions in a large-scale tube. *Fuel* **2010**, *89*, 329–335. [CrossRef]
137. Bai, C.; Gong, G.; Liu, Q.; Chen, Y.; Niu, G. The explosion overpressure field and flame propagation of methane/air and methane/coal dust/air mixtures. *Saf. Sci.* **2011**, *49*, 1349–1354. [CrossRef]
138. Kujala, P.; Arughadhoss, S. Statistical analysis of ice crushing pressures on a ship's hull during hull–ice interaction. *Cold Reg. Sci. Technol.* **2012**, *70*, 1–11. [CrossRef]
139. Kong, X.S.; Wu, W.G.; Li, J.; Chen, P.; Liu, F. Experimental and numerical investigation on a multi-layer protective structure under the synergistic effect of blast and fragment loadings. *Int. J. Impact Eng.* **2014**, *65*, 146–162. [CrossRef]
140. Pierre, L.S.; Deshpande, V.S.; Fleck, N.A. The low velocity impact response of sandwich beams with a corrugated core or a Y-frame core. *Int. J. Mech. Sci.* **2015**, *91*, 71–80. [CrossRef]
141. Niklas, K.; Kozak, J. Experimental investigation of Steel–Concrete–Polymer composite barrier for the ship internal tank construction. *Ocean Eng.* **2016**, *111*, 449–460. [CrossRef]
142. Moon, S.J.; Kwon, J.I.; Park, J.W.; Chung, J.H. Assessment on shock pressure acquisition from underwater explosion using uncertainty of measurement. *Int. J. Nav. Archit. Ocean Eng.* **2017**, *9*, 589–597. [CrossRef]
143. Jun, G.; Shichun, H.; Taranukha, N.; Mingqi, L. Vibration damping of naval ships based on ship shock trials. *Appl. Acoust.* **2018**, *133*, 52–57.
144. Li, Y.; Chen, Z.; Zhao, T.; Cao, X.; Jiang, Y.; Xiao, D.; Fang, D. An experimental study on dynamic response of polyurea coated metal plates under intense underwater impulsive loading. *Int. J. Impact Eng.* **2019**, *133*, 103361. [CrossRef]
145. Hsu, C.Y.; Chiang, C.C.; Hsieh, T.S.; Chen, T.H.; Chen, Y.H. A study of strain measurement in cylindrical shells subjected to underwater shock loading using FBG sensors. *Optik* **2020**, *217*, 164701. [CrossRef]
146. Jiang, X.; Zhang, W.; Li, D.; Chen, T.; Tang, Y.; Guo, Z. Experimental analysis on dynamic response of pre-cracked aluminum plate subjected to underwater explosion shock loadings. *Thin Walled Struct.* **2021**, *159*, 107256. [CrossRef]

147. Mieloszyk, M.; Majewska, K.; Ostachowics, W. Application of embedded fibre Bragg grating sensors for structural health monitoring of complex composite structures for marine applications. *Mar. Struct.* **2021**, *76*, 102903. [CrossRef]
148. Wang, H.; Fu, Z.; Zhou, J.; Fu, M.; Ruan, L. Cooperative collision avoidance for unmanned surface vehicles based on improved genetic algorithm. *Ocean Eng.* **2021**, *222*, 108612. [CrossRef]

MDPI AG
Grosspeteranlage 5
4052 Basel
Switzerland
Tel.: +41 61 683 77 34

Applied Sciences Editorial Office
E-mail: applsci@mdpi.com
www.mdpi.com/journal/applsci

www.ingramcontent.com/pod-product-compliance
Lightning Source LLC
Chambersburg PA
CBHW061516210926
51850CB00008B/1938
9783725815494